D1087685

Earthquake Protection

Earthquake Protection

Second Edition

Andrew Coburn

Risk Management Solutions Inc., USA

Associate of the Martin Centre for Architectural and Urban Studies, University of Cambridge, UK

and

Robin Spence

Reader in Architectural Engineering and Fellow of Magdalene College, University of Cambridge, UK

Director, Cambridge Architectural Research Ltd, Cambridge, UK

JOHN WILEY & SONS, LTD

Telephone (+44) 1243 779777

Email (for orders and customer service enquiries): cs-books@wiley.co.uk
Visit our Home Page on www.wileyeurope.com or www.wiley.com

Other Wiley Editorial Offices

John Wiley & Sons Inc., 111 River Street, Hoboken, NJ 07030, USA

Jossey-Bass, 989 Market Street, San Francisco, CA 94103-1741, USA

Wiley-VCH Verlag GmbH, Boschstr. 12, D-69469 Weinheim, Germany

John Wiley & Sons Australia Ltd, 33 Park Road, Milton, Queensland 4064, Australia

John Wiley & Sons (Asia) Pte Ltd, 2 Clementi Loop #02-01, Jin Xing Distripark, Singapore 129809

John Wiley & Sons Canada Ltd, 22 Worcester Road, Etobicoke, Ontario, Canada M9W 1L1

Library of Congress Cataloging-in-Publication Data

Coburn, Andrew (Andrew W.)
Earthquake protection / Andrew Coburn and Robin Spence. – 2nd ed.
p. cm.
Includes bibliographical references and index.
ISBN 0-471-49614-6
1. Earthquake hazard analysis. 2. Earthquake engineering. 3. Earthquakes – Safety measures. I. Spence, R. J. S. (Robin J. S.) II. Title.
QE539.2.S34 C63 2002
363.34′9563 – dc21

2002072616

British Library Cataloguing in Publication Data

A catalogue record for this book is available from the British Library

ISBN 0-471-49614-6 (Cloth)
ISBN 0-470-84923-1 (Paper)

Typeset in 10/12pt Times by Laserwords Private Limited, Chennai, India
Printed and bound in Great Britain by TJ International, Padstow, Cornwall
This book is printed on acid-free paper responsibly manufactured from sustainable forestry in which at least two trees are planted for each one used for paper production.

In memoriam:

Fred Cuny of Intertect, from whose ideas, vision and personal encouragement much of the work described in this book developed. Fred was murdered in Chechnya in 1995 in the course of a typically courageous and selfless effort to bring humanitarian support to the displaced people there.

Contents

ABOUT THE AUTHORS

Andrew Coburn MA, DipArch, PhD

Andrew Coburn is an executive of Risk Management Solutions, Inc., the world's leading insurance risk management and catastrophe modelling company, working with insurance clients to assist with the management of earthquake risk. Dr Coburn has many years of international experience of earthquake risk analysis and catastrophe modelling. He originally completed his PhD on earthquake risk at Cambridge University in the 1980s under the supervision of Dr Robin Spence and has studied many catastrophes and developed techniques for their analysis, modelling and quantification. For over 20 years, he has participated in the study of catastrophe events including 15 field damage surveys, ranging from the Italian earthquake in 1980 to the Gujarat earthquake in India in 2001. His research work has included research into human casualties in catastrophes, including Visiting Fellowships at Hokkaido University in Japan, Virginia Polytechnic Institute in Washington, DC, USA and University of Naples, Italy.

Robin Spence MA, MSc, PhD, CEng, FIStructE, MICE, Member EERI

Robin Spence is a structural engineer and Reader in Architectural Engineering in the Department of Architecture at Cambridge University. He has been active in the field of earthquake risk mitigation for over 20 years. During that time he has taken part in many field missions, and was one of the founders of EEFIT, the UK earthquake engineering team in 1983. He has also directed numerous research projects on earthquake vulnerability assessment, loss estimation and disaster mitigation, and is the author of many papers, reports and manuals on these subjects. He has frequently been a consultant to international agencies, national governments and insurance companies on the assessment and mitigation of earthquake and volcanic hazards.

After obtaining his PhD on the analysis of reinforced concrete structures, Dr Spence has been with the Department of Architecture at Cambridge University since 1975, and has been a Director and Joint Director of the Martin Centre since 1985. He has been a Visiting Professor at MIT and UCLA, at the University of Naples and at Macquarie University in Sydney. He is currently Director of the Cambridge University Centre for Risk in the Built Environment. He is also a Director of Cambridge Architectural Research Ltd, and a Fellow of Magdalene College, Cambridge.

Foreword

During the past ten years, since publication of the first edition of *Earthquake Protection*, the world has experienced many catastrophic earthquakes. These earthquakes have been in developed and developing countries and they have impacted urban as well as rural communities. Tens of thousands of lives and billions of dollars of economic losses have occurred due to these events. With every passing decade, the level of losses is increasing dramatically. Even in countries which are supposed to be at the cutting edge of science and technology in earthquake-related disciplines, we have seen devastating life and economic losses. It is then natural to inquire as to why our societies have not been able to mitigate the effects of such events.

The answer to such a query is very complex. The socio-economic conditions together with education and awareness of earthquake risk of the population at large play an important role in shaping the way an affected community will respond to an earthquake. Even amongst experts from seismology, geology, architecture, engineering and other technical disciplines related to earthquakes, there is relatively small appreciation of the complexities of earthquake risk mitigation issues. A holistic understanding and the application of strategies based on such an understanding are very urgently needed.

The second edition of *Earthquake Protection* by Coburn and Spence provides an excellent introduction to this complex problem. The book treats the problem of earthquakes from seismological, geological, architectural, engineering, social, and economic perspectives. The book provides a superb explanation about the role of insurance/reinsurance in earthquake risk management. It is important for all the earthquake risk management professionals to appreciate and understand the interactions of various disciplines to develop sustainable earthquake risk reduction strategies. This book by Coburn and Spence should be required reading for all these professionals.

Since the publication of the first edition of *Earthquake Protection* in 1992, the two authors have continued to build their reputations in this field, both in academic research and in commercial application. Their treatment of financial implications related to earthquakes is unique and provides state of the art knowledge to readers in the financial services industry. This book is also invaluable to students of architecture, engineering, and sciences who are trying to learn about earthquake-related loss estimation strategies.

In recent years, it has become obvious to all the professionals working towards reducing the losses due to earthquakes that a holistic approach in risk reduction is the only possible way of creating earthquake-resistant communities. Coburn and Spence in this book have provided excellent ideas towards that goal.

Haresh Shah
Obayashi Professor of Engineering, Emeritus, Stanford University
Chairman of the Board of Trustees, GeoHazard International, World Seismic Safety Initiative
March, 2002

Acknowledgements

We didn't write this book unaided. We used a lot of support, ideas and comments from our friends and colleagues in the earthquake field. It is always difficult to single out individuals, but we are grateful in particular to the following who assisted us in the development of the first edition and in this edition.

Professor Haresh Shah for his guidance and inspiration throughout the past 20 years and for kindly providing the foreword to this edition.

Professor Yutaka Ohta for his advice and encouragement over many years.

Dr Chris Bowitt, Dr Roger Musson and Dr Russ Evans from the British Geological Survey for their advice and reviews of Chapter 3.

Dr Herbert Tiedemann, formerly of SwissRe, for permission to include data from his Catalogue of Earthquakes and Volcanic Eruptions.

Dr Yasemin Aysan and Professor Ian Davis of the former Disaster Management Centre at Oxford Brookes University for their contributions to a number of the joint projects described here.

Antonios Pomonis for his invaluable and painstaking work compiling the statistics on earthquakes which formed the basis of the Martin Centre and Cambridge Architectural Research earthquake database.

Richard Hughes for his guidance on earthquake damage surveying on many joint field trips and for permission to use his drawings.

Jolyon Leslie and Dr Eric Dudley for their contributions to the discussion on upgrading rural construction, and for permission to use their drawings.

Dr Tom Corsellis and Michael Markus for suggestions based on their field experience in Gujarat which helped us to update Chapter 4.

Dr Peter Baxter for reviewing the medical aspects of Chapter 4 and suggesting many improvements.

Professor Mary Comerio of the University of California at Berkeley for many sources of inspiration, especially in improving Chapters 5 and 10.

Edmund Booth, Dr Dina D'Ayala, Domenico del Re and Giulio Zuccaro for contributions, comments and suggestions for the improvement of Chapters 8 and 9.

Colleagues at Risk Management Solutions for their support and input into this edition including Hement Shah, Dr Robert Muir Wood, Dr Gordon Woo, Dr Weimin Dong, Laurie Johnson and many others.

Paul Freeman for discussions about the cost of earthquakes to governments and for an advance copy of his PhD dissertation.

Nicholas Coburn, Nancy Peskett and Janet Owers for their help in preparing illustrations.

1 Earthquakes, Disasters and Protection

1.1 Earthquake Protection: Past Failure and Present Opportunity

In spite of the huge technical achievements of the last century – which have given us skyscraper cities, fast and cheap air travel and instant global telecommunications, as well as eradicating many major diseases and providing the potential to feed our burgeoning population – over much of the world the threat of earthquakes has remained untamed. As later chapters will show, the progress we have made in reducing the global death toll from earthquakes is modest, and at the beginning of the twenty-first century, we have become distressingly familiar with tragic media images of the total devastation of towns, villages and human lives caused by large earthquakes, for which their victims have been quite unprepared.

One possible reason for the lack of progress in saving lives from earthquakes is that although they are among our oldest enemies, it is only in the last quarter of the twentieth century that we have begun to understand how to protect ourselves against them. From time to time in our history, parts of the earth have apparently randomly been shaken violently by vast energy releases. Where these events have occurred near human settlements, the destruction has been legendary. Tales of destruction of ancient cities, like Troy in Greek mythology, and Taxila, have been attributed to the power of the earthquake. In more recent memory the cities of Messina in Italy, Tangshan in China, Tokyo and Kobe in Japan, and San Francisco in the United States have all been devastated by massive earthquakes. The apparent randomness of earthquakes, their lack of any visible cause and their frightening destructiveness earned them over the centuries the status of divine judgement. They were the instruments of displeasure of the Greek god Poseidon, the spiteful wriggling of the subterranean catfish *Namazu* in Japanese mythology, and punishment for sinners in Christian belief.

Only over the last century or so have we begun to understand what earthquakes are and what causes them. We have come to know that earthquakes are not random, but are natural forces driven by the evolutionary processes of the planet we live on. Earthquakes can now be mapped, measured, analysed and demystified. We know where they are likely to occur and we are beginning to develop predictive methods which reduce the uncertainty about where and when the next destructive events will happen. But in many of the parts of the world most at risk from large earthquakes, some aspects of the old attitudes live on; people are fatalistic, unwilling to believe that they have the means or ability to combat such destructive power, and thus they are reluctant to think in terms of planning, organising and spending part of their income – as individuals or as societies – on protection.

What makes matters worse is that the twenty-first century is experiencing an unparalleled explosion in the world's population growth, and an exponential growth in the size and number of villages, towns and cities across the globe. At the present time, unlike previous centuries, there is hardly a place on land where a large earthquake can occur without causing damage. As cities increase in size, so the potential for massive destruction increases. For this reason, the risk of earthquake disaster is higher than at any time in our history, and the risk is increasing. In the past few decades we have seen catastrophic disasters to cities and regions across the world on a scale unheard of a century ago. Unless serious efforts are made to improve earthquake protection worldwide, we can expect to see similar and greater disasters with increasing frequency in the years to come.

But the science and practice of how to protect ourselves, our buildings and our cities from earthquakes has also been developing rapidly during recent years. A body of knowledge has been built up by engineers, urban planners, financiers, administrators and government officials about how to tackle this threat. The approach to protection is necessarily a multi-disciplinary one, and one requiring a wide range of measures including well-targetted spending on protection, better building design and increasing quality of construction in the areas most likely to suffer an earthquake.

Earthquake protection involves everyone. The general public have to be aware of the safety issues involved in the type of house they live in and of earthquake considerations inside the home and workplace. The construction industry is involved in improving building design and increasing quality. Politicians and administrators manage risk by making decisions about how much to spend on earthquake safety and where public resources are most effectively allocated. Many other participants are involved either directly or indirectly, including urban planners in designing safer cities, community groups in preparing for future earthquakes and motivating their members to protect themselves, private companies and organisations in protecting themselves, their employees and customers, and insurance companies in assessing the risks and providing cover for people to protect themselves.

This book is for everyone interested in understanding, organising or participating in earthquake protection. It is intended to provide an overview of methods to reduce the impact of future earthquakes and to deal with earthquakes when they occur.

1.2 Earthquake Disasters

Earthquakes can be devastating to people as individuals, to families, to social organisations at every level, and to economic life. Unquestionably the most terrible consequence of earthquakes is the massive loss of human life which they are able to cause. The first task of earthquake protection is universally agreed to be reducing the loss of human life. The number and distribution of human casualties caused by earthquakes show the scale of the problem.

1.2.1 Casualties Around the World

Table 1.1 gives a list of confirmed or officially reported deaths in earthquakes in different countries around the world during the twentieth century. We know of at least 1248 lethal earthquakes during the twentieth century,[1] with a total of 1 685 000 officially reported deaths due to earthquakes. Over 40% of this total has occurred in a single country, namely China.

The total number of people actually killed by earthquakes is likely to be greater than the 1.7 million reported total. Small earthquakes causing only a few deaths may have gone unreported, and in 87 of the significant earthquakes reported this century, no figure for fatalities is officially available. Published estimates of fatalities may also be inaccurate, particularly in large events affecting many communities or in isolated areas. Some figures are also likely to be overestimates.

The risk to life from earthquakes is widespread. As Table 1.1 shows, at least 80 countries suffered life loss during the twentieth century. There also some other countries which are known to have suffered fatalities, sometimes on a large scale, in earlier centuries but which are not included in the list of countries suffering fatalities over the last 100 years. Future earthquakes may pose a significant threat in these countries. Large life loss is also widespread; half of all the countries which suffered any fatalities have had life loss running to thousands.

The extent of life loss in each country is primarily a function of the *severity* of life loss in individual earthquakes, rather than simply of the number of earthquakes experienced. Contrasting extreme examples from this list, the number of

[1] Authors' database of damaging earthquakes, 1900–2000.

Table 1.1 The world's earthquake countries: their loss of life, 1900–2000.

Rank	Country	No. of fatal earthquakes in 20th century	Total fatalities	No. of earthquakes killing more than 1000 people	No. of earthquakes killing more than 10 000 people	No. of earthquakes killing more than 100 000 people
1	China	170	619 488	21	7	2
2	Japan	84	169 525	10	1	1
3	Italy	45	128 031	6	2	
4	Iran	89	121 513	16	4	
5	Turkey	111	99 391	17	2	
6	Peru	62	76 016	3	1	
7	(former) USSR	44	75 813	8	3	
8	Pakistan	14	65 984	2	1	
9	Indonesia	66	43 992	5	2	
10	Chile	35	36 332	4	1	
11	India	21	33 329	3	3	
12	Venezuela	16	30 795	1	1	
13	Guatemala	16	25 345	2	1	
14	Afghanistan	15	23 312	4	1	
15	Mexico	48	17 625	3		
16	Nicaragua	4	13 718	3	1	
17	Morocco	2	12 013	1	1	
18	Nepal	3	11 853	1	1	
19	Taiwan	50	11 424	4		
20	Philippines	25	11 206	2		
21	Ecuador	22	9 303	4		
22	Greece	50	6 629	2		
23	Argentina	10	5 589	1		
24	Algeria	22	5 339	2		

25	Yemen	3	4 300	2
26	El Salvador	10	4 197	2
27	Colombia	33	3 734	1
28	Costa Rica	9	2 599	1
29	Romania	3	2 578	2
30	Papua New Guinea	8	2 329	1
31	Yugoslavia	16	2 008	1
32	Russia (since 1990)	1	1 989	1
33	USA	78	1 430	
34	Jamaica	2	1 003	1
35	Burma	7	675	
36	Egypt	4	576	
37	Albania	12	568	
38	Guinea	1	443	
39	Jordan	2	381	
40	Bulgaria	5	317	
41	Libya	1	300	
42	New Zealand	6	279	
43	Uganda	2	161	
44	Lebanon	1	136	
45	Portugal	3	122	
46	Puerto Rico	1	116	
47	Bolivia	3	111	
48	Dominican Republic	3	106	
49	Cyprus	4	94	
50	Turkmenistan	1	88	
51	Solomon Islands	4	81	
52	Ethiopia	3	72	
53	France	2	63	
54	Bangladesh	4	60	
55	Canada	3	57	
56	South Africa	7	53	

(continued overleaf)

Table 1.1 *(continued)*

Rank	Country	No. of fatal earthquakes in 20th century	Total fatalities	No. of earthquakes killing more than 1000 people	No. of earthquakes killing more than 10 000 people	No. of earthquakes killing more than 100 000 people
57	Sudan	2	52			
58	Zaire	3	33			
59	Azerbaijan	1	31			
60	Mongolia	1	30			
61	Ghana	1	22			
62	Iraq	1	20			
63	Tunisia	1	13			
64	Haiti	3	12			
65	Australia	1	11			
66	Malawi	1	10			
67	Cuba	2	9			
68	Fiji	1	8			
69	Honduras	2	8			
70	Spain	1	7			
71	Poland	2	6			
72	Croatia	1	5			
73	Brazil	2	4			
74	Former Czechoslovakia	1	3			
75	Tanzania	2	3			
76	Belgium	1	2			
77	Hungary	1	2			
78	The Netherlands	1	1			
79	Iceland	1	1			
80	Vanuatu	1	1			

Table 1.2 The twentieth century's most lethal earthquakes.

Rank	Fatalities	Year	Earthquake	Country	Magnitude
1	242 800	1976	Tangshan	China	7.8
2	234 120	1920	Kansu	China	8.5
3	142 807	1923	Kanto	Japan	8.3
4	83 000	1908	Messina	Italy	7.5
5	66 794	1970	Ancash	Peru	7.7
6	60 000	1935	Quetta	Pakistan	7.5
7	40 912	1927	Tsinghai	China	8.0
8	35 500	1990	Manjil	Iran	7.3
9	32 700	1939	Erzincan	Turkey	8.0
10	32 610	1915	Avezzano	Italy	7.5
11	28 000	1939	Chillan	Chile	7.8
12	25 000	1988	Armenia	USSR	6.9
13	23 000	1976	Guatemala	Guatemala	7.5
14	20 000	1905	Kangra	India	8.6
15	19 800	1948	Ashkhabad	USSR	7.3
16	17 118	1999	Kocaeli	Turkey	7.0
17	15 620	1970	Yunnan	China	7.5
18	15 000	1998	Afghanistan	Afghanistan	6.1
19	15 000	1917	Indonesia	Indonesia	N/A
20	15 000	1978	Tabas	Iran	7.4
21	15 000	1907	Tajikistan	USSR	8.1
22	12 225	1962	Buyin Zhara	Iran	7.3
23	12 100	1968	Dasht-e-Biyaz	Iran	7.3
24	12 000	1960	Agadir	Morocco	5.9
25	10 700	1934	Kathmandhu	Nepal	8.4

lethal earthquakes suffered by China is only double the number experienced by Greece, and yet the number of people killed is almost a thousand times greater.

The main contributors to the death toll are the small number of earthquakes which have caused large numbers of fatalities. Measured this way, the worst earthquakes of the twentieth century are listed in Table 1.2. The six worst events are responsible for almost exactly half of the total earthquake fatalities. A major reduction in the total number of people killed in earthquakes could be achieved if further repetitions of these extremely lethal events could be avoided. In order to avoid their repetition, it is first necessary to identify and understand the factors that made these events particularly lethal and then to work towards reducing these factors.

1.2.2 The Causes of Earthquake Fatalities

The statistics recording death due to earthquakes identify a wide range of earthquake-induced causes of death. Statistics include deaths from the fires following earthquakes, from tsunamis generated by off-shore events, from rockfalls, landslides and other hazards triggered by earthquakes. There are a wide range of other causes of death officially attributed to the occurrence of an earthquake,[2] ranging from medical conditions induced by the shock of experiencing ground motion, to accidents occurring during the disturbance, epidemic among the homeless and shootings during martial law. Any or all of these may be included in published death tolls from any particular earthquake.

It is clear from reports, however, that in most large-scale earthquake disasters, such as those in Table 1.2, the principal cause of death is the collapse of buildings. In earthquakes affecting a higher quality building stock, e.g. Japan and the United States, more fatalities are caused by the failure of non-structural elements or by earthquake-induced accidents than are killed in collapsing buildings, mainly because low proportions of buildings suffer complete collapse. Examples of failure of non-structural elements are pieces being dislodged from the exterior of buildings, the collapse of freestanding walls, or the overturning of building contents and equipment. Examples of earthquake-induced accidents include fire caused by the overturning of stoves, people falling from balconies or motor accidents.

Over the last century, about 75% of fatalities attributed to earthquakes have been caused by the collapse of buildings.[3] Figure 1.1 shows the breakdown of earthquake fatalities by cause for each half of this century. This shows that by far the greatest proportion of victims die in the collapse of masonry buildings. These are primarily weak masonry buildings (adobe, rubble stone or rammed earth) or unreinforced fired brick or concrete block masonry that can collapse even at low intensities of ground shaking and will collapse very rapidly at high intensities. These building types (one local example is shown in Figure 1.6) are common in seismic areas around the world and still today make up a very large proportion of the world's existing building stock.

Much of the increased populations in developing countries will continue to be housed in this type of structure for the foreseeable future. However, there are

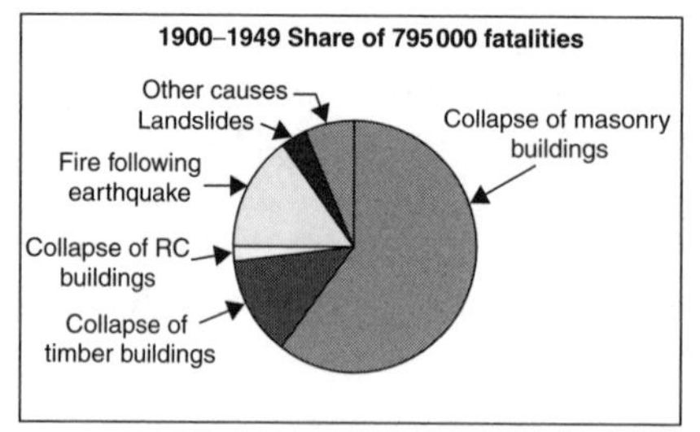

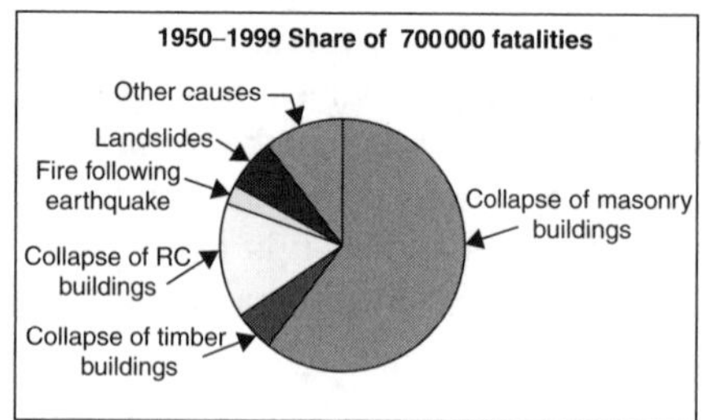

Figure 1.1 Breakdown of earthquake-related fatalities by cause

[2] See the list of causes of death due to the occurrence of an earthquake in Alexander (1984).

[3] Coburn *et al.* (1989).

Figure 1.2 The collapse of masonry buildings is the cause of most of the deaths in earthquakes around the world. The 1982 Dhamar Earthquake, Yemen Arab Republic

continuing changes in the types of buildings being constructed in many of the countries most at risk. Modern building materials, commercialisation of the construction industry and modernisation in the outlook of town and village dwellers are bringing about rapid changes in building stock. Brick and concrete block are common building materials in even the most remote areas of the world, and the wealthier members of rural communities who 20 or 30 years ago would have lived in weak masonry houses now live in reinforced concrete framed houses and apartment blocks.

Unfortunately, many of the reinforced concrete framed houses and apartment blocks built in the poorer countries are also highly vulnerable and, moreover, when they do collapse, they are considerably more lethal and kill a higher percentage of their occupants than masonry buildings. In the second half of the twentieth century most of the urban disasters involved collapses of reinforced concrete buildings and Figure 1.1 shows that the proportion of deaths due to collapse of reinforced concrete buildings is significantly greater than earlier in the century.

1.2.3 The World's Earthquake Problem is Increasing

On average, about 200 large-magnitude earthquakes occur in a decade – about 20 each year. Some 10% to 20% of these large-magnitude earthquakes occur in mid-ocean, a long way away from land and human settlements. Those that occur on land or close to the coast do not all cause damage: some happen deep in the earth's crust so that the dissipated energy is dispersed harmlessly over a wide area before it reaches the surface. Others occur in areas only sparsely inhabited and well away from towns or human settlements.

However, as the world's population grows and areas previously with small populations become increasingly densely settled, the propensity for earthquakes to cause damage increases. At the start of the century, less than one in three of large earthquakes on land killed someone. The number has gradually increased throughout the century, roughly in line with the world's population, until in the twenty-first century, two earthquakes in every three now kill someone. The increasing frequency of lethal earthquakes is shown in Figure 1.3.

But the annual rate of earthquake fatalities does show some signs of being reduced. Figure 1.1 shows that the total number of fatalities in the years 1950–1999 has averaged 14 000 a year – down from an average of 16 000 a year in the previous 50 years. And the number of earthquake-related fatalities in

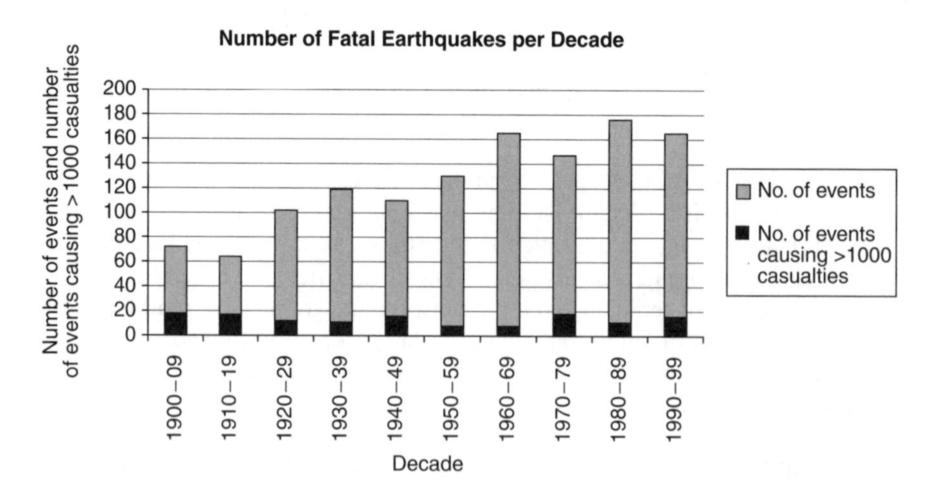

Figure 1.3 Number of fatal earthquakes per decade. This number has been increasing steadily over the last century. But the number per decade in which more than 1000 have been killed has remained roughly constant

the 1990s was 116 000, an average for the decade of 11 600 per year. Some of this reduction is undoubtedly due to beneficial changes: the reduction in fatalities from fire is largely due to changes in the Japanese building stock and successful measures taken by Japan to avoid conflagrations in its cities. And changes in building practices in some areas are making a significant proportion of buildings stronger than they used to be.

Nevertheless the present worldwide rate of reduction in vulnerability appears insufficient to offset the inexorable increase in population at risk. In the last decade the world's populationwas increasing by about 1.5% annually, i.e. doubling every 50 years or so, so the average vulnerability of the world's building stock needs to be falling at a reciprocal rate, i.e. halving every 50 years, simply for the average annual loss to be stabilised. The evidence suggests that although the average vulnerability of building stock is falling, it is not falling that quickly, so that the global risk of future fatalities is rising overall.

1.2.4 Urban Risk

Urban earthquake risk today derives from the combination of local seismicity – the likelihood of a large-magnitude earthquake – combined with large numbers of poorly built or highly vulnerable dwellings. A detailed analysis of the largest 800 cities in the world combining data on population, population growth rates, housing quality and global distribution of seismic hazard enables us to estimate the risks in all the large earthquake-prone cities, and compare them. Table 1.3 lists some of the world's most highly vulnerable cities and divides them into risk categories. Risk is here measured by the numbers of housing units which could be destroyed in the event of the earthquake with a 10% probability of exceedance in 50 years (approximately the once in 500 years earthquake). This assessment of loss is an indication of the overall risk, averaged out over a long period of time. The actual pattern of loss is likely to consist of long periods (a century or more) with small losses, with occasional catastrophic losses. Of the 29 cities in the three highest risk categories, only 8 cities (6 in Japan and 2 in the United States) are in the high-income group of countries; the 21 others are all in the middle- or low-income group of countries.

It is clear from both Table 1.1 and Table 1.3 that the risk today is polarising, with industrialised countries obtaining increasing levels of safety standards in their building stock while the increasing populations of developing countries become more exposed to potential disasters. This polarisation is worth examining in a little more detail.

1.2.5 Earthquake Vulnerability of Rich and Poor Countries

Earthquakes causing the highest numbers of fatalities tend to be those affecting high densities of the most vulnerable buildings. In many cases, the most vulnerable building stock is made up of low-cost, low-strength buildings. Some idea

Table 1.3 Cities at risk: the cities across the world with the highest numbers of dwellings likely to be destroyed in the '500-year' earthquake.

Name	Country	Population, 2002 (thousands)
Category A (over 25 000 dwellings destroyed in '500-year' earthquake)		
Guatemala City	Guatemala	1 090
Izmir	Turkey	2 322
Kathmandu	Nepal	712
Kermanshah	Iran	771
San Salvador	El Salvador	496
Shiraz	Iran	1 158
Tokyo	Japan	8 180
Yokohama	Japan	3 220
Category B (between 10 000 and 25 000 dwellings destroyed in '500-year' earthquake)		
Acapulco	Mexico	632
Kobe	Japan	1 517
Lima	Peru	7 603
Mendoza	Argentina	969
Mexicali	Mexico	575
Piura	Peru	359
San Juan	Argentina	439
Trujillo	Peru	600
Category C (between 5000 and 10 000 dwellings destroyed in '500-year' earthquake)		
Beijing	China	7 127
Bogota	Colombia	6 680
Chiba	Japan	902
Izmit	Turkey	262
Kawasaki	Japan	1 271
Manila	Philippines	10 133
San Francisco	USA	805
San Jose	USA	928
Sendai	Japan	1 022
Tehran	Iran	7 722
Tianjin	China	4 344
Valparaiso	Chile	301
Xi'an	China	2 656

The figures are derived from several sources of data. The '500-year' earthquake hazard for the city is based on the zoning of the 10% probability of exceedance in 50 years in the GSHAP map (http://seismo.ethz.ch/GSHAP/); this is combined with recent population figures from the world gazetteer (www.world-gazetteer.com), and average household sizes from UN data (UNCHS, 2001); estimates of the vulnerability of each city's building stock are based on information compiled by the authors from earthquake vulnerability surveys, recent earthquake loss experience and a variety of local sources of information. The resulting estimates are very approximate.

of the cost and quality of building stock involved in these fatal events can be obtained by comparing the economic costs inflicted by the earthquakes (chiefly the cost of destroyed buildings and infrastructure) with human fatalities. This is presented in Figure 1.4, for the countries most affected by earthquakes in the twentieth century.[4]

The highest casualties are generally those affecting low-cost construction. In Figure 1.4, the economic losses incurred range from $1000 of damage for every life lost (China) to over $1 million worth of damage for every life lost (USA). The location of individual countries on this chart is obviously a function of their seismicity as well as the vulnerability to collapse of their building stock and the degree of anti-seismic protection of their economic investment. The most earthquake-prone countries will be found towards the top right-hand corner of the chart, and the least towards the bottom left corner. Richer countries will lie above the diagonal joining these corners, poorer countries below it.

In general, high-seismicity countries want to reduce both their total casualties and their economic losses. In order to do this, those concerned with earthquake

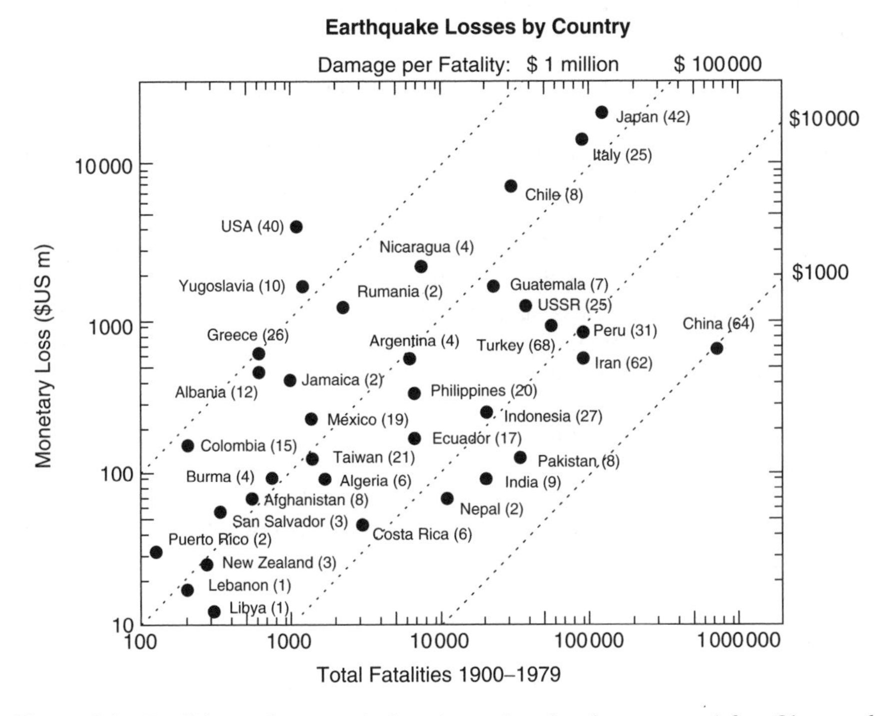

Figure 1.4 Fatalities and economic loss in earthquakes by country (after Ohta *et al.* 1986)

[4] After Ohta *et al.* (1986).

protection need first of all to understand some of the technical aspects of earthquake occurrence and the terminology associated with seismology, the study of earthquakes. There are a large number of books that explain earthquake mechanics in far greater detail than is possible here, and a number are listed in the suggestions for further reading at the end of the chapter. But some of the principles of earthquake occurrence are worth summarising here, to explain the terminology which will appear in later chapters.

1.3 Earthquakes

1.3.1 Geographical Distribution of Earthquakes

The geographical distribution of earthquake activity in the earth's crust is seen from the global seismic hazard map shown in Plate I. The map shows the distribution of expected seismicity across the earth's surface, measured by the expected intensity of shaking over a given time.[5] The concentration of seismicactivity in particular zones can be clearly seen. Two features of this map are worth elaborating.

1. Running down the western side of the Pacific Ocean from Alaska in the north to New Zealand in the south is a series of seismic *island arcs* associated with the Aleutian Islands, Japan, the Philippines and the islands of South East Asia and the South Pacific; a similar island arc runs through the Caribbean and another surrounds Greece.
2. Two prominent earthquake belts are associated with active mountain building at continental margins: the first is on the eastern shores of the Pacific stretching the length of the Americas, and the second is the trans-Asiatic zone running east–west from Myanmar through the Himalayas and the Caucasus Mountains to the Mediterranean and the Alps.

In addition to these major sources of earthquake activity, through the middle of each of the great oceans (but not shown on the map) there is a line of earthquakes, which can be associated with underwater mountain ranges known as *mid-ocean ridges*. Elsewhere, earthquakes do occur, but the pattern of activity is less dense, and magnitudes are generally smaller.

Tectonic Earthquakes

Seismologists explain this complex mosaic of earthquake activity in terms of *plate tectonics*. The continents on the earth's surface consist of large areas of relatively

---

[5] The expected intensity of shaking at each location is measured by the peak horizontal ground acceleration with a 10% probability of exceedance in 50 years.

cohesive plates, forming the earth's structure, floating on top of the *mantle*, the hotter and more fluid layer beneath them. Convection currents in the mantle cause adjoining plates to move in different directions, resulting in relative movement where the two plates meet. This relative movement at the plate boundaries is the cause of earthquakes. The nature of the earthquake activity depends on the type of relative movement. At the mid-ocean ridges, the plates are moving apart. New molten rock swells up from below and forms new sea floor. These areas are called *spreading zones*. At some plate boundaries, the plates are in head-on collision with each other; this may create deep ocean trenches in which the rock mass of one plate is thrust below the rock mass of the adjacent plate. The result is mountain building associated with volcanic activity and large earthquakes which tend to occur at a considerable depth; these areas are called *subduction zones*. The ocean trenches associated with the island arcs and the western shores of South America are of this type. Some collision zones occur in locations where subduction is not possible, resulting in the formation of huge mountain ranges such as the Himalayas.

There are also some zones in which plates are moving parallel and in opposite directions to each other and the relative movement is primarily lateral. Examples of these are the boundary between the Pacific plate and the North American plate running through California, and the southern boundary of the Eurasian plate in Turkey; in these areas large and relatively shallow earthquakes occur which can be extremely destructive.

Subduction Zones

The mid-ocean ridges are the source of about 10% of the world's earthquakes, contributing only about 5% of the total seismic energy release. By contrast, the trenches contribute more than 90% of the energy in shallow earthquakes and most of the energy for deeper earthquakes as well. Most of the world's largest earthquakes have occurred in subduction zones.

Intra-plate Earthquakes

A small proportion of the energy release takes place in earthquakes located away from the plate boundaries. Most of such *intra-plate* earthquakes occur in continental zones not very far distant from the plate boundaries and may be the result of localised forces or the reactivation of old fault systems. They are more infrequent but not necessarily smaller than inter-plate earthquakes. Some large and highly destructive intra-plate earthquakes have occurred. The locations of intra-plate earthquakes are less easy to predict and consequently they present a more difficult challenge for earthquake protection.

An important consequence of the theory of plate tectonics is that the rate and direction of slip along any plate boundary should on average be constant over a period of years. In any given tectonic system, the total energy released in earthquakes or other dissipations of energy is therefore predictable, which helps to understand seismic activity and to plan protection measures. Likely locations of future earthquakes may sometimes be identified in areas where the energy known to have been released is less than expected. This *seismic gap theory* is a useful means of long-term earthquake prediction which has proved valuable in some areas. Earthquake prediction is discussed further in Chapter 3.

1.3.2 Causes of Earthquakes

Earthquakes tend to be concentrated in particular zones on the earth's surface, which coincide with the boundaries of the tectonic plates into which the earth's crust is divided. As the plates move relative to each other along the plate boundaries, they tend not to slide smoothly but to become interlocked. This interlocking causes deformations to occur in the rocks on either side of the plate boundaries, with the result that stresses build up. But the ability of the rocks to withstand these stresses is limited by the strength of the rock material; when the stresses reach a certain level, the rock tends to fracture locally, and the two sides move past each other, releasing a part of the built-up energy by *elastic rebound*.

Once started, the fracture tends to propagate along a plane – the rupture plane – until a region where the condition of the rocks is less critical has been reached. The size of the fault rupture will depend on the amount of stress build-up and the nature of the rocks and their faulting.

1.3.3 Surface Faulting

In most smaller earthquakes the rupture plane does not reach the ground surface, but in larger earthquakes occurring at shallow depth the rupture may break through at the earth's surface producing a crack or a ridge – a *surface break* – perhaps many kilometres long. A common misconception about earthquakes is that they produce yawning cracks capable of swallowing people or buildings. At the epicentre of a very large earthquake rupturing the surface on land – quite a rare event – cracks in the earth do occur and the ground either side of the fault can move a few centimetres, or in very large events a few metres, up or along. This is, of course, very damaging for any structure that is built straddling the rupture. During the few seconds of the earthquake, the ground is violently shaken and any fault rupture is likely to open up several centimetres in the shaking. There is a slight possibility that a person could be injured in the actual fault rupture, but by far the worst consequences of damage and injury come from the huge amounts of shaking energy released by the earthquake affecting areas of hundreds of square kilometres. This energy release may well cause landslides and ground cracking in areas of soft or unstable ground anywhere in the affected area, which can be confused with surface fault traces.

1.3.4 Fault Mechanisms; Dip, Strike, Normal

According to the direction of the tectonic movements at the plate boundary the fault plane may be vertical or inclined to the vertical – this is measured by the angle of dip – and the direction of fault rupture may be largely horizontal, largely vertical, or a combination of horizontal and vertical.

The different types of source characteristic do produce recognisably different shock-wave pulses, notably in the different directional components of the first moments of ground motion, but in terms of magnitude, intensity and spatial attenuation the different source mechanisms can be assumed fairly similar for earthquake protection planning.

1.3.5 Earthquake Waves

As the rocks deform on either side of the plate boundary, they store energy – and massive amounts of energy can be stored in the large volumes of rock involved. When the fault ruptures, the energy stored in the rocks is released in a few seconds, partly as heat and partly as shock waves. These waves are the earthquake. They radiate outwards from the rupture in all directions through the earth's crust and through the mantle below the crust as compression or *body* seismic waves. They are reflected and refracted through the various layers of the earth; when they reach the earth's surface they set up ripples of lateral vibration or seismic waves which also propagate outwards along the surface with their own characteristics. These *surface waves* are generally more damaging to structures than the body waves and other types of vibration caused by the earthquake. The body waves travel faster and in a more direct route so most sites feel the body waves a short time before they feel the stronger surface waves. By measuring the time difference between the arrival of body and surface waves on a seismogram (the record of ground motion shaking some distance away) seismologists can estimate the distance to the epicentre of a recorded earthquake.

1.3.6 Attenuation and Site Effects

As the waves travel away from the source, their amplitude becomes smaller and their characteristics change in other complex ways. Sometimes these waves can be amplified or reduced by the soils or rocks on or close to the surface at the site. Theground motion which we feel at any point is the combined result of the source characteristics of the earthquake, the nature of the rocks or other media through which the earthquake waves are transmitted, and the interaction with the site effects.

A full account of earthquake waves and their propagation is outside the scope of this book, but is well covered elsewhere.[6] The effect of site characteristics on the nature and effects of earthquake ground motion is further discussed in Chapter 7.

---

[6] See e.g. Bolt (1999).

Not all earthquakes are tectonic earthquakes of the type described here. A small but important proportion of all earthquakes occur away from plate boundaries. These include some very large earthquakes and are the main types of earthquakes occurring in many of the medium- and low-seismicity parts of the world. The exact mechanisms giving rise to such intra-plate earthquakes are still not clearly established. It is probable that they too are associated with faulting, though at depth; as far as their effects are concerned they are indistinguishable from tectonic earthquakes.

Earthquakes can also be associated with volcanic eruptions, the collapse of underground mine-workings, and human-made explosions. Generally earthquakes of each of these types will be of very much smaller size than tectonic earthquakes, and they may not be so significant from the point of view of earthquake protection.

1.3.7 Earthquake Recurrence in Time

Given the nature of the large geological processes causing earthquakes, we can expect that each earthquake zone will have a rate of earthquake occurrence associated with it. Broadly, this is true, but as the rocks adjacent to plate boundaries are in a constant state of change, a very regular pattern of seismic activity is rarely observed. In order to observe the pattern of earthquake recurrence in a particular zone, a long period of observation must be taken, longer in most cases than the time over which instrumental records of earthquakes have been systematically made. A statistical study of earthquake occurrence patterns, using both historical data and recent data from seismological instruments, can enable us to determine average return periods for earthquakes of different sizes (see Figure 1.5). This is the approach which has been used to develop the global seismic hazard map shown in Plate I and is discussed further in Chapter 7.

1.3.8 Severity and Measurement of Earthquakes

The size of an earthquake is clearly related to the amount of elastic energy released in the process of fault rupture. But only indirect methods of measuring this energy release are available, by means of seismic instruments or the effects of the earthquake on people and their environment.

The terms magnitude and intensity tend to be confused by non-specialists in discussing the severity of earthquakes. The *magnitude* of an earthquake is a measure of its total size, the energy released at its source as estimated from instrumental observations. The *intensity* of an earthquake is a measure of the severity of the shaking of the ground at a particular location. 'Magnitude' is a term applied to the earthquake as a whole whereas 'intensity' is a term applied to a site affected by an earthquake, and any earthquake causes a range of intensities at different sites.

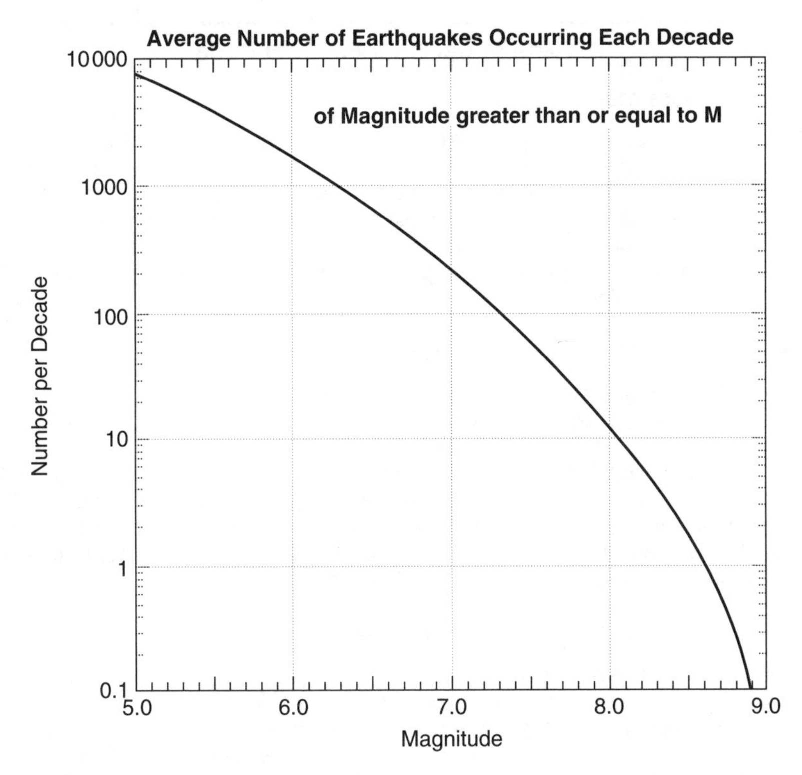

Figure 1.5 Average recurrence rate of earthquakes of different magnitudes worldwide (after Båth 1979)

1.3.9 Earthquake Magnitude

A number of magnitude scales are in use. The oldest is the *Richter* magnitude (M_l) scale, defined by Charles Richter in 1936. It is based on the logarithm of the amplitude of the largest swing recorded by a standard seismograph. Because earthquakes of different types cause different forms of seismic wave trains, more detailed measurements include *body wave magnitude* (m_b) and *surface wave magnitude* (M_s), based on the amplitudes of different parts of the observed wave train. In general, the definition of magnitude which best correlates with the surface effects of earthquakes is the surface wave magnitude M_s, since it is the surface waves which are most destructive to buildings. There are a number of correlations between the different magnitude definitions.

Because magnitude scales are derived from the logarithm of the seismograph amplitude, the amount of energy released in an earthquake is not a simple function of the magnitude – each unit on the Richter scale represents a 32-fold increase in the energy released.

A guide to earthquake magnitude

Magnitude less than 4.5

Magnitude 4.5 represents an energy release of about 10^8 kilojoules and is the equivalent of about 10 tonnes of TNT being exploded underground. Below about magnitude 4.5, it is extremely rare for an earthquake to cause damage, although it may be quite widely felt. Earthquakes of magnitude 3 and magnitude 2 become increasingly difficult for seismographs to detect unless they are close to the event. A shallow earthquake of magnitude 4.5 can generally be felt for 50 to 100 km from the epicentre.

Magnitude 4.5 to 5.5 – local earthquakes

Magnitude 5.5 represents an energy release of around 10^9 kilojoules and is the equivalent of about 1000 tonnes of TNT being exploded underground. Earthquakes of magnitude 5.0 to 5.5 may cause damage if they are shallow and if they cause significant intensity of ground shaking in areas of weaker buildings. Earthquakes up to magnitudes of about 5.5 can occur almost anywhere in the world – this is the level of energy release that is possible in normal non-tectonic geological processes such as weathering and land formation. An earthquake of magnitude 5.5 may well be felt 100 to 200 km away.

Magnitudes 6.0 to 7.0 – large magnitude events

Magnitude 6 represents an energy release of the order of 10^{10} kilojoules and is the equivalent of exploding about 6000 tonnes of TNT underground. A magnitude 6.3 is generally taken as being about equivalent to an atomic bomb being exploded underground. A magnitude 7.0 represents an energy release of 10^{12} kilojoules. Large-magnitude earthquakes, of magnitude 6.0 and above, are much larger energy release associated with tectonic processes. If they occur close to the surface they may cause intensities at their centre of VIII, IX or even X, causing very heavy damage or destruction if there are towns or villages close to their epicentre. Some of these large-magnitude earthquakes, however, are associated with tectonic processes at depth and may be relatively harmless to people on the earth's surface. There are about 200 large-magnitude events somewhere in the world each decade. A magnitude 7.0 earthquake at shallow depth may be felt at distances 500 km or more from its epicentre.

Magnitudes 7.0 to 8.9 – great earthquakes

A magnitude 8 earthquake releases around 10^{13} kilojoules of energy, equivalent to more than 400 atomic bombs being exploded underground, or almost as much as a hydrogen bomb. The largest earthquake yet recorded, magnitude 8.9, released 10^{14} kilojoules of energy. Great earthquakes are the massive energy releases caused by long lengths of linear faults rupturing in one break. If they occur at shallow depths they cause slightly stronger epicentral intensities than large-magnitude earthquakes but their great destructive potential is due to the very large areas that are affected by strong intensities.

Very sensitive instruments can record earthquakes with magnitudes as low as −2, the equivalent of a brick being dropped from the table to the ground. The energy released from an earthquake is similar to an explosive charge being detonated underground, with magnitude being the measure of the energy released.

In the guide to magnitude (see box), an explosive equivalent of each magnitude level is given as a rough guide. The destructive effects at the earth's surface of the energy released are also affected by the depth of the earthquake: energy released close to the surface will be more destructive on the area immediately above it, and a deep energy release will affect a wider area above, but the energy will be more dissipated and the effects weaker.

1.3.10 Limits to Magnitude

The larger the area of fault that ruptures, and the bigger the movement that takes place in one thrust, the greater the amount of energy released. The length of the fault and its depth determine the area of its rupture: in practice the depth of rupture is constrained by the depth of the earth's solid crust, so the critical parameter in determining the size of earthquake is the length of the fault rupture that takes place. The tectonic provinces where long, uninterrupted fault lengths exist are limited, and are by now fairly well defined. The limits to magnitude appear to be the sheer length of fault that could possibly unzip in one single rupture. The largest magnitude earthquake yet recorded measured 8.9, rupturing over 200 continuous kilometres down the coast of Chile.

Because of this tendency for magnitude scales to saturate at about 9, seismologists have developed a new measure of the magnitude of an earthquake which derives more directly from the source characteristics. *Seismic moment* is defined by the rigidity of the rocks, multiplied by the area of faulting, multiplied by the amount of the slip. Seismic moment can be inferred from instrument readings, and for larger earthquakes checked by observations of the surface fault trace. Based on seismic moment, a *moment magnitude* (M_w) has been defined which correlates well with other measures of magnitude over a range of magnitudes.

1.3.11 Intensity

Intensity is a measure of the felt effects of an earthquake rather than the earthquake itself. It is a measure of how severe the shaking was at any location. For any earthquake, the intensity is strongest close to the epicentre and attenuates away with distance from the source of the earthquake. Larger magnitude earthquakes produce stronger intensities at their epicentres. Intensity mapping showing *isoseismals*, or lines of equal intensity, is normally carried out after each damaging earthquake by the local geological survey. Isoseismal maps of

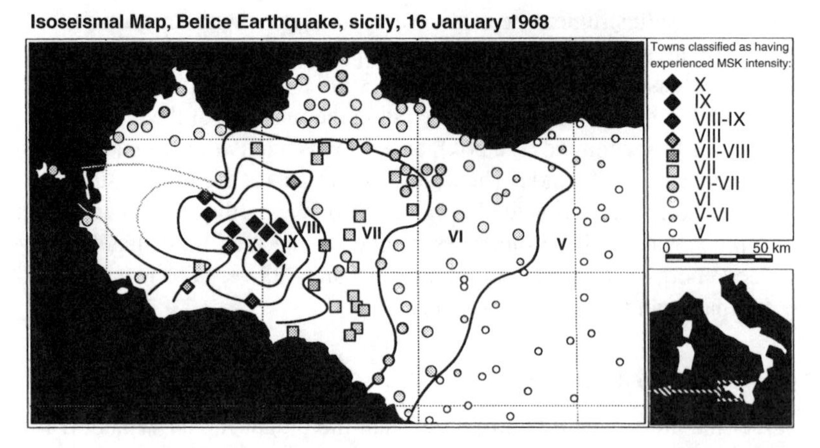

Figure 1.6 An example of an isoseismal map: the Belice earthquake, 1968, Sicily, Italy, using the MSK intensity scale (after Cosentino and Mulone, in Barbano *et al.* 1980)

past events play an important part in the estimation of the probable occurrence of future earthquakes. An example of an isoseismal map is shown in Figure 1.6.

Intensity is assessed by classifying the degree of shaking severity using an intensity scale. The intensity level is assigned for a particular location from the visible consequences left by the earthquake and from reports by those who experienced the shaking. The level of intensity is identified by a Roman numeral commonly on a scale from I to X (or even up to XII), indicating that the scale describes a succession of states but is not numerical. An example of an intensity scale, the definitions of the EMS 98 intensity scale, are given in the box. It may be worth noting that intensities of degree X are rare, and the higher degrees, XI and XII, have rarely, if ever, been scientifically verified.

The European Macroseismic Scale 1998: definitions of intensity[7]

Note: the arrangement of the scale is: (a) effects on humans, (b) effects on objects and on nature, (c) damage to buildings.

Intensity I: Not felt

(a) Not felt, even under the most favourable circumstances.
(b) No effect.
(c) No damage.

[7]Based on Grünthal (1998).

Intensity II: Scarcely felt

(a) The tremor is felt only at isolated instances (<1%) of individuals at rest and in a specially receptive position indoors.
(b) No effect.
(c) No damage.

Intensity III: Weak

(a) The earthquake is felt indoors by a few. People at rest feel a swaying or light trembling.
(b) Hanging objects swing slightly.
(c) No damage.

Intensity IV: Largely observed

(a) The earthquake is felt indoors by many and felt outdoors only by very few. A few people are awakened. The level of vibration is not frightening. The vibration is moderate. Observers feel a slight trembling or swaying of the building, room or bed, chair, etc.
(b) China, glasses, windows and doors rattle. Hanging objects swing. Light furniture shakes visibly in a few cases. Woodwork creaks in a few cases.
(c) No damage.

Intensity V: Strong

(a) The earthquake is felt indoors by most, outdoors by few. A few people are frightened and run outdoors. Many sleeping people awake. Observers feel a strong shaking or rocking of the whole building, room or furniture.
(b) Hanging objects swing considerably. China and glasses clatter together. Small, top-heavy and/or precariously supported objects may be shifted or fall down. Doors and windows swing open or shut. In a few cases window panes break. Liquids oscillate and may spill from well-filled containers. Animals indoors may become uneasy.
(c) Damage of grade 1 to a few buildings of vulnerability class A and B.

Intensity VI: Slightly damaging

(a) Felt by most indoors and by many outdoors. A few persons lose their balance. Many people are frightened and run outdoors.
(b) Small objects of ordinary stability may fall and furniture may be shifted. In a few instances dishes and glassware may break. Farm animals (even outdoors) may be frightened.
(c) Damage of grade 1 is sustained by many buildings of vulnerability class A and B; a few of class A and B suffer damage of grade 2; a few of class C suffer damage of grade 1.

Intensity VII: Damaging

(a) Most people are frightened and try to run outdoors. Many find it difficult to stand, especially on upper floors.
(b) Furniture is shifted and top-heavy furniture may be overturned. Objects fall from shelves in large numbers. Water splashes from containers, tanks and pools.

(c) Many buildings of vulnerability class A suffer damage of grade 3, a few of grade 4. Many buildings of vulnerability class B suffer damage of grade 2, a few of grade 3. A few buildings of vulnerability class C sustain damage of grade 2. A few buildings of vulnerability class D sustain damage of grade 1.

Intensity VIII: Heavily damaging

(a) Many people find it difficult to stand, even outdoors.
(b) Furniture may be overturned. Objects like TV sets, typewriters, etc., fall to the ground. Tombstones may occasionally be displaced, twisted or overturned. Waves may be seen on very soft ground.
(c) Many buildings of vulnerability class A suffer damage of grade 4, a few of grade 5. Many buildings of vulnerability class B suffer damage of grade 3, a few of grade 4. Many buildings of vulnerability class C suffer damage of grade 2, a few of grade 3. A few buildings of vulnerability class D sustain damage of grade 2.

Intensity IX: Destructive

(a) General panic. People may be forcibly thrown to the ground.
(b) Many monuments and columns fall or are twisted. Waves are seen on soft ground.
(c) Many buildings of vulnerability class A sustain damage of grade 5. Many buildings of vulnerability class B suffer damage of grade 4, a few of grade 5. Many buildings of vulnerability class C suffer damage of grade 3, a few of grade 4. Many buildings of vulnerability class D suffer damage of grade 2, a few of grade 3. A few buildings of vulnerability class E sustain damage of grade 2.

Intensity X: Very destructive

(c) Most buildings of vulnerability class A sustain damage of grade 5. Many buildings of vulnerability class B sustain damage of grade 5. Many buildings of vulnerability class C suffer damage of grade 4, a few of grade 5. Many buildings of vulnerability class D suffer damage of grade 3, a few of grade 4. Many buildings of vulnerability class E suffer damage of grade 2, a few of grade 3. A few buildings of vulnerability class F sustain damage of grade 2.

Intensity XI: Devastating

(c) Most buildings of vulnerability class B sustain damage of grade 5. Most buildings of vulnerability class C suffer damage of grade 4, many of grade 5. Many buildings of vulnerability class D suffer damage of grade 4, a few of grade 5. Many buildings of vulnerability class E suffer damage of grade 3, a few of grade 4. Many buildings of vulnerability class F suffer damage of grade 2, a few of grade 3.

Intensity XII: Completely devastating

(c) All buildings of vulnerability class A, B and practically all of vulnerability class C are destroyed. Most buildings of vulnerability class D, E and F are destroyed. The earthquake effects have reached the maximum conceivable effects.

Definitions of quantity

Few means less than about 15%; many means from about 15% to about 55%; most means more than about 55%.

Classification of damage to masonry buildings[8]

Grade 1: Negligible to slight damage (no structural damage, slight non-structural damage)
Hair-line cracks in very few walls. Fall of small pieces of plaster only. Fall of loose stones from upper parts of buildings in very few cases.

Grade 2: Moderate damage (slight structural damage, moderate non-structural damage)
Cracks in many walls. Fall of fairly large pieces of plaster. Partial collapse of chimneys.

Grade 3: Substantial to heavy damage (moderate structural damage, heavy non-structural damage)
Large and extensive cracks in most walls. Roof tiles detach. Chimneys fracture at the roof line; failure of individual non-structural elements (partitions, gable walls).

Grade 4: Very heavy damage (heavy structural damage, very heavy non-structural damage)
Serious failure of walls, partial structural failure of roofs and floors.

Grade 5: Destruction (very heavy structural damage)
Total or near total collapse.

Classification of damage to buildings of reinforced concrete

Grade 1: Negligible to slight damage (no structural damage, slight non-structural damage) (Figure 1.8b)
Fine cracks in plaster over frame members or in walls at the base. Fine cracks in partitions and infills.

Grade 2: Moderate damage (slight structural damage, moderate non-structural damage) (Figure 1.8c)
Cracks in columns and beams of frames and in structural walls. Cracks in partition and infill walls; fall of brittle cladding and plaster. Falling mortar from the joints of wall panels.

Grade 3: Substantial to heavy damage (moderate structural damage, heavy non-structural damage) (Figure 1.8d)
Cracks in columns and beam column joints of frames at the base and at joints of coupled walls. Spalling of concrete cover, buckling of reinforced rods. Large cracks in partition and infill walls, failure of individual infill panels.

Grade 4: Very heavy damage (heavy structural damage, very heavy non-structural damage) (Figure 1.8e)
Large cracks in structural elements with compression failure of concrete and fracture of rebars; bond failure of beam reinforced bars; tilting of columns. Collapse of a few columns or of a single upper floor.

Grade 5: Destruction (very heavy structural damage) (Figure 1.8f)
Collapse of ground floor or parts (e.g. wings) of buildings.

---

[8] Damage grades 1 to 5 as defined in this scale are referred to elsewhere in this text as damage levels D1 to D5.

Classification of typical vulnerability classes

Class A: rubble stone, fieldstone, adobe
Class B: simple stone, unreinforced masonry with manufactured masonry units
Class C: massive stone, unreinforced masonry with RC floors; RC frame or walls without ERD
Class D: reinforced or confined masonry, RC frame or wall with moderate ERD, timber structure
Class E: RC frame or wall with high ERD, steel structure

But vulnerability could be one class higher or one or two classes lower according to standard of construction.[9]
Note: ERD = earthquake-resisting design.

There are a large number of intensity scales, most of which have been modifications or adaptations of previous scales, and originate from the attempts of early seismologists to classify the effects of earthquake ground motion without instrumental measurements. The most common ones in use today include the Modified Mercalli (MM) scale, a 12- point scale mainly in use in United States; the European Macroseismic Scale (EMS), a development from the MM scale now used more in Europe and given as an example in the box; the Japanese Meteorological Agency (JMA) scale, a seven-point scale used in Japan; and other scales similar to the MM scale are used in the former USSR and in China for their own building types. The evolution of these various intensity scales is summarised in Figure 1.7.

Nowadays, intensity scales are primarily used to make rapid evaluations of the scale and geographical extent of a damaging earthquake in initial reconnaissance, to guide the emergency services.

1.4 Earthquake Protection

The term *earthquake protection*, as used in this book, refers to the total scope of all those activities which can be taken to alleviate the effects of earthquakes, or to reduce future losses, whether in terms of human casualties or physical or economic losses. The term is similar in meaning to the more widely used expression *earthquake risk mitigation*, although this usually refers primarily to interventions to strengthen the built environment, whereas earthquake protection is taken to include the human, financial, social and administrative aspects of reducing earthquake effects.

[9]For a more detailed definition of the vulnerability classes, see the vulnerability table and the guidelines given in the European Macroseismic Scale document (Grünthal, 1998).

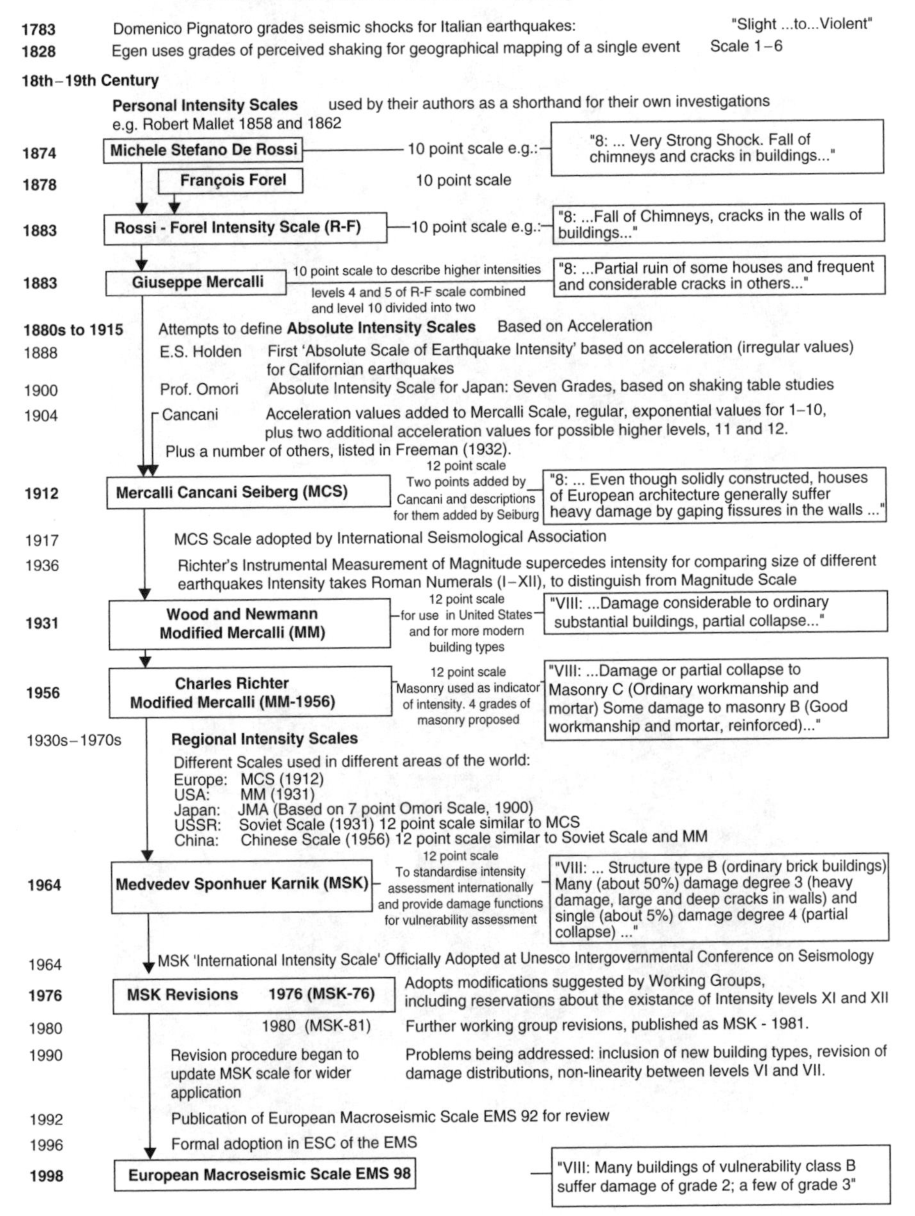

Figure 1.7 The genealogy of intensity scales

Figure 1.8(a) EMS damage state D0 (undamaged)

Figure 1.8(b) EMS damage state D1 (slight damage)

Figure 1.8(c) EMS damage state D2 (moderate damage)

Figure 1.8(d) EMS damage state D3 (heavy damage)

Figure 1.8(e) EMS damage state D4 (very heavy damage/partial collapse)

Figure 1.8(f) EMS damage state D5 (destruction)

Figure 1.8 Damage to mid-rise reinforced concrete frame buildings in the 1999 Kocaeli earthquake in Turkey, in relation to the EMS damage states defined on p. 25

1.4.1 Self-protection Measures

There is no doubt that in some areas of the world where earthquakes are a common occurrence, people do take some basic actions to protect themselves without any external prompting. They build their houses more robustly than elsewhere, using materials which are able to resist some degree of ground motion without damage, and they avoid sites which previous disasters have shown to be dangerous because of landslides, rockfalls or tsunamis. The culture and traditions of such areas are often full of references to past disasters which help to maintain present-day earthquake awareness. Earthquake damage surveys from many parts of the world have often reported unexpectedly good performance by vernacular structures, and it has been suggested that the awareness of the earthquake risk has been incorporated into the traditional form of construction of these buildings.

There are a number of reported examples of traditional construction techniques that may have evolved within certain communities as a response to repeated occurrences of earthquakes. Such examples include:

- The construction of energy-absorbing timber frame joints in traditional Japanese construction.
- Traditional timber reinforcement in weak masonry construction in the Alpine–Himalayan seismic belt.[10]
- Roof systems supported on a dual structure of walls and posts, allowing posts to keep the roof up when walls collapse in earthquakes thereby preventing injury to the occupants.[11]
- Composite earth-and-timber vernacular structures in a number of earthquake-prone areas that combine heavy mass for thermal insulation with the resilience and ductility of a timber frame structure.[12]
- The use of arches, domes and vaults which appear to suffer less earthquake damage by transmitting lateral forces safely.[13]

---

[10] The use of horizontal timber-strengthening elements in traditional masonry construction along the Alpine–Himalayan seismic belt from Southern Europe through the Middle East (*hatil* construction) to the Indian Subcontinent (Arya and Chandra 1977) has been attributed to the earthquake-resisting properties of this construction type in Ergünay and Erdik (1984). It also has other attributes, including adding general stability to the construction, which may also encourage its widespread adoption in these regions.

[11] The safe collapse of walls in earthquakes while roofs are supported on extra posts has been noted in a number of earthquake reports, including Ambraseys *et al.* (1975) report of the Patan earthquake in Pakistan, and the characteristics of the traditional Bali Balinesian hut, described in LINUH (1976) which allows a thatched roof to shed its mud walls in an earthquake without collapsing.

[12] For example, the *quincha* construction in Peru and other parts of Latin America and the use of *Bagdadi* construction in Iran and elsewhere.

[13] Several earthquake reports from Iran and elsewhere have noted that traditional dome construction, particularly quasi-spherical domes, and arches have remained relatively undamaged in regions of heavy destruction; an example is in Ambraseys *et al.* (1969) reporting the Iran, Dasht-e-Bayaz, earthquake in 1968.

There are also many examples of ancient earthquake engineering knowledge for more monumental structures, including the construction of pendulum-like central posts in pagodas in China,[14] anti-seismic engineering for temples in Ancient Greece[15] and earthquake-resistant reinforcement of monuments, mosques, minarets and other structures of Ottoman architecture[16] throughout the Middle East. Other historical accounts of protection measures include the legislation measures enacted by the Neapolitancourt during the seventeenth century[17] and the numerous attempts in the nineteenth century by the City Fathers of San Francisco to protect the city against future earthquakes.[18]

This evolution of construction techniques by communities increasingly adopting the building types that perform well in successive earthquakes has been dubbed 'Architectural Darwinism', the survival of the fittest building methods.[19] There is no doubt that earthquakes and other disasters can act as powerful prompts, causing a community to change its construction practices, adopt new and safer building types and to pass new legislation to protect itself. It is even argued that change *only* comes about as a result of a major disaster, with most of the advances in disaster protection in a community attributable to a major disaster in the past.[20]

But many of the most damaging earthquakes of the last few decades have occurred in locations where there is no general public awareness of the earthquake risk, either because they have been recently settled, or because the interval since the last large earthquake is many centuries. In these cases[21] the earthquake tends to be particularly disastrous.

Thus, where self-protection happens, it can make some contribution to providing an adequate level of protection, and it is useful to be aware of the extent of earthquake awareness and self-protection which exists. But self-protection cannot always be assumed to take place, and even where it does, it is very unlikely that self-protection alone will provide adequate protection. Some degree of action by

---

[14] Needham (1971) has suggested that the knowledge of the superior earthquake resistance of timber was learned early by Chinese craftsmen.

[15] Excavations and reconstructions of classical Greek temples reveal iron cramping of stone blocks and pre-loading of foundations to create monolithic foundations that would withstand earthquake waves.

[16] Mosque design by the famous sixteenth-century Ottoman architect Sinan included chain reinforcements around domes and towers to resist earthquake forces.

[17] Tobriner (1984).

[18] Tobriner in NCEER (1989).

[19] Wood (1981).

[20] Davis (1983).

[21] Cases of earthquakes recurring unexpectedly and disastrously include Tangshan in China 1976, the Leninakan region of Armenia in 1988, the Dhamar area of Yemen in 1982, and the 1995 Kobe earthquake in Japan.

national, regional and local authorities can be assumed to be necessary wherever earthquakes are a known or a potential hazard.

1.4.2 Vulnerability and Protection

Any discussion of earthquake protection must attempt to identify the distribution of vulnerability in any society, and across the world. It is clear from the earlier discussion that earthquake vulnerability is heavily concentrated in the poorer developing countries of the world. Consequently the book will place particular emphasis on earthquake protection policies which can be of application in countries with limited resources. In such countries it rarely makes sense to attempt to implement earthquake protection as an activity separate from other measures to improve the general living conditions of the most economically vulnerable groups.

Likewise, there is evidence that even in the wealthiest countries, there is significant earthquake vulnerability among the poorest members of society, who are forced to live in old weak buildings because this is the only accommodation they can afford. Methods of upgrading these buildings are becoming available and better understood, and they will be discussed in later chapters. But it is essential not to overlook the political dimension of allocating priorities for earthquake protection within a society in which all members feel vulnerable, and recent experiences in implementing protection policies will be described.

One of the key questions for any society to determine is what level of protection it should attempt to provide. Earthquake protection is costly and must compete for limited resources with other priorities for individual and public expenditure, such as health care and environmental protection. In common with many other areas of expenditure it is very difficult to define with any precision what benefits are purchased by any given expenditure. Often earthquakes are seen as a remote threat, unlikely to occur within the planning timescale of governments, adult taxpayers or corporations, and even then very unlikely to be fatal; and it is difficult to raise public enthusiasm for spending money on protection except in the immediate aftermath of an earthquake. Overspending on protection will waste resources, restricting economic development and economic growth, and these opportunity costs are easier to perceive. The question of setting the right level of protection and how to evaluate alternative protection strategies is therefore one of the topics which the book will discuss.

Another matter which will be considered is whose responsibility it is to take initiatives and to pay for protection. Apart from the individual or corporate property owner, concern for the effects of earthquakes is also experienced by local community groups, local government, and regional and national governments. International agencies are also involved, particularly in the activity of post-earthquake relief. The community at each of these levels will benefit from improved earthquake protection, and needs to be drawn into a comprehensive and effective protection strategy. At the lowest end, individuals and community groups have the smallest probability of experiencing a disaster, and the least resources to implement a protection strategy; on the other hand, perhaps only community-based groups can effectively determine priorities for protection. At the upper end national governments at the same time both face the greatest risk of a disaster and have potentially the greatest financial and legislative resources to implement protection, but without the active support at the level of individual or community-based action, earthquake protection cannot succeed. Strategies and actions appropriate to all levels of decision-making are described in this book.

1.5 Organisation of the Book

The following chapter, Chapter 2, discusses the costs of earthquakes: what is lost, who pays, and how risks are being measured and shared in the newly developing international risk transfer market. Each of the following five chapters then deals with a separate aspect of earthquake protection.

Chapter 3 deals with earthquake preparedness. The evidence shows that if the public can be made aware of the risk of an impending earthquake, and trained to know how to act when an earthquake strikes, the casualties will be considerably smaller than if the earthquake strikes an unprepared community, regardless of any additional action that might be taken to strengthen buildings. The chapter discusses the present state of earthquake prediction and how this might be used to improve public preparedness. Actions which can be taken in advance of expected earthquakes such as training in emergency procedures and the role of evacuation are also considered. Developing an earthquake safety culture is the key to success.

Chapter 4 looks at the earthquake emergency itself, and examines what can be done to reduce losses by the operation of effective disaster plans and by facilitating speedy search and rescue operations. Detailed aspects of the way buildings are designed are shown to have a crucial influence on the survival chances of those caught in damaged or collapsed buildings.

Chapter 5 deals with post-earthquake recovery and reconstruction. It is clear that the immediate aftermath of one earthquake provides the best opportunity for building-in protection from future earthquakes, in the damaged area itself and in adjacent areas – an opportunity which is often lost through lack of awareness of the appropriate actions. The appropriate and politically acceptable response is likely to be different in areas where the interval between damaging earthquakes is a few decades from that where the expected interval is measured in centuries, as a number of reconstruction case studies show.

Chapter 6 is concerned with defining the roles and strategies appropriate to the different groups acting to protect themselves and society as a whole. Measures suitable for individuals, households and neighbourhood community groups are discussed first, then suitable measures for private companies or organisations are itemised. The role of urban authorities in developing earthquake protection programmes at a city level is considered. Then national government activities and priorities for implementing protection measures are presented and it is argued that it is necessary for government to take a lead role in instigating a safety culture. Finally measures for international and national aid and development organisations are considered.

Chapter 7 presents the effects of siting and location on earthquake risk. It describes the use of seismic hazard maps to support decisions on earthquake protection, especially building design regulations, and it discusses the use of microzoning techniques for earthquake protection in urban areas.

Chapter 8 considers the means available for improving the earthquake resistance of buildings. It discusses the manner in which buildings resist earthquakes and the choice of appropriate structural form and materials for new buildings is considered. The approaches for engineered buildings designed to codes of practice will be very different from those for non-engineered buildings. Older existing buildings constitute the greatest source of earthquake vulnerability almost everywhere and the chapter concludes by describing some of the techniques for strengthening existing buildings which have been developed in particular locations.

Chapter 9 deals with loss estimation and seismic risk assessment techniques. As the techniques of risk analysis develop, it becomes an increasingly important part of the earthquake protection strategy for any organisation or community to be able to assess the extent of losses, of all types, which it faces. The methods available to carry out loss assessment and the way in which the uncertainties involved can be dealt with are the subject of Chapter 9.

Chapter 10 follows from the arguments of the previous chapter, identifying the range of strategies which have been adopted which could make measurable reductions in future earthquake risk, mainly through building improvement programmes. It also considers how such alternative earthquake protection strategies can be evaluated, and how comparisons can be made in a situation where avoiding human death and injury is the primary goal of protection policies, and in which simple monetary evaluation of losses is consequently inadequate. It concludes by reviewing the progress in earthquake protection which has been made so far. And it considers the potential for progress through international action during the years ahead.

Further Reading

Bolt, B.A., 1999. *Earthquakes* (4th edition), Freeman, New York.
Cuny, F., 1983. *Disasters and Development*, Oxford University Press, Oxford.
Hadfield, P., 1991. *Sixty Seconds That Will Change the World: The Coming Tokyo Earthquake*, Sidgwick & Jackson, London.
Richter, C.F., 1958. *Elementary Seismology*, Freeman, San Francisco.

2 The Costs of Earthquakes

2.1 The Costs of Earthquakes in the Last Century

During the last century, from 1900 to 1999, earthquakes caused damage estimated to be worth more than $1 trillion ($1 000 000 000 000) at modern values. This represents the loss estimates from the historical repair costs of each of the 1248 destructive earthquakes recorded around the world during the century, and adjusted to the value of money in the year 2000.

2.1.1 Costs of Earthquakes are Increasing Rapidly

These costs if averaged over the century represent a loss of over $10 billion a year. But costs are rising dramatically and during the century the average annual rate of earthquake cost increased by an order of magnitude. In the last decade of the century the loss rate averaged more than $20 billion a year. This is because there is more property to be affected by earthquakes and property is more valuable. The historical costs of earthquakes much earlier in the century were lower because population densities were lower and property cost less to build and repair.

Historical Earthquakes would Cost more if they Occurred Today

For example, the contemporary estimates of the 1906 earthquake in San Francisco put the costs of rebuilding the ruined and burnt city at over $300 million. At today's prices, this represents a sum of over $50 billion. But at that time, San Francisco was a city of around 340 000 people, with much less sophisticated infrastructure, less expensive buildings and much simpler personal possessions compared with the city that exists there today. Today San Francisco is one of the world's leading and richest cities, with a population of 7 million people in the Bay Area and a gross product of over $100 billion a year. A similar-magnitude earthquake to the 1906 event affecting today's San Francisco would almost certainly not cause the same extent of fires as occurred in 1906, and most of the modern buildings built to earthquake codes would not suffer major damage, but the damage that would be caused would cost hundreds of billions of dollars to repair. One analysis puts the total economic loss from a repeat of the 1906 earthquake on modern San Francisco at $170 to $225 billion.[1]

2.1.2 Different Types of Damage Cost Estimates

But the estimation of costs from all these earthquakes is very imprecise. The historical data is only approximate and comes from many different sources of different quality. People estimating earthquake losses use different terminology and different components of the cost when they produce a damage cost estimate. Definitions of different expressions of earthquake loss costs commonly encountered are given in Table 2.1.

2.1.3 Difficulties of Costing an Earthquake

Quantifying the costs of an earthquake is difficult. Loss figures given for historical earthquakes are usually estimates, based on aggregates of approximate information. Often assessors make their estimate by modelling likely loss levels against approximate information about the numbers and values of property in the affected area. In only a few cases have detailed studies been made to collect the actual costs incurred by all of the many people, businesses and stakeholders affected by the earthquake, and to compile an overall assessment once the repairs and reconstruction costs are known. Such studies show that it can take many months and years for the true costs to be recognised.

Losses come from many Stakeholders

There are many different components of loss, and many different people and organisations suffer losses, so establishing a definitive inventory of losses across all the various stakeholders is complex. Table 5.1 provides a framework of the major categories of likely loss-sufferers and various economic sectors impacted by an earthquake, based on checklists used for initial loss scoping for United Nations disaster reconnaissance missions.

Losses become more Apparent over Time

There is usually an urgency to establish an early loss assessment, and earthquake event reports that publish a loss estimate quickly after an event sometimes remain

[1] Risk Management Solutions (RMS) (1995).

Table 2.1 Definitions of different earthquake loss costs.

Physical loss	Costs of repairing the physical environment, including repairing damaged buildings, rebuilding infrastructure and replacing destroyed possessions
Economic loss	The total costs of repairing damaged property, the costs of the emergency operations and relief efforts, and the costs of lost economic production arising from the disturbance caused by the earthquake. Economic loss estimation is usually an attempt to aggregate the losses from all the stakeholders directly affected, such as the population, the commercial businesses, the public sector and the insurance industry
Insured loss	The loss to the insurance industry, arising from claims made by policyholders covered for earthquake. Insurance may cover repair to damaged buildings, replacement of damaged possessions and compensation for business interruption, and additional living expenses for people made homeless by earthquake damage. Only a proportion of people and private companies affected are likely to have earthquake insurance cover, and policies may have deductibles and limits, so insurance repayments cover only part of the costs incurred by the private sector
Shock loss	The cost of damage arising from the initial shaking, but excludes any subsequent losses, such as damage caused by fires triggered by the earthquake, or damage caused by landslides, sprinkler leakage or other secondary hazards
Historical loss	The value of the actual costs at the time of the earthquake. To compare the costs of earthquakes that occurred in different years, some account needs to be taken of the change in purchasing value over time, such as using a retail price index or inflation index
Value-adjusted loss for a reference date (e.g. loss at 2000 value)	The cost of an historical earthquake, adjusted to a standardised value, such as the value at a reference year, to account for change in purchasing value over time. In this book, earthquake loss costs are adjusted to values for the reference year 2000
Local currency loss	The value of costs in the currency of the country affected. To compare the costs of earthquakes that occurred in different countries, currencies are converted, usually standardised on US dollars
$ loss	The cost converted from the local currency to US dollars at the exchange rate prevailing at the time of the earthquake. Fluctuations in exchange rates over time can distort comparisons of costs between countries
Recurrence loss or 'as-if' loss	The loss that would be caused by a historical earthquake as if it were to recur on a modern population and building stock. The loss is modelled, calculating the effects of the known ground motion severity of the historical earthquake on the built environment or insured portfolio that exists in those locations today. Such studies should provide the benchmark year assumed for the infrastructure

authoritative and are quoted in subsequent catalogues when better estimates may have been made some time later. Sometimes estimates are never reviewed even when accurate data finally becomes available many months later, as by this time interest may have waned.

Loss estimates change very significantly over time, as more information becomes available. It takes time for people to discover and to provide accurate estimates of the costs of repair and replacement of goods. In earthquakes a large proportion of the cost is in repair of buildings, which is notoriously difficult to estimate accurately. Costs of carrying out repair work initially estimated from normal construction rates can escalate when the local demand from the disaster causes price inflation. Damage can prove more complex and costly once construction work starts. As buildings and machinery return to use, the recommissioning process can also reveal more damage and complexities.

Insurance Loss Development over Time

Insurance claims are only a proportion of the losses suffered in an earthquake, but they indicate how loss development can take time to occur and for the true nature of the loss to take many months to be finalised. Figure 2.1 shows how the estimates by the US insurance industry of insured losses from the Northridge earthquake were revised over time. Initial surveys of insurance companies compiled in the first two months after the earthquake estimated that insurance claims

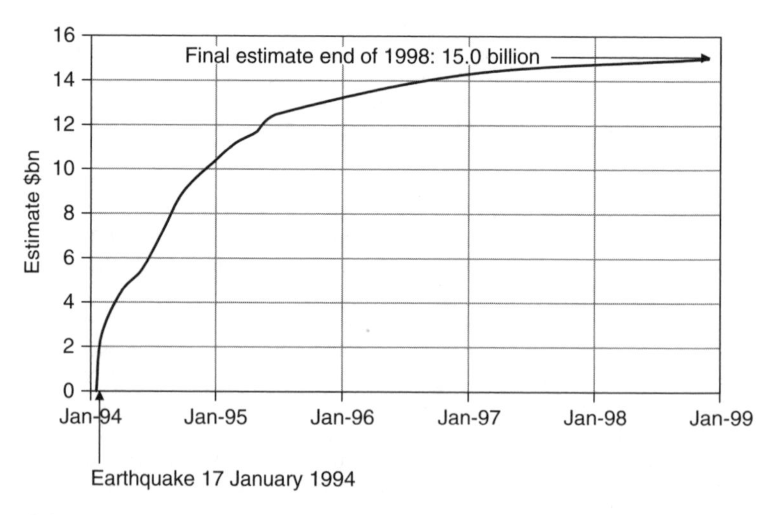

Figure 2.1 Estimates over time of the insured losses from the Northridge earthquake in California (from CDI 1998)

costs were likely to be around $4.5 billion, but the final inventory of losses compiled nearly five years later showed that the total payouts reached more than three times that, at $15 billion.

Estimating Lost Economic Activity

A complete assessment of the financial impact of an earthquake is only possible by including the losses caused by lost economic production as a result of the damage to facilities and disruption to the infrastructure. Some of the biggest uncertainty in estimating earthquake losses arises in quantifying the costs of lost production. Lost production is an abstract quantity and cannot be physically surveyed in the way that it is possible to count the number of damaged buildings. It arises from companies being forced to suspend their activities for a period of time, losing revenues or incurring manufacturing shortfalls or some other financial loss. In some cases companies are unable to pay wages and parts of the working public also suffer reduced incomes at a time when they have suffered economic losses. This is further discussed later in the chapter.

Many estimates of economic loss either ignore these losses, assuming they are marginal, or add notional amounts to the costs of physical damage. Several analysts believe that losses from lost economic production are considerably underestimated in earthquake accounting. Losses from economic activity may take many months to become apparent and as time goes by, if businesses do not resume production, these losses get larger. Loss estimates made in the first few months after an earthquake are rarely able to assess losses from economic downtime with any accuracy and so most earthquake cost statistics are likely to underestimate this component.

Use of Earthquake Loss Estimates

For all these reasons, earthquake loss estimates are highly approximate data, limiting their use for statistical analysis or detailed comparisons between individual events. The compilation of this information in a catalogue such as the database built by the authors can be used to show the scale of economic loss in general and broad trends.

2.1.4 Why Is It Important to Know about the Costs?

While loss of human life and injury are the most tragic and intolerable consequences of earthquakes, their social and economic losses consequences are far reaching, and provide a powerful argument for earthquake protection. The scale of economic loss from earthquakes is considerable and, for many of the people and organisations affected, their individual loss is financially devastating. Their collective losses have an impact on broader society and, as is shown later in this chapter, earthquake losses reduce economic growth and make for a less plentiful society for everyone.

Efforts are increasingly focusing on reducing the losses from earthquakes, and accurate information is needed to assess how much good would be done by different approaches. Assessment of the costs and benefits of protection measures is only possible if we can assess the likely losses with some precision. The financial management of risk requires an accurate analysis of costs. An individual or a company ultimately has to decide how to manage their own risk – how much they can afford to lose if an earthquake or other catastrophe occurs, and whether to buy insurance or to carry out other risk transfer and risk mitigation measures. Better information on losses helps with these decisions. Professional risk managers, like insurance companies and financiers, need to set rates to sell insurance policies and to assess the risk that they assume when they accumulate a portfolio of property at risk from loss. The financial models that they use to quantify and assess their risk all rely on good input data about the losses from earthquakes.

2.1.5 Intangible Losses

Estimates of earthquake loss as described above are derived by looking at the measurable aspects of cost, sometimes referred to as the *tangible* losses. There are also losses that cannot be formally quantified, the *intangible losses*; these can be significant and have important financial consequences.

Intangible losses include the human misery and the deprivation caused by the earthquake, and its effects on morale and confidence. Earthquakes destroy historical heritage and culture that contribute to the quality of our lives and our identity as a community. Earthquakes disrupt communication networks, and disrupt social activities, which means that people lose contact with friends, customers and business suppliers.

Effects on Culture and Heritage

Damage and destruction of the historical buildings and cultural artefacts of a region are a major loss that cannot be quantified. The historical buildings of a community are one of the major ways in which it defines its own cultural identity; they and the contents of its museums help it to connect with its past. They are irreplaceable, and their loss is beyond economic valuation. Attempts are sometimes made to rebuild destroyed town centres to look the same as they were before the earthquake, but this, costly though it is, does not replace them. The costs of repair of historical buildings are much higher than other buildings, and some regions of the world cannot afford the costs of repair involved. An example is the Gujarat earthquake in India in 2001 that damaged several hundreds of monuments, temples and palaces, many of which are beyond the resources of the local communities or national heritage organisations to restore.[2] International appeals raised funds for some of the major monuments, but many of the other fine buildings could not be saved.

Effects on Long-term Economic Development

The social consequences of large-scale destruction can be wide ranging and can last a long time. People are rendered homeless, jobs and services are disrupted, communications fail, and many elements of day-to-day administration are likely to be suspended. The extent of this social disruption depends both on the scale of the earthquake damage and on the robustness and degree of preparedness of the community. There are many positive examples where earthquake destruction has acted as a spur for an affected community to respond constructively, rallying round in adversity and reinvigorating the economy through its reconstruction efforts. However, the psychological effects of living in a devastated town or village can be profound, and there are examples where an earthquake has demoralised a community that has lost families, friends, housing and jobs. Examples have been documented where the economic potential or competitiveness of a region has been permanently shaken by severe destruction.[3] In an area where the economy is already marginal, the destruction caused by an earthquake may be enough to cause an irreversible decline: the immediate loss of employment forces the young and economically active to leave the area, damaged industry is not replaced, and the resulting stagnation is never reversed.

The psychological dimension of living through earthquake destruction is commonly recognised in relief operations where organisers often provide some element of counselling and morale support to the worst affected communities, but the effects can be long term.

Effects on Consumer and Investor Confidence

Consumer and investor confidence can also be casualties of an earthquake. Consumer purchasing can be hit, leading to loss of economic demand. The affected community may be forced to channel its resources into replacing its losses from the earthquake, reducing the disposal income it may have to buy other goods. This has an impact on the sales of goods and if this happens on a large scale it can depress whole regions and economic sectors. A period of economic stringency caused by the losses in one city or region may have a wider effect on

[2] Booth and Vasavada (2001).

[3] D'Souza (1984).

the population further afield: there have been cases where unaffected populations elsewhere in a country affected by a serious earthquake have exhibited restraint in purchasing during a period of national solidarity, having a marked impact on retail trade generally.

Major earthquakes can also cause a loss of confidence in the national and international investment community, causing the stock market to plunge. Most analysts believe that such effects are short term and that in general, the risk of stock market investment losses being correlated with catastrophe losses is very small. Investor confidence in such cases has also tended to be polarised across sectors selling insurance stock and buying into the construction sector. However, when large catastrophe losses have coincided with other trends, such as recession or political instability, large loss events have caused value losses on stock exchanges,[4] and some analysts have described scenarios where a major earthquake catastrophe in a financial centre like Los Angeles or Tokyo could have widespread repercussions across the world's financial markets.[5]

The reality of earthquake loss is that it is suffered individually by a large number of different stakeholders. Each has a perspective and a different view of their risk. This is developed further in the next sections of this chapter.

2.2 Who Pays?

2.2.1 Stakeholders in the Loss of the Kocaeli Earthquake, Turkey 1999

In the boxes running throughout this chapter, examples are given of the losses suffered by several different types of stakeholder in one selected earthquake, the Kocaeli earthquake in Turkey in 1999. The examples given are all fictional, but based on real case studies. They illustrate the way that different groups of people are impacted financially by the earthquake loss and how those losses are interrelated between the various 'stakeholders' in the loss. Circumstances vary considerably between any individual examples of loss, and in other earthquakes in different parts of the world situations are quantitatively quite different, but these examples are given to illustrate a process of risk sharing that is common in many earthquakes, from the comparatively richest nations in the world to some of the poorest.

---

[4] The multi-billion-dollar losses caused by the World Trade Center destruction from terrorist attack on 11 September 2001 caused large losses on the New York Stock Exchange, where stocks lost 13% of their value within a week. This was not directly comparable with a natural catastrophe loss, as it was linked to fears of future terrorism and military reprisals, but shows how major shocks can cause investor reactions.

[5] Hadfield (1991).

The loss stakeholders in the Kocaeli earthquake, Turkey, 1999

1. The homeowner

Selim and Basak Birgoren, homeowners, salvaging possessions from their damaged apartment in Yalova

Selim and Basak Birgoren lived in an apartment in a five-storey building in the centre of Yalova. The building was badly damaged in the earthquake, judged to be beyond structural repair and condemned. Some of the contents of the apartment were damaged in the shaking; most were recovered. The couple moved in to stay with relatives in another town nearby. Selim returned to his work at the factory, which reopened after three weeks closure during which it paid him half-time rates. The Birgorens had bought their apartment by buying into an association with a group of 20 other apartment owners in the five-storey building. The association of apartment owners is, however, unable to afford the rebuilding cost of the building, as most of the owners cannot afford their share of the reconstruction. The building was not insured. Selim is applying for a government rehousing grant which would provide him with a secured loan at preferential interest rates. Until they find a new apartment they are renting a room in a friend's house.

Estimated loss: \$55 000 *Annual earnings:* \$6000 *Loss/earnings:* 9 years

2.2.2 A Wide Range of Loss

In an earthquake catastrophe, destructive shock waves ripple across several thousand square kilometres of land, shaking and damaging hundreds of thousands of buildings. The earthquake damages farmhouses and homes, office buildings and factories, schools and law courts. All the owners of these buildings suffer

The loss stakeholders in the Kocaeli earthquake, Turkey, 1999

2. The small business

Tunc Tunali, restaurant owner closes for business for several months

Tunc Tunali, 62, ran a restaurant business serving the travellers on the main Izmit road. The restaurant was heavily damaged in the earthquake and had to close. Tunc Tunali had contents insurance covering fire and theft, but not earthquake cover, so was not eligible to recover any money from his insurer. Fortunately his home nearby was undamaged, but the cost of the demolition, rebuilding and re-equipping of the restaurant was considerably more than he could afford. After considering initially whether to retire, he decided to rebuild the restaurant, using some savings, and negotiated a bank loan, repayable over 10 years. The repayments are costly, so the income he is able to take home from the restaurant is considerably less than before the earthquake. He expects ultimately to sell the business as a going concern when he retires and to pay off the outstanding loan at that time.

Estimated loss: \$70 000 *Annual earnings:* \$24 000 *Loss/earnings:* 3 years

losses and need to find the funds to repair the damage or replace the facility. The farmer, the householder, the office block landlord, the factory owner and the government education and legal authorities are just a few of the many different stakeholders who have to look to their resources to overcome the loss. Some may receive external assistance, in terms of grants or loans. Some may be able to cover the costs of repair from their own resources. Others may have purchased insurance to help them cover the costs of this eventuality.

The loss stakeholders in the Kocaeli earthquake, Turkey, 1999

3. The corporate business

Huseyin Ceran, Plant Manager, major textile factory examines damage to production equipment

The factory is the main production facility of a company producing polymer textiles. It operates continuously in three shifts a day, producing about $5 million worth of output a week. In the earthquake the factory suffered light damage, with some storage tanks ruptured and many leaks in the pipe runs. The damage cost $12 million to repair. The factory was insured for earthquake damage, with a total insured value of $200 million, which is probably an underestimate of what it would cost to completely rebuild the factory. The standard earthquake insurance conditions include a 5% deductible, which on the $200 million factory means that the owners pay the first $10 million of the loss. The insurance claim on the $12 million repair bill meant that the owners recovered $2 million from their insurer. The factory was closed for six weeks, from a combination of waiting for the water company to reconnect the supply, carrying out repairs, obtaining critical replacement equipment and an extensive period of recommissioning the plant after repair. The lost production cost the company a further $30 million. The company was not insured for business interruption for earthquake. The losses were announced to the shareholders of the company and a rights issue made. The losses will affect the earnings and valuation of the company and possibly make it a takeover target for a competitor.

Estimated loss: $40 million *Annual earnings:* $34 million
Loss/earnings: 1.5 years

The loss stakeholders in the Kocaeli earthquake, Turkey, 1999

4. The government

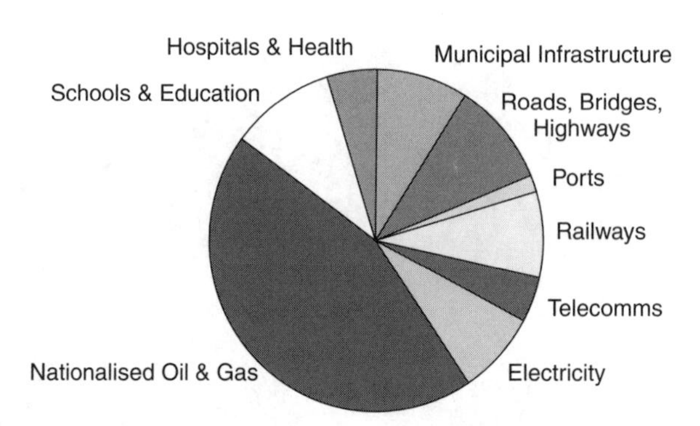

The breakdown of $1 billion infrastructure costs borne by the local administrations of Izmir, Yalova, Adapazari and Istanbul and the national government of Turkey

Many different government departments were involved in responding to the earthquake emergency and in managing the reconstruction. The emergency aid exercise and military mobilisation cost several millions of dollars. The earthquake damaged a lot of public infrastructure, from water supply to transport. The overall costs across damaged infrastructure are over $1 billion. The Turkish government has long had a proactive involvement in providing rehousing for citizens made homeless by natural disasters, under Disaster Law 7269. The government instigated a major housing reconstruction programme, budgeted at over $3 billion, in which some new housing will be built by government-appointed contractors, and other assistance and subsidised loans will be made to other sectors of the affected community. The government loss will be funded from the national exchequer, which means that it will ultimately be paid for by the Turkish taxpayer. In the past the government has funded previous disasters with an additional purchase tax levied on tobacco and consumer goods. Government revenues are about 25% of gross national product (GNP) and the earthquake will add to the budget deficit, already running at 13% of GNP. The government raises the funds it needs from the international financial markets by issuing treasury bonds and other debt instruments. Part of the funds it needs is to be borrowed from the World Bank International Bank for Reconstruction and Development, which allows a three-year period before starting the loan repayments. The other financial implication of the earthquake for the government is that its impact is forecast to affect the national economy, causing slower economic growth by as much as two to three percentage points of GNP for the next two years. Slower economic growth will result in lower revenues to the Turkish Treasury, reducing the budget available for public services.

Turkey 1999 GDP: $184 billion
Estimated loss: $4 billion
Annual government revenues: $47 billion *Loss/earnings:* 1 month

The loss stakeholders in the Kocaeli earthquake, Turkey, 1999

5. International aid

The European Community Humanitarian Office distributes relief supplies to earthquake-affected families

Many non-governmental organisations (NGOs), volunteer and charity organisations were involved in providing emergency assistance and aid in the aftermath of the earthquake. Volunteers helped with emergency operations, including feeding and sheltering victims in the immediate aftermath. The European Community Humanitarian Office (ECHO) was one of many providers of international aid that contributed to the Turkish earthquake. ECHO is the main coordinating aid agency of the 15 countries of the European Union and is one of the major aid providers worldwide, contributing an estimated 25–30% of global humanitarian aid. Thirteen of the member states contributed to ECHO's Turkey earthquake fund, the main donors being Germany ($17 million), The Netherlands ($6 million), Sweden ($4 million) and Greece ($4 million). These funds were disbursed in a series of tranches in the months following the earthquake through partner organisations – NGOs that carry out the logistics and implementation of aid. In the first two months, $2 million was given to the Red Crescent, and another $200 000 to Médecins du Monde, to assist with emergency medical care. Funds were also given to United Nations High Commissioner for Refugees and World Food Program, the United Nations hunger relief agency, who in turn worked with local NGOs to provide tents and basic shelter for the homeless, and to provide water and sanitation during the emergency period. Later aid was used to set up a number of reconstruction projects, such as rebuilding schools, re-equipping health centres and providing community assistance to the families worst affected by the earthquake.

Many foreign governments sent aid in the form of medical supplies, blankets, tents and other contributions of materiel. A number of international aid agencies set up appeals in countries such as the United States, Japan and many other countries, where private donors contributed important funds to the recovery effort.

Aid provided: $32 million *Annual budget 1999:* $870 million
Loss/budget: 2 weeks

The loss stakeholders in the Kocaeli earthquake, Turkey, 1999

6. The insurance company

Mehmet Avci, Loss Adjuster, Izmir Insurance Company, assesses an insurance claim from a damaged property

The Izmir Insurance Company provides both homeowner and commercial property insurance nationally across Turkey. It has a portfolio of many hundreds of thousands of buildings that it insures. Following the earthquake it received 18 000 insurance claims. The total value of the claims it paid out was over $120 million. The Izmir Insurance Company has a reinsurance programme where it buys catastrophe 'excess of loss' reinsurance from a panel of London market reinsurers, negotiated through a reinsurance broker. Its reinsurance contract provides for reimbursement of a loss in excess of a certain amount. The programme consists of two layers: one is a layer 50 xs 50 – that is, for one catastrophe, a group of reinsurers will cover the loss above $50 million and up to $100 million. The second layer, 25 xs 100, is covered by another, smaller group of reinsurers who will cover a loss above $100 million and up to $125 million. The insurer has lost $120 million so the first group of reinsurers pay a claim for $50 million and the second group pay a claim for $20 million. The total reinsurance recovery was $70 million. The net loss to the insurance company was $50 million. The Izmir Insurance Company funded this by selling some of the investment holdings it has in the stock market. The capital it uses to pay its claims cost is provided by its shareholders and it builds up its investment portfolio by investing its premium income profits over several past years. Regulations set the minimum amount of capital that the company needs to have available to pay claims.

Estimated loss: $50 million
Annual property insurance premiums: $22 million *Loss/earnings:* 2.5 years

The loss stakeholders in the Kocaeli earthquake, Turkey, 1999

7. The reinsurance company

Hamilton Syndicate at Lloyd's of London, processes reinsurance claims from the earthquake in Turkey

The Hamilton Syndicate is a reinsurance company providing several different types of reinsurance and special risk cover to insurance companies in many different countries around the world. Its portfolio includes two Turkish insurance companies and three 'facultative' contracts (taking a share in the insurance of large factories and oil refineries) that incurred loss in the Turkey earthquake. The total losses from these contracts to the syndicate was $39 million. This was a bad year for the syndicate who also had substantial losses from Typhoon Bart in Japan, Hurricane Floyd in the United States and Windstorm Lothar in France. Its total payouts this year amount to $180 million, making a loss of around 13% – about average for the catastrophe reinsurance industry in 1999. But the reinsurance business is cyclical and the syndicate expects to increase the price of reinsurance generally across its global business next year to recoup the losses over several years. All insurers worldwide that buy reinsurance will ultimate fund the cost of the losses through paying higher reinsurance premiums.

Estimated loss: $39 million *Annual earnings:* $160 million
Loss/earnings: 3 months

2.2.3 Stakeholders Who Share the Loss

Other stakeholders share in the loss. Insurance companies, reinsurance companies and the capital markets investing in public companies meet some of the cost.

The loss stakeholders in the Kocaeli Earthquake, Turkey, 1999

8. The capital markets

Kiyoshi Kanbe, Catastrophe Bond Trader, Tokyo Stock Exchange, buys and sells financial instruments that fund catastrophe losses

There were no catastrophe bonds issued that could be affected by the Turkish earthquake, but in other parts of the world, e.g. the United States and Japan, a major earthquake could trigger a payment of bonds offered directly by insurance companies to obtain capital from the capital markets around the world. The total capital available in the capital markets around the world is very much larger than that in the insurance industry. The capital markets trade in the stocks of reinsurance companies, insurance companies and major corporate companies that are affected by earthquake damage. The bond markets also deal in treasury bonds issued by national governments to raise money for their national borrowings. Investors in the stock exchanges around the world share in the losses of all of the various companies and help fund the national government debt that is used to pay for the reconstruction.

Total economic loss of Kocaeli earthquake: $20 billion
Annual daily variation on worldwide stock markets: $5 billion
Loss/earnings: 4 days

They fund the losses from the premiums they charge to their broader customer base of insurance policyholders. Private investors and shareholders will share the loss of the commercial operations affected. National governments provide many of the resources to help local authorities rebuild their infrastructure and may also provide resources for social assistance programmes for the worst-affected

The loss stakeholders in the Kocaeli earthquake, Turkey, 1999

9. So who paid what?

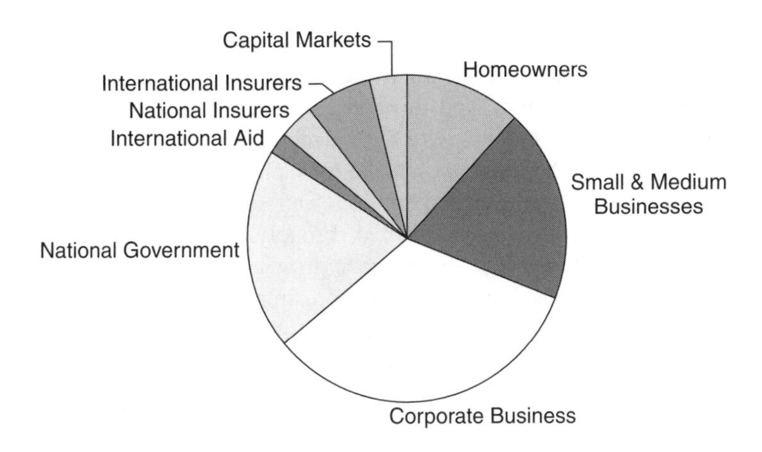

Share of $20 billion economic loss to each stakeholder

In the Kocaeli earthquake, over 500 000 households suffered some level of material loss, with around 130 000 dwellings destroyed or uninhabitable. An estimated 150 000 small businesses suffered significant damage, but less than 12% had any insurance cover. Corporate business was badly affected – more than 5000 industrial facilities in the area were damaged, and almost three months of production was lost from the region. The national government reported its reconstruction costs at over $3 billion, with some additional losses from local administrations and the emergency mobilisation. Aid was received from over 40 countries and more than 200 NGOs were active in the relief and reconstruction effort. The insurance industry in Turkey shared parts of the losses from homeowners, small businesses and corporate businesses – over 100 000 insurance claims were paid. Almost two-thirds of the insurance loss were recovered from international reinsurance companies in London, Germany, the United States and other financial centres.

The proportions of losses between these different stakeholders vary in different earthquakes and in different countries.

citizens. Governments meet these costs from their exchequers, financed from public taxes, so ultimately the taxpayers of earthquake-prone countries are all stakeholders in the loss. International aid may also assist when the resources needed are more than can be raised by the national government of a disaster-struck country, providing development banking assistance or subsidised loans. Development banking assistance is funded by the richer countries, making the citizens of the developed world also minor stakeholders in the catastrophe loss of a developing country.

2.2.4 Spreading the Risk across a Broad Network of Stakeholders

The network of stakeholders in the loss from an earthquake spreads out from the immediately impacted owners of the damaged property to broader groups of shareholders and investors, and to other insurance policyholders, through to other taxpayers, and in some cases to the citizens of countries on the other side of the world far from where the earthquake took place. Many stakeholders make small contributions to fund parts of the loss and some losses are so large that they can only be met by sharing the risk broadly. As commerce and finance become increasingly global, so the number grows of the people who have an interest in the losses caused and, perhaps more importantly, in the steps needed to implement earthquake protection to reduce those losses. People all over the world are now stakeholders in the earthquake risk of places they may not even have heard of.

The next sections describe some of the main stakeholders and describe how losses are commonly dealt with and how risks are shared and approached.

2.3 The Private Building Owner

Private building owners are the main risk stakeholders. As the dust settles after the earthquake, the property that they own has suffered damage and is worth less than it was previously. It will require cash expenditure to repair the damage or to rebuild the property and to replace its contents.

2.3.1 Homeowners

The largest number of private building owners are homeowners. In modern cities there may be three or more times as many residential buildings as buildings for other uses. Houses are smaller and cheaper than most civic or commercial buildings and owned in their millions across the earthquake zones of the world. A large earthquake near populated areas can damage tens of thousands of houses, and there are several cases documented where hundreds of thousands of houses have been made uninhabitable by damage.[6] For each of these individual households this is a personal and financial tragedy.

[6] Several earthquakes in China have had over 100 000 residential units destroyed, and in the 1999 Shanxi earthquake in China 600 000 houses were reported destroyed. In the 1995 Kobe earthquake in Japan 200 000 buildings were destroyed. In the 1994 Northridge earthquake in California, officials estimated that over 60 000 residential units had damage costing over $5000 and were therefore uninhabitable.

Complete Loss of a Home is Rare

The house represents a key asset that is typically the majority of the assets a household possesses. If the building has collapsed it may also have destroyed the contents and possessions of the house. Fortunately the number of cases where a house collapses and destroys contents is much more rare than the house being damaged. Damage ranges from complete collapse through to minor cracking and in most earthquakes many more buildings suffer minor damage than severe damage. Some householders are faced with minor repair costs for cosmetic damage and broken fixtures. Others need to find the resources to carry out major repairs on structural damage. Where the building is uninhabitable, the household needs to find alternative accommodation and this can be a significant additional expense. A small number of affected households may have lost the entire house and all its contents. In a major earthquake in a developing country with vulnerable buildings, this could be as high as 5 or 10% of the affected households, but in lesser earthquakes and in more developed countries with more robust building stock, it is likely to be a very small percentage, less than 1%, of the affected households that suffer total loss of their property. Figure 2.2 shows some typical distributions of repair costs across the thousands of householders affected in different earthquakes. This distribution will differ according to the strength of the earthquake, how many people are living close to the earthquake epicentre and the vulnerability of their homes, but it shows some examples of the ratios between losses of different levels. The majority of homeowners affected by an earthquake are faced with a repair cost of less than 30% of the rebuilding cost of their house, but this can still be a major loss to fund.

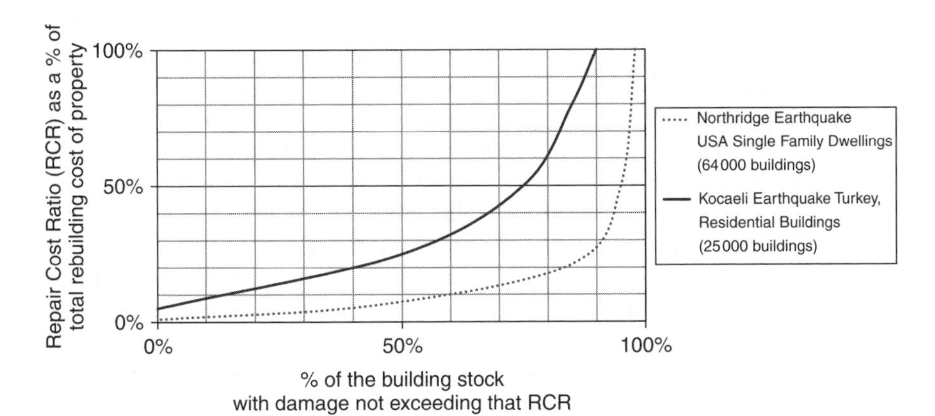

Figure 2.2 Distribution of repair cost ratios for residential buildings damaged in earthquakes

Finding the Resources for Repair

Repairing or replacing a damaged home can consume a lot of a family's remaining resources.

In most of the developed and developing world the average house value represents between three and eight times the average annual income of the households that live in them. Individual situations of course vary enormously. The reconstruction cost of a property is usually slightly less than the market value (which includes the land) but a typical repair bill of 30% of the rebuilding cost of the house is likely to represent more than a year's income for the average household.

A wealthy household may have savings or other assets it can draw upon to cover this cost. Individuals may have savings, investments or pension schemes that can be used to fund sudden losses of this type. It may be possible to sell other assets to raise funds, but if an earthquake has affected many people similarly, then if many people are selling assets the 'firesale' value of assets could be a lot lower than they are normally worth. Other households may have other family members or friends who are prepared to help, if the earthquake has not affected them also. However, many people find it understandably difficult to meet the costs of repairing their house. Some analysts recommend a rule of thumb that no more than 10% of the liquid assets (cash, savings, pension schemes and investments) should be at risk from loss. It can be seen that in an earthquake, many people will have losses far greater than 10% of their liquid assets. The more prudent householders may have insurance or other financial reserves to assist with the loss.

Homeowner Insurance

In many parts of the world, homeowners can purchase insurance against earthquake. The policyholder pays an insurance company a premium each year and if an earthquake strikes, they can make a claim for part of the repair costs. Insurance policies have many different terms, conditions and coverages. Some may include cover for earthquake within a standard fire insurance policy, others offer it as a separate product, for payment of an additional premium. Most insurance requires a deductible – a portion of the cost of the claim that the householder pays. For example, in California the usual deductible is 15% of the insured value of the home,[7] so if a home was valued at $100 000, and it suffered earthquake damage of $40 000, the insurers would pay the sum above the deductible of $15 000, so would reimburse the policyholder for $25 000. Even in relatively wealthy seismic countries, homeowner insurance has a relatively low penetration, with only

[7] The standard deductible offered by the California Earthquake Authority (CEA) in 2002 was 15%. Many companies offer the CEA or a similar policy in California.

a proportion of householders (perhaps only 10 or 20%) having earthquake insurance.[8] This is partly due to the cost, as earthquake insurance can be expensive, and may be expensive relative to the perception of the risk of earthquake damage held by the homeowner.

Where a house is purchased on a home loan or mortgage, the lending institution usually insists on having insurance. This ensures that if the house is damaged, the homeowner has sufficient funds to repair, and the lender secures its collateral – the house – on the loan.

2.3.2 Companies

Earthquake losses suffered by private companies have to be funded from the assets of the company and ultimately by its owners or shareholders and investors. If it cannot meet its losses, the company will become insolvent and its owners and creditors will suffer some level of loss. Companies operating in earthquake areas consider insurance a vital component of their strategy for managing their business risks and larger companies have dedicated risk managers who decide how much insurance to buy and other aspects of managing their risk.

Damage to the buildings that a company operates in may well not be the most important element of the loss the company may suffer in an earthquake. Equipment, plant and inventory stored in warehouses may be far more valuable than the costs of the sheds they occupy. Interruptions to the production carried out by the company, either production of manufactured goods or the business it does servicing customers, may be more costly than the buildings. The compensation payments to workers who are injured in an earthquake at the workplace, or for liability payments to third parties, may be the largest component of all.

Earthquakes are only one of many Business Risks

The executives of large commercial operations manage a wide range of financial risks in running their business, of which an earthquake impact is one scenario. They may anticipate currency exchange rate fluctuations, the risk of a major creditor defaulting, public relations scandals and other events that could damage their revenues or profitability. They balance the capital they have in reserve with the likely calls and costs from a multitude of potential causes. Managers look to produce a return on their capital while being able to manage the variations of cost. Financial risk analysis may show them how much they can afford to lose in accidents and how often they can face a loss above a certain level. Managers

[8] The numbers of people insured against earthquake vary considerably from country to country and place to place. In California, where earthquake risk is a daily fact of life, the CEA estimates that less than 25% of all housing is insured (and less than 17% for people who own their own homes). In Japan, the Japan Earthquake Reinsurance Company estimates that the number of households with residential earthquake insurance policies is around 16%.

use financial risk modelling to decide how much to keep in reserve against the vagaries of their operations and how much insurance to buy for the more extreme losses. They may decide that they can afford a minor loss themselves and only buy insurance cover for a very large loss that exceeds their capacity to cover it.

Some companies are so large that they can diversify their risk and fund their own losses without needing insurance – indeed many of the largest companies are bigger than insurance companies – and some set up their own 'captive' insurance company that manages the capital they need for a rainy day.

Buying Commercial Insurance Cover

Insurance brokers help companies decide how much insurance they need and the best deals that provide them the cover they need.

A company may commonly buy one or more of the following types of insurance cover for earthquake risk:

- buildings insurance
- contents insurance
- business interruption insurance
- workers compensation or personal accident insurance.

Small-business Operators

Most commercial earthquake insurance is sold on large commercial buildings. Small commercial businesses tend to have very low take-up of earthquake insurance, possibly because of the cost of the premiums, but also because of the short time horizons of small businesses and often because the amount of assets they have at risk is comparatively low. Disaster recovery for small businesses is complex and important. After the 1994 Northridge earthquake, the downtown areas of the towns of Northridge, Whittier and Santa Cruz in Los Angeles County took a long time to recover, partly because so many people and businesses moved out of the damaged towns. The issue of speeding up economic recovery in badly affected areas is one that continues to challenge planners and researchers.

2.4 The Insurance Industry

2.4.1 *Insurance is Profit Driven*

Insurance companies take on risk in return for a premium payment – when the policyholder suffers a loss within the terms of the insurance contract, the insurer pays the claim of the policyholder. This is a commercial transaction and the insurance company makes its business and profits out of successfully selling and managing insured risks. In addition to the costs of paying claims, insurance companies have their operating expenses to cover and a profit margin to achieve, to provide returns for their shareholders. The economics of the insurance business is driven by successfully understanding the costs and designing insurance 'products' that customers want to buy – the sales and marketing of insurance products are like any other product and depend on matching the content of the product with the demand of the customers, setting the price at an attractive level to maximise sales and ensuring that the product is promoted and sold in sufficient numbers through appropriate distribution channels.

2.4.2 Catastrophe Perils

An insurance company calculates the costs of its products from the frequency and severity of its claims. Most of the day-to-day cost of general lines of property business is driven by the more routine perils of fire and accidental damage (high-frequency, low-severity claims). Thousands of claims from these causes occur every year and the average rate and the likely variation from one year to another can be costed by statistical analysis. Earthquake coverage is more difficult to cost because earthquakes occur so rarely that insurance companies do not have a statistical dataset of recent claims. Earthquakes are one of several 'catastrophe perils' that cause low-frequency, high-severity events – they occur rarely but can cause severe damage when they do occur. Other catastrophe perils offered by insurance companies include hurricanes, tornadoes, windstorm, hailstorm and floods in countries where these natural hazards occur.

2.4.3 Insurance Markets

Insurance markets are largest in the richer countries. The United States is the single largest insurance market, with over 40% of the world's non-life insurance premiums. Within the United States, the insurance markets are by state, with each state having its own regulatory framework and standard practices. Here, the biggest market for earthquake insurance is California, where earthquake insurance premiums totalled more than $386 million in 1998 (more than half the US national earthquake insurance total of $738 million and almost six times the business done anywhere else). The state with the second-highest premiums was Washington with just under $66 million, and third was Missouri at $38 million. Some earthquake cover is sold in all 50 states.[9] The second-largest insurance market is Japan, which has a long history of destructive and lethal earthquakes.

Figure 2.3 shows the world with the size of the countries drawn in proportion to the size of their non-life insurance industry.

---

[9] California Earthquake Authority (1998).

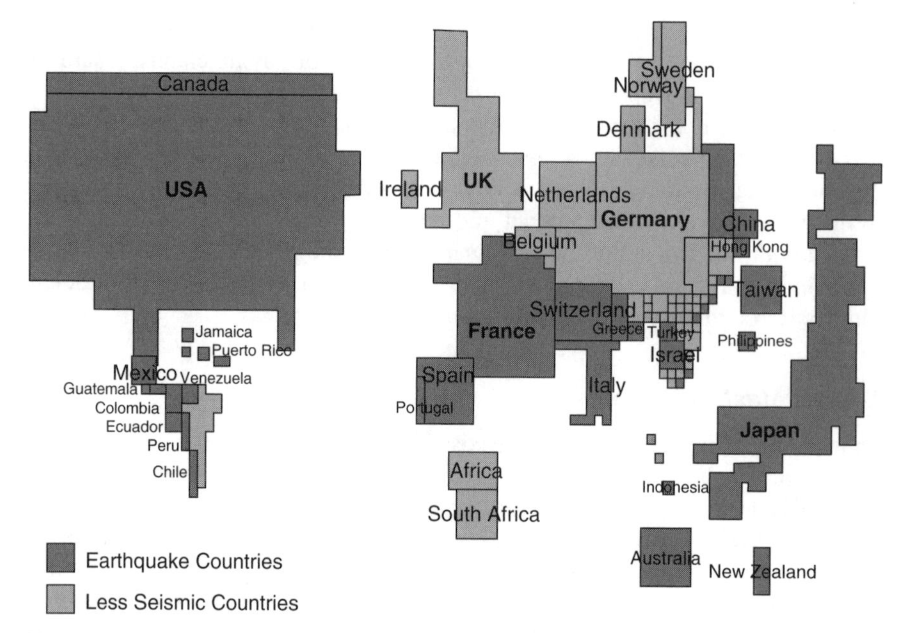

Figure 2.3 How the world looks to the insurance industry: the area of each country is shown in proportion to the size of the insurance industry (non-life insurance premiums in 1999). From data provided by Swiss Re (2000)

2.4.4 Catastrophe Reinsurance

The insurance company may in turn make a claim from its reinsurer. Reinsurance companies provide protection to insurance companies against sudden peaks of claims costs – so-called ‘rapid onset’ disasters, like earthquake. In many of the major insurance markets, catastrophe reinsurance is dominated by the risk of earthquake or hurricanes. Different perils drive the catastrophe risk in each of the main markets of the international insurance industry.

Risk management for earthquakes works by diversifying a portfolio of risk across different geographical areas. As different markets for earthquake and other catastrophe risks grow, the opportunities for risk transfer increase.

One of the most Destructive Natural Perils

When an earthquake does strike, it has the potential to be one of the most destructive perils that an insurer covers. The insurance industry paid out more than $1 billion after the Loma Prieta earthquake in northern California in 1989. Five years later, the insurance industry paid out another $15 billion for the 1994 Northridge earthquake.

A large earthquake can cause strong shaking capable of causing very high levels of destruction over a large region. In 1999, a magnitude 7.4 earthquake in

Turkey caused shaking of at least intensity VIII over an area of some 2000 square kilometres. The levels of damage that intensity VIII can cause are dependent on the quality of the building stock it affects, but weaker property types can suffer over 50% loss. This represents a massive potential loss if the earthquake strikes in a densely insured region of weaker property.

Insurance coverage varies considerably in products from country to country and across different lines of business. In a number of countries the level of the deductible is relatively high – this reduces losses to insurers from the widespread but small-scale damage likely from small events and from those on the periphery of large events. But where large earthquakes cause high damage levels, the deductible is of only marginal protection. There are also variations in how different countries deal with fire following earthquakes (some include it as a standard cover, others do not) and business interruption. Business interruption can be a very major component of earthquake loss in commercial and industrial risks.

2.4.5 Catastrophe Losses

The trend of rapidly growing economic losses from earthquakes is even more pronounced in the insurance industry. Monitoring of catastrophe losses shows that insured losses are increasing rapidly worldwide. Industry analysts show that natural catastrophe losses in the 1990s grew to 15 times as large as those in the 1960s.[10]

The frequency and severity of insurance losses from natural hazards are increasing. Records are constantly being broken in each country of the world for the cost of a natural disaster. This recognition has had wide-ranging implications for the reinsurance industry and has brought to prominence new techniques of risk management, successive waves of new capital being brought into the insurance industry, and a growing role for the capital markets in the transfer of catastrophe risk.

Earthquakes account for about 20% of insured catastrophe losses (and over a third of all economic losses from natural hazards).[11] Figure 2.4 shows the increasing insured losses from catastrophe insurance and how earthquake losses contribute to the growth. The statistics are difficult to generalise from because the losses are dominated by a small number of individual catastrophes, such as Hurricane Andrew in 1992 and the Northridge earthquake in 1994, and yet the trend is clear that these individual extreme events are occurring more frequently.

In general, there have not been any more hurricanes or earthquakes during the past quarter of a century than have happened during other 25-year periods in history. The evidence suggests that the main driver for the increased cost is that the natural hazards that occur are causing more losses than they did previously.

[10] Munich Re (1999).

[11] Munich Re (1999) estimates earthquake losses accounted for 18% of insured losses and 35% of economic losses from 1950 to 1999.

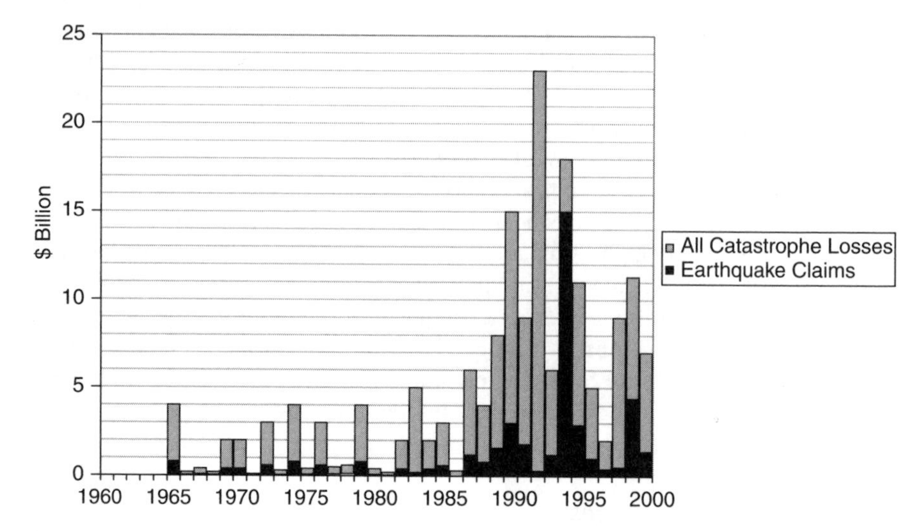

Figure 2.4 The growth of insured losses from catastrophe in the last four decades of the twentieth century (from Munich Re 1999, Swiss Re 2000 and authors' earthquake database)

The number and value of insured property in the paths of the events that occur are very much larger today than was the case a generation ago. The values at risk are increasing. The population of the planet has doubled in a generation. Increasing numbers of people have their assets or industries insured. The pattern of insured assets is changing across the areas where hazards occur. The severity, locations and types of losses suffered in the past are no longer a very good guide to the losses that will occur in the future.

Those who live in the countries where insurance is a way of life are becoming wealthier and expect to have their increasing assets covered, largely within existing insurance arrangements. Although there is little growth in premium income from property insurance in the OECD countries (averaging less than 2% growth during the price-competitive 1990s) there is little doubt that the insured values at risk are increasing rapidly. Economists show that more wealth was created in the United States in the last 10 years of the twentieth century than in the first 60. Large amounts of this wealth turn into property and find their way into the insurance industry's portfolio. The average householder is far better off than their parents' generation and today owns houses and contents of far greater value. Commercial operations have more (and different) property, liabilities and dependencies than ever before.

The demographics of risk have also changed – population movement has meant that the population of the state of California and the earthquake-prone regions of the United States have grown by 50% since the 1970s.

Increasing numbers of developing countries are developing an insurance industry. Insurance premium growth in the newly emerging economies is averaging 10% a year. India and China, representing a third of the world's population, ended the 1990s more than twice as rich as they started it. As countries become more prosperous, they buy insurance to protect that wealth. The types of property in these regions are more vulnerable to the prevailing hazards, being built to less demanding construction standards, so relative to the developed world they suffer higher proportional losses when disaster does strike.

The Roaring 90s

The early 1990s saw a sequence of catastrophe events that put unprecedented pressure on the reinsurance industry. Even though the second half of the decade proved less eventful and catastrophe reinsurance pricing slumped to low levels afterwards, the sequence of sizeable losses in the first half of the 90s had significant consequences for catastrophe reinsurance as an industry. Of the worst 20 catastrophe events in history, ranked by insured loss, 15 occurred in the early 1990s. These included Hurricane Andrew in Florida, 1992 ($16.5 billion), the Northridge earthquake in California, 1994 ($15 billion); Typhoon Mireille in Japan, 1991 ($5.2 billion) and Storm 90A in Northern Europe, 1990 ($3.2 billion).

Super-cats

The major events of the 1990s put unprepared insurance companies and reinsurers out of business. Catastrophe capital was depleted and people began to recognise that a potential existed for even larger loss events. If Hurricane Andrew had tracked across Miami, they realised, the total losses could have been far higher. If a major earthquake occurred closer to San Francisco, there could be even larger losses. A major earthquake in Tokyo would have a severe impact on the global reinsurance industry. Such events became termed 'super-catastrophes' or super-cats. During the 1990s analysts talked about a shortage of capital in the reinsurance industry. The analysis of potential catastrophes became important in understanding the needs for capital. The management of portfolios became an issue – how to spread the risk and balance the capital allocated to business in different regions. Insurance companies learned how to measure and model catastrophe risk. The use of catastrophe models, simulating the effects of an earthquake on an insured city, became a standard part of risk management.

Alternative Risk Transfer

The risk of a super-catastrophe causing capital shortages in the insurance industry has caused people to look for other sources of capital. New insurance and reinsurance companies were set up to provide new capital – many of them in

Bermuda to take advantage of the favourable tax regime – creating the Bermuda insurance market. The whole insurance industry, although large, is much smaller than the capital markets. Daily variation in the value of stock markets exceeds the losses from a major insurance catastrophe. A number of ways have been devised to access the capital markets with financial instruments based on catastrophe risk. These are alternatives to traditional reinsurance treaties, grouped under the term alternative risk transfer or non-traditional reinsurance. The securitisation of catastrophe risk has grown in significance each time the reinsurance price cycle hardens and costs of risk transfer rise. A typical catastrophe bond is issued by an insurance company (or sometimes even a large corporation, bypassing the insurance market) offering to pay a certain rate of return. Investors who purchase the bond receive the rate of return, but if a catastrophe occurs and the insurance company suffers a certain level of loss, the investors may lose some or all of their investment. The value of this arrangement is that the bond is a tradable commodity.

2.5 The Public Sector

2.5.1 Government Costs

Damage to Publicly Owned Infrastructure

The physical destruction from an earthquake hits the infrastructure and the public services organisations as much – and sometimes more than – it affects the individuals and businesses in the stricken region. Community facilities such as schools, hospitals and leisure may be destroyed. The centres of administration and public buildings are likely to suffer. The equipment, personnel and buildings that make up the police service, the fire service and even the military facilities in the earthquake area can suffer loss. Transport networks suffer from ground deformations, ground shaking and landslides that cut roads, damage railways, destroy bridges and close tunnels. Public utilities are publicly owned in many countries and these can be badly damaged, cutting supplies of power and water to large proportions of the population. Electricity generators and substations are vulnerable to earthquake forces and power lines are easily cut. Water and gas supplies, sewers and sanitation are difficult and expensive to repair when underground pipe networks are damaged by ground deformation. In some countries the telecommunications networks are in public ownership, and damage to telephone lines and switching stations needs to be paid for from the public purse.

Funding the Emergency Operations

In addition to the costs of the damage, the emergency operations involved in managing an earthquake disaster are largely paid for from government budgets. Major mobilisations of the emergency services, including police, fire services, hospitals and the military, can cost millions of dollars in salaries and equipment costs.

Assistance to Citizens

Governments are also likely to provide assistance to the worst-affected individuals, particularly in housing the homeless. Governments may set up social housing programmes or loans or credit schemes for those who otherwise would be unable to find the resources to house themselves. Similarly, there may be government-backed loan schemes or subsidies for small businesses to revitalise the economy in worst-affected areas. Social programmes, welfare and unemployment benefit schemes may all increase as a result of the earthquake causing increased deprivation and job losses.

2.5.2 Impact of the Losses

The impact of such economic losses can be severe and have national and international repercussions. Government assets do not tend to be insured – governments usually bear their own risks. Costs of building national infrastructure are met through the government treasury, ultimately funding capital investment from tax revenues. Governments raise money through borrowing to fund major capital projects. Management of the national debt is an important function of the treasury. Most earthquake losses are funded in the short term by increasing the national debt. Borrowing is made from the capital markets, through instruments such as treasury bonds. Developing countries may be eligible to obtain loans from international development banks, such as the World Bank, providing loans at commercial rates of interest but with initial repayment periods of grace. Some losses may be offset by reconstruction aid provided by wealthier countries to the developing countries, through bilateral or multi-lateral aid arrangements.[12]

2.5.3 Revenue Losses

In addition to the direct costs of replacing damaged infrastructure, an earthquake that has a major impact in reducing the economic productivity of a region also reduces the revenues to the government through reduced taxes on the production. The example of the Kocaeli earthquake shows that if the impact of the earthquake reduces the economic growth of the country by two percentage points, the net difference to the treasury the following year would be around a billion dollars, a quarter of the government's direct cost of the earthquake. And a loss of economic

[12] The difficulties of financing catastrophe loss for developing countries and new ways currently being explored for financing are described in Freeman (2000).

growth in one year can cause shortfalls in government budgeted revenues for several years.

2.5.4 Effects of Earthquake Economic Impact

The cost of reconstruction after a major earthquake can greatly increase a country's national debt, set back economic development and cripple local and national economies. In severe cases, the severity of the economic problems caused by an earthquake can cause long-term reductions in the growth of a nation's economy, trigger inflation and unemployment rises. For example, economists observed a number of effects on the national economy of the Philippines after the Luzon earthquake of 1990. They identified that the earthquake caused a reduction in GNP growth of nearly a third from the pre-earthquake forecast, inflation increased several percentage points and there was a major decline in the balance of payments, directly due to earthquake effects.[13] In extreme cases, the economic impact of a sudden downturn may even contribute to the destabilisation of a country's administration. The decline of the Nicaraguan economy under the Sandanista government during the 1970s and 1980s can be traced back to the initial national debt created by the 1972 Managua earthquake, according to economic analysts.[14]

A comparison of earthquake losses with GNP of various countries (Table 2.2) shows how serious such losses can be for the national economy. GNP is an indicator of the country's own potential for recovery and in many cases earthquake losses constitute a significant proportion of GNP. The poorer nations, with lower GNP, tend to be more vulnerable to the economic impact of a costly earthquake, even though in absolute terms, the cost of the damage may not be as high as elsewhere. This gives an indication of the greater relative vulnerability of the smaller or poorer nations to an earthquake disaster.

The high costs of national reconstruction may have international repercussions with economic assistance being provided by international finance and multinational aid. In severe cases of earthquake destruction, reconstruction and full recovery can take decades. In addition to the costs of damage replacement and lost production, economists also recognise that costs include 'opportunity costs', the other things that the money could have been used for if it had not been needed to recover from an earthquake. For a nation, the opportunity costs of earthquake losses are the investments that would otherwise have been made improving the quality of life and the economic conditions of its citizens. Money spent on rebuilding damaged hospitals is money that could have been used to build roads to attract new industry, create jobs and promote more economic growth.

---

[13] NEDA (1990).

[14] Brooking Institute, Washington, DC, reported in *The Independent*, London, 28 February 1990.

Table 2.2 Economic losses from earthquakes in the late twentieth century, as a proportion of GNP.

Country	Earthquake	Year	Loss ($bn)	GNP that year ($bn)	Loss (% GNP)
Nicaragua	Managua	1972	2.0	5.0	**40.0**
El Salvador	San Salvador	1986	1.5	4.8	**31.0**
Guatemala	Guatemala City	1976	1.1	6.1	**18.0**
Greece	Athens	1999	14.1	110.0	**12.8**
Yugoslavia	Montenegro	1979	2.2	22.0	**10.0**
Iran	Manjil	1990	7.2	100.0	**7.2**
Italy	Campania	1980	45.0	661.8	**6.8**
Romania	Bucharest	1977	0.8	26.7	**3.0**
Mexico	Mexico City	1985	5.0	166.7	**3.0**
USSR	Armenia	1988	17.0	566.7	**3.0**
Japan	Kobe	1995	82.4	2900.0	**2.8**
Philippines	Luzon	1990	1.5	55.1	**2.7**
Greece	Kalamata	1986	0.8	40.0	**2.0**
China	Tangshan	1976	6.0	400.0	**1.5**
Quindio	Colombia	1999	1.5	245.0	**0.6**
USA	Los Angeles	1994	30.0	7866.0	**0.3**
USA	Loma Prieta	1989	8.0	4705.8	**0.2**
Turkey	Kocaeli, Izmit	1999	20.0	184.0	**0.1**
Taiwan	Chichi	1999	0.8	N/A	

2.6 Interrelated Risk

This chapter has shown how many different stakeholders are involved in the losses from an earthquake. This was illustrated with a case study of the losses from the Kocaeli earthquake in an industrial region of Turkey in 1999. In this case study, the entire 'food chain' of earthquake risk is shown to be shared between individual houseowners, corporate businesses, government, insurers and global financiers, and ultimately the citizens and insurance premium-payers in many different countries around the world.

Other Earthquakes are Different

The Kocaeli earthquake in Turkey was only one of several earthquakes that occurred in 1999. An earthquake in Greece near Athens, an earthquake in Taiwan and an earthquake in Colombia also caused many deaths and major economic losses that year. Each earthquake was quite different in the type of region it affected – an urban area, a rural agricultural and tourist region. The levels of losses and the distribution of the losses between the various players affected are different in every earthquake. How the loss is shared between the different stakeholders depends on the number and value of homes and industry, the level of infrastructure and the relative levels of wealth in each of the sectors. In other earthquakes in different parts of the world, there are different ratios of wealth and loss, different levels of take-up of insurance, different participations by government in social loss and different international involvement by financiers.

No Winners, Only Losers

However, the overall picture has some similarities wherever it occurs. There are no winners when an earthquake occurs, only losers. When earthquakes occur, the damage they cause is a financial cost to the householders, companies and governments affected. Financial losses damage economies and hinder development. In this way, the losses of the different stakeholders are all linked.

There are a number of interactions between the various stakeholders and their losses. The losses of corporate businesses are linked to the losses of the general population and homeowners – when the workforce is made homeless the manufacturers have to stop production, and when the workplace is destroyed, the employees lose their jobs. When the population is destitute, restaurant owners lose their customers. The government shares in the losses of its citizens. Insurance companies take on large losses on behalf of their policyholders. And ultimately, an increasing global financial structure spreads losses among many shareholders, investors, insurance premium-payers and taxpayers around the world.

Risk Transfer

One important interrelationship between stakeholders is risk transfer – when one party buys an insurance policy from another they are transferring risk from the policyholder to the insurance company who spread the risk across many other similar policy holders. Increasingly this is becoming an important method of providing protection, and other methods of risk transfer, and aggregating risk to share it, swap it, or spread it across other people who have risk, both implicitly and explicitly, are increasingly being explored.

Co-interest in Risk

Where people share in a single loss, e.g. when a homeowner loses their house and it falls to the government to provide a new house or housing loan, both the government and the homeowner suffer a loss as a result. Both parties have an interest in reducing that loss.

Regulatory Environments

Sometimes when the risks are shared, or are more societal, legislative or regulatory measures are adopted to ensure that socially responsible actions are taken to protect against unacceptable losses. Regulatory frameworks ensure that insurance companies meet capital adequacy tests, so that they can meet their claim obligations in the event of a major catastrophe.

2.6.1 A Shared Interest in Earthquake Protection

There are major differences between the levels of risk faced by the individual stakeholders in the earthquake. The potential loss to an individual homeowner may represent decades of income, and as a proportion of their total assets it can be overwhelming. However, the probability of it occurring to any one individual is very small. By comparison, the losses to an insurance company are a much lower proportion of their total assets. However, because insurance companies spread their risk and insure many people in different parts of the company, and perhaps in different parts of the world, the probability of them experiencing a loss is much larger – they experience more frequent losses.

Increasingly the losses from earthquakes are being scrutinised and researched. Economic loss and the hardship that results is a major penalty resulting from earthquake activity. Risk can be spread from those who can least afford it to the larger community capable of shouldering a smaller share of loss. The reduction of losses is a major priority for all concerned and an area of mutual interest between stakeholders in the loss. Throughout the rest of this book, strategies and measures to provide earthquake protection are explored.

Further Reading

Bronson, W., 1986. *The Earth Shook, The Sky Burned: A Photographic Record of the 1906 San Francisco Earthquake and Fire*, Chronicle Books, San Francisco.

Comerio, M., 1998. *Disaster Hits Home*, University of California Press, Berkeley, CA.

EQE, 2002. *The EQE Earthquake Home Preparedness Guide* (available from www.eqe.com).

Freeman, P.K. and Kunreuter, H., 1997. *Managing Environmental Risk Through Insurance*, The AEI Press, Washington, DC.

Munich Re, 1999. *Topics 2000: Natural Catastrophes – the Current Position*, Munich Re Group, Koniginstrasse 107, 80802 Munchen, Germany.

RMS, 1995. *What if the 1906 Earthquake Strikes Again? A San Francisco Bay Area Scenario*, Topical Issue Series, May 95; Risk Management Solutions, 7015 Gateway Boulevard, Newark, California 94560, USA.

3 Preparedness for Earthquakes

3.1 Earthquake Prediction

One of the obvious methods to reduce the loss of life and injury that occur in major earthquakes would be to predict the earthquake and evacuate the occupants of buildings before it arrives. Short-term prediction is unlikely to reduce the damage to property which is the main economic loss in an earthquake, but the benefits in reducing human injury and some of the secondary hazard effects of earthquakes (fires, industrial accidents and others) mean that some degree of warning would be immensely valuable.

Efforts to predict earthquakes successfully have been made since the 1950s when seismology provided a new theoretical framework for the process of earthquake occurrence. Rapid developments in the science in the early 1960s led to an optimism that prediction would be routine within a few years. Since then it has been realised that the development of the conditions giving rise to an earthquake and the process by which an earthquake is triggered are far more complex than was at first thought. Routine, short term prediction remains elusive, although a number of individual events have in the past been predicted with varying degrees of success.

There are a number of methods for earthquake prediction that are continually being researched and further developed which may in the future offer increased reliability and usefulness for the earthquake protection planner.

3.2 Long-term Prediction (Years)

Earthquakes are large-scale phenomena occurring on a geological timescale and in three-dimensional space within the earth's brittle crust. Trying to determine their exact time and position in relation to our tiny towns and settlements is like trying to predict the crack patterns on a slowly melting sheet of ice.

However, there are patterns in earthquake activity and reasons for their occurrence that can give us clues to future earthquakes. By studying how, why and where they occur the chances of successfully predicting future events is greatly increased.

3.2.1 Integrated Earthquake Hazard Studies

Identifying possible locations for future earthquakes depends on gaining a thorough understanding of the processes that cause them. This means understanding the detailed local deformations and geological processes going on in a region as well as the overall global tectonic evolutions that drive them. For this, geologists, seismologists and historians need to put together what is known about the geological structure of a region or country, its current seismic activity and deformation processes and a detailed history of earthquakes from as far back as written records exist.

Seismologists need earthquake catalogues of instrumentally recorded earthquakes over a significant length of time. Large-magnitude earthquakes can be recorded a long distance away at international seismic monitoring stations, but the smaller magnitude earthquakes and the micro-tremors that build up the full picture of seismic activity in a region can only be recorded by sensitive monitoring networks in and around the area of study. An effective seismic monitoring network, adequately staffed and resourced over a long period, is of course a primary prerequisite for earthquake protection planning in any region.

Geologists need a series of sources of information including aerial survey, field surveys, borehole data, profiling of geological strata and mapping of the surface geology. From these a skilled geologist can interpret the geomorphological processes that have formed the region and that are still continuing. The geologist may be able to identify the location and extent of some individual active fault along which the shaping processes are taking place. Unfortunately most of the main structural faults lie deep below the surface geology and are likely to be undetectable without very detailed and expensive studies[1] but the geological investigations may well suggest probable locations for them. Slow, geological deformations occurring across the region can be monitored with a *geodetic survey*, accurate, triangulated measurements across the countryside. From these, repeated every few years, ideas of the deformation rates and directions can be obtained, giving important information on the geological processes occurring and identifying potential areas for earthquake occurrence. Satellite surveying techniques can also be used for geodesy using *global positioning systems* (gps).

[1] Such as *seismic profiling*, i.e. setting off explosive charges and monitoring shock-wave reflections from geological strata.

The geological timescale being investigated and the long return period of earthquake catalogues mean that the long-term timescale of earthquake occurrence is of great importance in understanding the earthquake hazard. Where faults are known to exist, *paleoseismology* can provide information on prehistoric earthquakes from detailed examination of the faults, by digging trenches and dating the offsets in the strata that ruptures have produced. Historians can help build up a picture of past earthquake patterns over a number of centuries if the region is one that has had any length of literate tradition. Historical studies require years of painstaking study, locating, reading and indexing documents from the past. Newspapers, official logs, diaries, books, letters to friends and travellers' tales are all likely to contain references to any earthquakes that have occurred within the region. Earthquakes are remarkable events, and few writers who have experienced one do not record it. By a systematic logging and cross-referencing of earthquake reports, a seismic history can be built up. This shows the frequency, size and location of past earthquakes and is invaluable in locating fault systems, estimating *return periods* and identifying possible *seismic gap* that may be the location of the next damaging earthquake (see below).

3.2.2 Probabilistic Seismic Hazard Assessment (PSHA)

One of the most important patterns in earthquake occurrence is the recurrence of earthquakes at the same approximate locations over a long enough period of time. This gives an idea of activity rates and when, approximately, it might be reasonable to expect another one. The *average return period* for an earthquake in any region can be estimated from past records. To have an estimate that is at all reliable, it is necessary to have a long and accurate historical record. Unfortunately, return periods are the average of widely differing time intervals between the recurrence of earthquakes at any particular place. An average return period of, say, 100 years, means that we could expect approximately five earthquakes within 500 years, but one might occur only 20 years after its predecessor and another perhaps 250 years later. Except where there is accurate knowledge of the past behaviour of known faults, the seismic source area over which it is sensible to discuss the return period of earthquakes is large, so the use of PSHA is useful only for earthquake prediction in the longest time frame and for regional preparedness planning. Measurement of average return periods and their variation is further discussed in Chapter 7.

3.2.3 Characteristic Earthquakes

Statistical return periods are often associated with a general level of energy released over an area in which earthquakes occur, rather than on individual faults. The energy released can take place through small or large earthquakes or as a seismic movement and occurs on geological faults extending some kilometres below the ground which may not be visible at all on the earth's surface.

Sometimes earthquakes connected with a single fault or fault system recur on different parts of the same fault or hundreds of kilometres away on a related part of the fault system. But some well-known, surface-identified faults have been found to be associated with more regular, and possibly more predictable, *characteristic earthquakes*. Reports indicate that these faults release their energy in relatively regular, similar-sized bursts of energy. It may be that with intensive study of earthquakes areas, seismologists may be able to identify many more individual faults that can be associated with characteristic earthquakes, making the task of predicting earthquakes on these faults far more straightforward.

3.2.4 Seismic Gaps

Over a long enough time period, the relative motion of tectonic plates gives fairly constant energy release all along a certain seismic belt. Plotting past earthquakes along a geographical zone of slippage shows that earthquakes occur in between each other and eventually add up to a continuous rupture along the zone. Gaps between recorded earthquakes may indicate the positions of future earthquakes. *Seismic gaps* represent one of the best methods of intermediate-term prediction because it can identify candidate locations for future earthquakes within a seismic zone with known return period or energy release characteristics. Seismologists are then able to concentrate their monitoring equipment in a small area to await shorter term precursory phenomena. Where there is sufficient information on slip rates and past earthquakes, hazard analysis uses *time-dependant fault rupture models* to assess how overdue an earthquake may be on a particular fault.

3.2.5 Creep Freeze

Similarly, in areas of aseismic deformation between two tectonic plates, the relative deformation or *fault creep* can be measured by accurate land surveying. If movement is not occurring in one area between other areas of deformation then that area may be locked, and the accumulating stresses within it may be suddenly released in the form of an earthquake. This *creep freeze* may indicate the position of a future earthquake for more intensive fault monitoring.

3.3 Short-term Prediction (Days/Hours)

3.3.1 Precursory Phenomena

Many reports of past earthquakes include descriptions of unusual events noticed in the days or hours beforehand that may have been due to the imminence of the earthquake. Reports include dropping water levels in wells, strange glows in the sky, peculiar behaviour by animals and unusual sounds. Some of these reports may be fanciful – part of the mythology surrounding earthquakes – but a number of *precursory phenomena* have been authenticated for individual earthquakes and if reliably detected in future events could form the basis of short-term earthquake prediction. Not all of the recorded precursory phenomena have a good scientific explanation, but it is plausible to think that the geological conditions of the crustal material just before a major rupture could cause effects on the surface above. One of the major features of pre-rupture crustal material is the build-up of stress in the rocks at depth. This high stress level, it is argued, could have a number of characteristic features which could be detected beforehand and acted upon. Very high stresses in rocks produce heat, cause deformation and expansion of rocks, release gases from within the rocks, align crack formations within the rock crystals, and may have other geomagnetic effects. These in turn may cause some phenomena at or near the surface. Over the last 30 years many research projects have been devoted to investigating different precursory phenomena in the hope of finding a reliable short-term predictor of a destructive earthquake.

3.3.2 Foreshock Activity

One of the most likely indicators of a big earthquake is the occurrence of a number of small earthquakes beforehand, building up to a big event. As stresses build up, smaller fractures are likely to occur before the main rupture. Unfortunately less than half of big earthquakes are preceded by a significant *foreshock*. And the vast majority of small earthquakes that could be interpreted as foreshocks are not followed by a big earthquake. If, statistically from past earthquake records, 2 out of the 100 earthquakes of magnitude 4.0 recorded in a region were followed by an earthquake of magnitude greater than 6.0, then if a seismologist records another magnitude 4.0 earthquake, there is a 2% probability of a magnitude 6.0 or greater occurring soon. The seismologist can qualify it further by saying a magnitude 6.0 within two days, and its epicentre less than 10 km away from the foreshock, and so on, according to past patterns.

From a single small earthquake, the probability that it will be followed by a larger event is generally very low. A sequence of small earthquakes, however, in an area that is usually quiet, or has been identified as a seismic gap, does significantly increase the probability that it could be followed by a damaging earthquake. If the ratio of the numbers of small and large earthquakes in a region changes significantly over a short time period – that is, if there are suddenly more smaller earthquakes than there used to be – this marks a sudden *stress drop* which may well indicate that a larger event is about to be triggered. Complex patterns of multiple small earthquakes occurring may well not have been recorded in this particular area before (earthquakes have only been instrumentally recorded during the past 100 years), so the numerical probability of it indicating a future large-magnitude event may not be able to be derived from a statistical study of past earthquake records. In this case seismologists may make a probabilistic estimate based on their judgement and case studies elsewhere. An administrator or politician, who will have to take the responsibility for the decision whether to call a full-scale evacuation or not, should be aware of the process and uncertainties in the predictions presented by the seismologists. The use of probabilistic estimates for public safety is discussed below, in Section 3.5.

3.3.3 Ground Deformation

It has been observed that the rupture of a fault does not occur completely as a brittle fracture, instead it yields a bit before it breaks. The yielding of ground is only microscopic – deformations of a few millimetres over a few days – and it only happens close to the fault, but if it can be detected it could serve as a warning of an impending fault rupture. This could be useful in the case of a well-known fault suspected of being ripe for rupture where permanent instruments can be installed to detect small deformations. These instruments are expensive; for example, tilt-meters used to detect small rotations in the ground have to be carefully fitted into precisely drilled, deep boreholes, and exact geodetic survey measuring points have to be mounted across the suspect area. This limits the possible survey area to perhaps a few tens of square kilometres (depending on the resources available). The instruments do not require constant attention as they can transmit steady readings to a centralised control point and sound an alarm if significant deformations begin.

3.3.4 Anisotropy

When rock is under stress, close to its breaking point, the pressures cause a number of changes to occur in the characteristics of the rock. One characteristic is that the pressure causes the micro-crack formations within the rock to become aligned parallel to the stressing forces. The rock becomes *anisotropic*. These crack patterns cause any shock waves travelling through the stressed rock to become polarised, a characteristic that can be detected by sophisticated seismographic instruments. It may be possible to locate areas of highly stressed rock within crustal material by detailed monitoring of the background 'noise' of small shock waves from deep micro-tremors, and detecting polarisation in certain directions. Studies of anisotropy are continuing[2] but it is not yet clear how much information would be derived or the level of instrumentation (and therefore cost) required to pinpoint the possible sources of future earthquakes this way.

3.3.5 Radon, Chemical and Water Table Monitoring

The micro-crack patterns that develop in highly stressed rocks also appear to soak up ground water. In severe cases this can lead to a localised drop in the level of the water table nearby. Monitoring of water tables, by measuring the depth of wells, has shown that lowering of water tables has occurred in the vicinity of major earthquakes shortly beforehand. Unfortunately water tables vary from day to day for many reasons and a drop in the water table cannot on its own

[2] Crampin and Zatsepin (1997).

be used as a predictor. Water that permeates into the stressed rock also appears to absorb certain chemicals from the rock, possibly released owing to the stress. Monitoring the chemical content of the water table in deep wells shows that concentrations of radon and other chemicals have increased shortly before major earthquakes. Chemical and water-level monitoring could become part of an integrated prediction programme in areas already identified by other means as likely locations for a future earthquake, and despite being rather labour intensive, could add information to other prediction studies.

3.3.6 Abnormal Animal Behaviour

For centuries, there have been reports of strange behaviour of animals, fish and birds shortly before a large earthquake. Tales include horses bolting, rats climbing telegraph wires, birds flocking and fish jumping. The catfish became the symbol for earthquake in ancient Japan because, it is reported, they would leap from the water in the hours before an earthquake. There is no scientific explanation for these reports, although rationalisations have included possible changes in geomagnetic forces or gaseous chemical releases. The aggregation of reports of animal abnormalities has become a standard part of Chinese civil protection planning for an earthquake and, it is claimed, contributed to the successful prediction of the Haicheng–Yingkou earthquake in 1975 (discussed below). It is quite possible that reports of strange animal behaviour are illusory, and are part of the mythology of the terrifying event after it has taken place. Studies of animal behaviour have been carried out inconclusively and no doubt further studies will continue. There is, however a danger that if animal behaviour is not a genuine predictor of an earthquake and it is included in the signs considered by a decision-maker, it could lead to a false alarm, or worse, that if in the absence of animal anomalies an earthquake prediction could be mistakenly discounted.

3.3.7 Short-term Earthquake Prediction – an Illusory Goal

Despite half a century of work on short-term earthquake prediction, the prevailing mood among scientists is rather pessimistic. To date no reliable and widely accepted precursors have been found. In 1991 IASPEI (The International Association for Seismology and Physics of the Earth's Interior) set out a set of guidelines with regard to the data accuracy and validation of suggested precursory phenomena; of the 31 precursory phenomena proposed none satisfied all the guidelines. Of the many short-term predictions of earthquakes that have been made, none (except possibly that in Haicheng in 1975) have been both precise enough to lead to public action and subsequently proved correct. Claims for success tend to rest on the prediction of events expressed in a rather imprecise way.[3] Two particular predicted

[3] The VAN group in Greece claim to be able to anticipate forthcoming earthquakes based on geoelectric currents. However, their predictions have mainly been imprecise or incorrect in relation to magnitude, location or time window (Geller 1997).

earthquakes, the anticipated Tokai earthquake in Japan and the anticipated Parkfield earthquake in California, based on the idea of a characteristic earthquake, have both failed to materialise within their anticipated time period. And there is a growing belief among seismologists that the nature of the earthquake triggering process is inherently unpredictable. Nevertheless, there are scientists and groups who still have confidence in their ability ultimately to provide useful predictions, given sufficient instrumentation to generate the data needed.

3.4 Instantaneous Warning (Seconds)

One sure way to predict an earthquake is after it has happened. In a number of specialised cases, the danger from earthquakes comes from the shock waves arriving from an earthquake with its epicentre some distance away. This is the case for many of the deep earthquakes off the coast of Japan, the coastal earthquakes of Central and Latin America and elsewhere, affecting the towns and regions some distance inland. These earthquakes occur some 20 or 30 seconds before their shock waves hit the towns inland. Japanese Railways have pioneered an alarm to register the occurrence of a large coastal earthquake and signal an automatic braking system for the high-speed *Shinkansen* bullet trains operating in the vicinity inland. The 20 seconds or so gained from advanced instantaneous warning means that the trains can be slowed to a much safer speed by the time the ground starts to shake.

Similar warning systems have been tested in other areas[4] and may be useful in locations a long distance from likely earthquake epicentres for many factories, power stations and other mechanical operations that would be safer if shut down by the time the ground starts to shake. Unfortunately, they are probably of limited use in evacuating people, as the warning period is much shorter than the time needed to recognise the warning, react and evacuate buildings – evacuation times and mobilisation requirements are discussed in Section 3.5 below. Indeed it may only add to the danger of panic, stampede and injury. However, in conjunction with a well-considered earthquake drill, such a warning might be thought worthwhile to let people carry out rapid preparation measures and brace themselves in a safe position.

3.5 Practicalities of Prediction and Evacuation

In spite of the more pessimistic attitude to short-term prediction which currently prevails, groups who have confidence in their predictive methods will continue to anticipate earthquakes (although often in terms which are rather imprecise

[4] For example, a warning system for Mexico City, described in Rosenblueth (1991).

about location and timing), and those responsible for earthquake protection in the region affected will have to decide how to respond in these situations.

Given the uncertainty of any prediction or anticipation the scientist who provides the prediction information should not have to determine what decisions are made on the basis of the uncertain information. The hard decisions remain the responsibility of an elected politician or an appointed administrator who is ultimately answerable to the people affected. The penalties for getting a prediction wrong are heavy. The evacuation of a city or region and putting emergency countermeasures into operation are highly disruptive, costly and likely to be unpopular. Economic production is lost, factories closed down, wages missed and all aspects of life are disrupted. A false alarm not only causes economic losses comparable to the effects of a (smaller) earthquake, but also destroys the confidence of the public in prediction for a future event. The insurance situation of being able to claim for business interruption caused by false prediction has not yet been tested. Legally there may be grounds for companies to recover their losses from the government through lawsuits.

The politician or administrator is faced with an unenviable decision: that of ordering a highly disruptive and costly exercise on the basis of probabilistic advice. There are no rules to determine how probable an event has to look before it should be acted upon. There will not be time to take a cool look at the costs and benefits of a possible false alarm against the possible life saving that could result.

In places where monitoring equipment is in place, such as Japan and the United States, an exhaustive procedure has been proposed for reviewing the data, consulting teams of specialists, agreeing on a prognostication and passing their advice to a government committee. The final authority for declaration of emergency countermeasures is the responsibility of the Prime Minister (in Japan) and the President (in the United States). The period envisaged for full consultations and checks is several hours or even days.[5] To prevent public panic, news that the earthquake warning committee is meeting is generally not made public. Predictions of greater urgency than the hours or days required may not be able to be authenticated quickly enough to act on, so the prediction science needs to be geared to give accurate time and place forecasts with at least several days' warning.

3.5.1 Notification

In addition to the problems of making predictions and deciding whether to act on them, there are the logistics of how to notify, convince and motivate large numbers of people to leave their buildings at short notice. After the precursory signs are first recorded, analysed, a scientific committee has authenticated them

[5] NLA (1987).

and passed its advice on to the government and the government has declared a state of emergency, the time left before the occurrence of the earthquake may be very short. The success of prediction relies on a wide-scale and effective warning system reaching the entire population within the likely affected area.

Communication systems needed for successful evacuation must be in place well before any prediction can be acted on. It has to be possible to get a warning to many thousands of people within a few minutes, and this cannot be improvised. It requires considerable investment in warning systems, perhaps for sirens in every neighbourhood, mobile public address systems and radio and television newsflashes. But it also requires the public themselves to be prepared for the emergency, to know what the warning means and what to do, and most importantly to participate as part of the warning system. They have to check that their neighbours have heard the warning, they have to look after sick, elderly or deaf people, and they must take the actions to make the warnings effective. This requires considerable education of the public by the government well before any prediction and the creation of a climate of public awareness. Public education for earthquake safety is discussed later.

The time taken to carry out an evacuation very much depends on the readiness and alertness of the general public. A very significant factor is the time taken for *alarm*. A distant siren is much less easy to recognise as an alarm than the interruption of a TV program – but at any particular time only a very small proportion of the population are watching TV or listening to the radio. Studies of fire drills and evacuation show that regular fire drills and person-to-person warning greatly speed up alarm recognition – without them, alarms can ring loudly for hours without building occupants taking notice.

3.5.2 Evacuation

After the alarm has been recognised, and people believe it is authentic, they then have to get to safety. The time taken to get out of a building depends on its size, the number of storeys, the complexity of the plan layout, and the number of people trying to get out at the same time. This is well known in building planning for fire regulations. Figure 3.1 shows the evacuation time for plan layout size and number of storeys of different height.

Note that if the evacuation starts only when the earthquake begins, there is insufficient time for most people to get out of buildings before a major earthquake has finished (and the building will have collapsed if it is going to) – even if it is still possible to walk downstairs while the shaking is going on.

3.5.3 The Population Outdoors

Once outside, the population must be able to find sufficient open space for everyone to stand safely away from the possibility of falling glass or collapsing

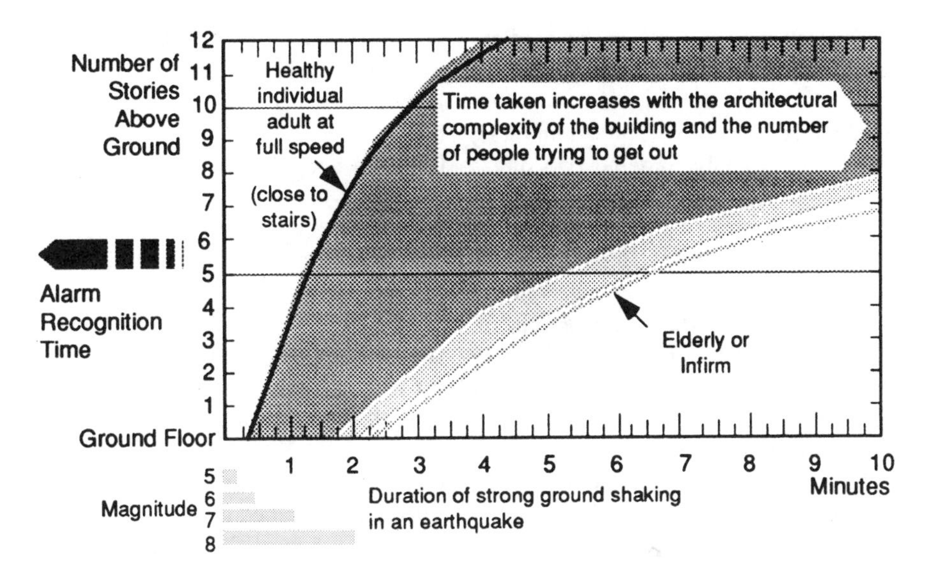

Figure 3.1 Evacuation times from multi-storey buildings (after Georgescu 1988)

buildings. It is necessary to designate congregation points and refuge areas close to each person's home or workplace where facilities, organisation and information can be provided. These refuge areas should be well publicised and provide rallying points for post-earthquake roll-calls and community emergency mobilisation. In dense areas of tall buildings this may not always be easy. There are not many downtown areas in large cities that have sufficient open space to give a square metre to every person who works there.

Refuge areas need to be equipped for people to remain out of doors for many hours. It may be difficult to prevent people returning to their homes or workplaces after some time – particularly in cold weather. Facilities will have to be provided outside the buildings for any extended evacuation period, including rain shelters, public toilets, food and possibly tents and blankets for sleep. After a day or so, people will want to return to their homes for fresh clothing, washing facilities and other needs, unless these are also provided. It is probably impossible to prevent people returning to their buildings if they finally lose confidence in the prediction and are suffering discomfort. People will have to believe strongly in the prediction to maintain their evacuation – something which can be reinforced by keeping everyone well informed of developments with regular news announcements and well-established communication systems to the congregation points.

3.5.4 Past Examples – Successes and Failures

There have been a few occasions on which a major damaging earthquake has been successfully predicted and an evacuation effected before it struck. The most celebrated of these is the prediction of the Haicheng–Yingkou earthquake of 4 February 1975 in China.[6] Predictions from June 1974 onwards led to a slow build-up of preparation work for the earthquake and in December many inhabitants were moved into temporary shelters. The evacuation of three large cities was officially ordered two days beforehand and a quake warning issued in the morning of 4 February. The magnitude 7.3 earthquake occurred in the afternoon damaging 22 million square metres of property. It killed 1328 and injured 16 980 people, claimed as the lowest mortality rate in China (0.02% of the population within intensity greater than VII) in recent decades.

There are also occasional reports of individual actions before an earthquake saving lives, such as the personal decision by the Mayor of Perahora in Greece to evacuate his town after a major foreshock.[7] The subsequent earthquake, the 1981 Corinth earthquake, caused 157 buildings to collapse in the town but resulted in the death of only three people.

There have also been a number of occasions when predictions have not worked. A review of earthquake prediction has identified and described nine further unsuccessful, publicly announced predictions between 1974 and 1995.[8] One example is in Tuscany, Italy, when a large-scale evacuation was ordered after a small tremor. 56 000 people living in the towns of Lucca and Modena were evacuated. The police and fire brigades were mobilised, 13 000 hospital beds were freed and railway carriages brought in to house evacuees. The alert caused wide-scale traffic jams and petrol shortages as people tried to drive out of the area, and closure of shops and businesses for two days. When the alert was called off, recriminations from angry businesses and townspeople brought the eventual resignation of Lucca's mayor and administration.[9]

3.5.5 Pros and Cons of Evacuation

The organisation and pre-planning needs for a successful evacuation should not be underestimated. The scientific investment necessary to make earthquake prediction a practicality will be pointless unless there is complementary investment in warning systems, and organisation to make a prediction usable.

If an earthquake is correctly predicted and a warning issued then it is likely to save the lives of many of the people who would otherwise have been inside buildings when they collapsed. A warning also gives time to shut down vital industries, prevent fire outbreak, stabilise furniture and building contents, stop

[6] Zhang (1987).
[7] Ambraseys and Jackson (1981).
[8] Geller (1997).
[9] Alexander (1984).

trains, assemble and prepare emergency services and many other useful preparations to help alleviate the impact of the earthquake. But physical damage to buildings cannot be prevented, infrastructure cannot be protected, the main cause of economic loss in earthquakes is not removed and not all people reach safety.

Even with a successful evacuation, the complete elimination of casualties may not be achieved. Despite eight months' preparation and a full-scale official evacuation, the Haicheng–Yingkou earthquake in 1975, described above, still killed over 1300 people. The problems of successfully isolating the 6.5 million people in the region away from the possibility of earthquake-induced hazards were obviously immense.

Even in very successful evacuations, some people will remain unwarned, perhaps in isolated communities or out of communication, and some are unable or unwilling to be moved. Often the most vulnerable members of the community, the old or the critically ill, may be the most difficult to move and there are others for whom rapid evacuation may pose particular problems, like miners, workers in vital services, prisoners, etc.

The difficulties of prediction mean that most damaging earthquakes will continue to occur unannounced for the foreseeable future. Earthquakes will continue to occur outside those areas where they are expected to be imminent, so actions to reduce the possible impact of earthquakes must also be carried out outside those targeted for short-term prediction. The fact that short-term prediction cannot reduce the heavy economic impact of destruction to property means that even within the areas targeted for prediction, proper protection for the community must range wider than a few days' warning, towards full damage mitigation policies.

3.5.6 Invest in Mitigation Rather than Prediction

In the light of these difficulties, the value of large-scale investment in short-term prediction and evacuation organisation should be carefully considered. If resources are scarce, then investment may be better justified in the essential long-term prediction studies necessary (building up a better seismic monitoring network rather than installing prediction tiltmeters, for example) and investment in long-term mitigation actions to reduce building damage rather than spending money to prepare for evacuation. An alternative policy to life saving by evacuation is life saving by preventing building collapse. This has the added justification that money spent on improving building strength will save damage cost as well as reducing the chances of human injury. These policies are dealt with in the following chapters.

What does an earthquake feel like?

Most earthquakes are small and, if felt at all, consist of a rattling vibration and the sound of a low rumble. Window panes may rattle and crockery chinks because they are more sensitive to high-frequency vibration that people do not feel. Some very distant earthquakes feel like the gentle swaying on board a ship. Small earthquakes do not last very long – only a few seconds.

Large earthquakes that occur some distance away may also begin like a small earthquake, but quickly build up to strong, violent shaking. One of the most frightening things about an earthquake is its noise. The noise of the ground vibrating – the energy travelling through the earth – is deep and loud. The earthquake also sets in motion hundreds of items around you that also clatter and groan. Sometimes the sound of the earthquake can be heard approaching, rolling over you and finally disappearing away into the distance. The motion of the ground can build up to a violent shaking, preventing you from standing or walking, and throwing items off tables, books off shelves and overturning furniture. The motion itself is likely to be disorientating and the violent motion may make you feel dizzy and sick. The more severe the shaking and the longer the earthquake lasts, the more likely the building is to be damaged. The larger the earthquake is the longer it lasts. Duration of shaking mainly depends on how long the fault is and the length of time taken for the rupture motion to travel along it. An earthquake of around magnitude 6 might cause strong shaking lasting 30 seconds or more. An earthquake of magnitude 7 or more may last minutes. The earthquake may also last longer on soft ground as this continues to vibrate after the earthquake has stopped. At levels of shaking where buildings are being damaged, it is difficult to walk, so running out of a building after it starts to develop damage is difficult. It is best to find protection within the building.

If you are close to the source of a large earthquake, the shaking does not build up in strength at all, but begins immediately with a loud clap of thunder-like noise and strong violent shaking. This may feel like a giant punch upwards into the air, or sideways, followed by powerful waves of shaking. Few people have time to understand what is happening. Fortunately, the chances of being caught right in the centre of a powerful earthquake are much less than the possibility of being in the surrounding area. The chances are that if you are aware of the fact that you are in an earthquake, you are some way away from the epicentre.

3.6 Getting the General Public Prepared

If everyone knows what to do in an earthquake, some of the injuries will be avoided and the aftermath will be more organised. Moderate earthquakes will be less disruptive if people take simple precautions. People can be encouraged to protect themselves. If the general public understand about earthquakes, then they will understand and support efforts made on their behalf to protect them.

What to do in an earthquake

If you are:

Near an exit

Or can get to it easily, leave the building as soon as possible. Do not stay to collect belongings or valuables. As you go outdoors, put your arms over your head to protect yourself against possible objects falling from above and move as far away from nearby buildings as possible. Do not look up until you are well clear of the buildings, in case objects hit you in the face. Do not rush straight out into the middle of the road: watch out for traffic.

Upstairs

If you cannot get to the exit quickly, look for protection within the building. Stay away from balconies, parapets, low windows and balustrades in case a sudden jolt throws you off balance or the rail gives way. Keep away from bookshelves, wardrobes or tall furniture that could topple over on you. Find a strong piece of furniture (like a table or a steel-framed bed) and sit or lie down beside it or underneath it. If you are in bed, roll out of bed and lie next to or underneath the bed. Brace yourself against the furniture and hold or cover your head to reduce the disorientation produced by vibration. Pull a cloth, sheet or piece of clothing over your head to protect yourself from breathing the thick dust that may be thrown up if the building suffers any damage. When the shaking has stopped, go straight outdoors.

In a high-rise building

Sit or lie down on the floor, next to or underneath a strong piece of furniture (like a strong table or filing cabinet). When the shaking has finished, get up and evacuate the building. Do not use the elevators.

In a car

Slow down and stop the car when safe to do so. Keep the car away from roadside structures, bill-boards, tall buildings or any other structures that could fall onto the car. Stay inside the car until the shaking has finished.

Cooking, working with machinery, or near a fire or naked flame

Shut down the machinery, switch off your cooker and extinguish any flames. If you cannot do so quickly, stay away from the machinery or flame and shut it down as soon as the shaking has stopped.

Some types of advice may be useful for the general public and volunteer groups. The following pages include examples of the types of public information and advice that might be useful (see boxes).

After an earthquake

- If your building has been damaged do not re-enter it. Another earthquake can come at any time. Even if your building has not been damaged, stay outdoors for an hour or so.
- Do not use the telephone unless somebody has been injured, or a building is damaged or burning. The emergency services may need all the lines they can use.
- Try to see that all heaters, boilers, cookers and naked flames are extinguished. If you can turn off gas or fuel supplies from outside a building, do so. Do not re-enter the building.
- If a fire breaks out, you may have to organise a civilian group to fight it; it is possible the fire department may be overloaded with emergency calls.
- Have a look at the other buildings nearby: some buildings may have suffered collapse or heavy damage in the middle of an undamaged neighbourhood. Their occupants may need your help.

If a building nearby has collapsed

- Get somebody to notify the authorities, either the fire department or the police. If it is not possible to contact the emergency services by telephone, get a volunteer to run to the nearest station. Give the exact address, estimated number of people inside and type of building.
- At the same time, try to help the occupants of the collapsed building out of the ruins. This requires organising a team of people to lift and move heavy pieces of rubble or beams off trapped people.
- Do not use lifting machinery, bulldozers or mechanical diggers, even if they are available, without the direction of skilled rescue advisors. Machinery is more likely to kill trapped victims than help them. Use manual labour to excavate rubble.
- Get some idea where the people are inside the rubble before you start digging. Stop the digging and other noises, form rescuers into a circle around the rubble, and call out. Listen for somebody inside the rubble replying or making a noise. Get everyone to point at where they think the noise is coming from. Place a flag or marker at places where it appears noises are coming from and continue digging. Repeat after some time to confirm the direction of digging.
- Do not expose yourself or your team to unnecessary risks. A collapsed building is highly unstable and dangerous. Do not crawl into confined spaces or stand underneath damaged walls without putting in proper shoring, and preferably only under the direction of a skilled rescue advisor. Take a few moments to prop a beam against damaged walls in the vicinity of rescue efforts, and dig rubble from the top of a slope before excavating at the base.

Public awareness can be raised in a number of ways, from short-term, high-profile campaigns using broadcasts, literature and posters, to more long-term, low-profile campaigns that are carried out through general education.

3.6.1 Awareness of Earthquakes: Familiarisation and De-sensationalising

Everybody who lives in a seismic area should know about earthquakes – they are a fact of life. If people know and understand about the threat of earthquakes they can take actions to protect themselves. Their understanding should include being aware of what to do in the event and being conscious, even at a low level, that their choice of house, the placing of that bookcase or stove, and the quality of construction of the garden wall around their children's play area all affect their own safety.

If public education is handled well, there should eventually develop a climate of everyday practicality to earthquake safety – a *safety culture* – where people take conscious, automatic precautions through being conscious of, but not terrified of, the possibility of an earthquake. People are unfamiliar with earthquakes because they happen very rarely – even in the most seismic areas few places are damaged more than once or twice in a lifetime. So earthquake risk is not like traffic risk or fire risk in the home that can be learned through experience. Earthquake risk has to be taught through abstract images and concepts.

The first part of creating a safety culture is familiarisation with earthquakes. Regular reporting of earthquakes in other parts of the world on TV and in the media is a help, together with occasional mentions of them (in less disastrous forms) in everyday contexts, such as stories, TV soap operas, novels, press newspapers and other common media.

Information about earthquake hazard should be part of the standard curriculum of all children at school, all professional training and part of the briefing of officials and administrations.

The second part is to de-sensationalise the effects of earthquakes. Only one perceptible earthquake in a thousand causes a disaster. Reporting only the catastrophic earthquakes causes fear and fatalism: 'If an earthquake lays waste a town, what difference does it make where I put my bookcase?' Fear is a well-known barrier to learning. If somebody is afraid of something their mind shuts out the valuable information. If a child is shown a film of a garden wall falling on a woman, the child will not learn that garden walls are dangerous, it will simply fear for the life of its mother. The treatment of fictional earthquakes in the common media should be aimed at showing how a household copes or otherwise with a disruptive tremor, not the annihilation of the soap opera family through cataclysm.

Formal programmes of posters, lectures and public information films will be a useful addition to a public that have already developed a sense of earthquake public awareness. If a climate of earthquake safety awareness has not been created, then public programmes are meaningless and may appear ridiculous – out of context a warning against earthquakes may seem as relevant as one against alien invaders. There have been a number of examples in different countries of 'public education programmes' consisting of colourful bill-boards in the streets or public information broadcasts, but alone, without a coordinated strategy for raising awareness, these efforts have often been wasted.

Awareness of risk locally is aided by reminders of past events: the preserved ruins of a building damaged in a past earthquake can be a useful reminder of earthquake hazard as well as a memorial or symbol of reconstruction. Involvement of the community in earthquake protection plans may involve public meetings and consultations, public inquiries and full discussion of decisions in the normal political forum.

Further awareness is developed through drills, practice emergencies and anniversary remembrances. In hospital, schools and large buildings it is often common to have evacuation practices to rehearse what the occupants should do in the event of fire, earthquake or other hazard. In schools children may practise earthquake drills by getting under desks. This reinforces public awareness and develops behavioural responses.

In some countries, the anniversary of a major disaster is remembered as Disaster Awareness Day – 1 September in Japan, 20 September in Mexico, and the month of April in California (Figure 3.2). On these occassions drills are performed, ceremonies and activities held to promote disaster mitigation.

3.6.2 Selling Safety

Earthquake protection will only come about when there is a consensus that it is desirable. In many places, the individual hazards that threaten are not realised, the steps that people can take to protect themselves are not known and the mandate of the community to have itself protected is not forthcoming. Earthquake preparedness planning should aim to develop the 'safety culture' in which the

Figure 3.2 The California public awareness programme involves an 'Earthquake Preparedness Month'. Images used in publicising the 1992 event

general public are fully aware of the hazards they face, protect themselves as fully as they can and fully support efforts made on their behalf to protect them.

The concept of earthquake safety has to be advertised and sold to the general public in the same way as any other marketable product: by educating the market to understand that the product is more desirable or has a higher priority than rival claims to their resources. Somebody building or buying a house has a choice over whether to invest their money in a stronger structure or more expensive finishes; it is important for their own safety that they choose the stronger structure. A good protection promotion campaign should make people consider safety features on a building as an asset, in the same way they might be sold on a car.

Community groups can help by educating their members, promoting public awareness and giving out information about earthquake protection. One of the greatest pressures that shape attitudes towards safety is the opinion of colleagues and friends. If it is generally accepted by the community, particularly by community leaders and opinion-formers, that it is sensible and beneficial to be protected ('safe is smart') then many people will conform.

In the end, only the communities and individuals affected can turn preparedness for a future earthquake into a force for safety.

Further Reading

EERI, 1984. *The Anticipated Tokai Earthquake: Japanese Prediction and Preparedness Activities*, Publication No. 84-05 (ed. C. Scawthorn), Earthquake Engineering Research Institute, 2620 Telegraph Avenue, Berkeley, California 94704, USA.

FEMA publications (from www.fema.org).

Geller, R.J., 1997. 'Earthquake prediction: a critical review', *Geophysical Journal International*, **131**, 425–450.

NLA, 1987. *Earthquake Disaster Countermeasures in Japan*, National Land Agency, Prime Minister's Office, Government of Japan, Tokyo.

Publications on earthquake preparedness prepared by Bay Area Regional Preparedness Project (BAREPP), Metrocenter, 101 8th Street, Suite 152, Oakland, California 94607, USA, include:

A Guide to Marketing Earthquake Preparedness: Community Campaigns That Get Results

Local Incentive Programs: Case Studies (Examples of community programmes for earthquake preparedness in Southern California)

Resources for School Earthquake Safety Planning (Teaching and curriculum materials, videos, instruction books and guidelines)

Earthquake Media to Public: Guidelines for Department Managers (The role of the media in earthquake preparedness).

4 The Earthquake Emergency

4.1 Emergency Management

A well-coordinated response to an earthquake is likely to save many lives, greatly reduce the disruption to the population, and prevent earthquake-triggered hazards escalating the magnitude of the disaster. Poor emergency response or a follow-on disaster can double, treble or multiply 10-fold[1] the death toll of an earthquake.

In the immediate aftermath of a major earthquake, the situation can rapidly become chaotic, with many uncoordinated activities, poor communications between groups and a general ignorance by the population of what to do. Time is essential: most people trapped in collapsed buildings who are not rescued within a few hours will die. They have to be found, retrieved and given adequate medical attention. People are out on the streets without shelter. Society has been disrupted, communications are knocked out, aftershocks are frequent, and normality is suspended.

There may be no overall authority in charge and ad hoc groups of people, organisations and local administration are likely to be dealing with the emergency in a number of different localities. Each of these groups may have to rely on their own resources and ingenuity for several hours or days. Containment of the emergency is the first priority, preventing any possibility of the disaster escalating, followed by establishment of order and a gradual return to normality. This requires an urgent and efficient organisation of labour and resources, prioritisation of actions with time, and an understanding of the likely consequences of the disaster. In most cases this has to be carried out with imperfect information, perhaps even in the absence of any idea of the extent of the catastrophe.

[1] Failure to suppress major fires that can follow earthquakes in Japan has been shown to multiply death tolls by a factor of 10 (Coburn *et al.* 1987).

Pre-earthquake emergency planning is one of the best ways to ensure that the earthquake can be handled effectively. If, before the event, there has been an emergency plan drawn up, public information has been given out and people have been trained in what to do, the emergency can be handled effectively and the effects of the earthquake will be reduced. However, if no emergency plan exists, or for some reason the plan fails to be appropriate, a good understanding of the issues and priorities can enable an effective emergency response to be improvised.

This chapter deals with the issues involved in dealing with an earthquake emergency, both to help in the preparation of an emergency plan in preparation for some future event and to structure an improvised emergency response should it ever prove necessary.

4.1.1 Reinforcement of Volunteer Groups

It can be assumed that, in a large, recognisable disaster like a major earthquake, participation of the general public, the normal emergency services and volunteer groups will occur spontaneously. If buildings have collapsed or have caught fire in a neighbourhood, people nearby will be attempting to help. If people are injured, they will be attended to by other people on the scene.[2] They do not wait for instructions from higher authorities before starting to help. It is often incorrectly assumed that the best model for emergency management by central authorities is a military 'command and control' response, because disaster impact has certain similarities with a war situation. The difference with a disaster is that response activities are spontaneously underway without a command from a centralised control. However, the very definition of a disaster is that the emergency exceeds the capability of normal, local resources to deal with it. Disaster management by central authorities in the first instance is the procurement and distribution of additional resources to reinforce the local response where it is most needed. Emergency services must be geared to operate independently without centralised control or coordination.

4.1.2 Agency Coordination

In the emergency response a very large number of agencies, organisations and individuals may become involved. Many of them are likely to be autonomous or not under the direct control of any single central agency. Examples of agencies involved in emergency response are given in Figure 4.1.

It can be seen from Figure 4.1 that many of the agencies likely to be involved in the response to a large-scale emergency are not under the direct control of

[2] After the Kobe earthquake, for example, 630 000 volunteers worked in the area during the first month (IFRC 1996).

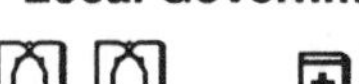

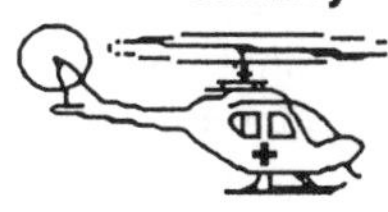

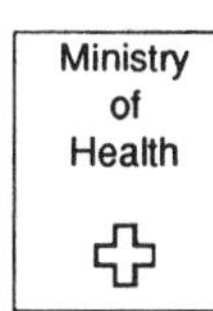

Figure 4.1 Organisations likely to be involved in emergency response after an earthquake

any single, central agency but are independent or answerable to other authorities outside central government. During the emergency period these groups may well agree to be directed by a central disaster committee, but each will effectively be working towards their own perceived objectives and with different criteria.

Effective disaster management requires the coordination of these disparate groups. It requires integrating a large number of parallel agencies towards a common goal. A primary requirement is information – both to and from the organisations. True coordination between different groups, however, goes far beyond the exchange of information to include standard operational policies, response doctrines, standards of practice and compatible specifications of equipment.

Ultimately, of course, the final decisions on declarations of emergency, scale of response, request for international assistance and strategic decisions on recovery and reconstruction rest with the national government and the presidential or cabinet administration. Most structures of disaster management are topped by a premier (Prime Minister/President) or a presidential or cabinet committee. The hierarchy of how this committee relates to the large number of agencies involved in the response is a matter for the disaster plan of the individual country.

Structures of disaster management administration in government have been categorised[3] into presidential, (a coordinating office within the office of the prime minister, cabinet or presidential administration), ministerial (a specific ministry for disaster issues), multi-ministerial units (disaster units within a number of ministries) and voluntary council (a disaster coordinating council formed of many different bodies within and outside government). The presidential model of disaster management administration is thought by many to be most effective as it outranks other ministries and centralises power for obtaining resources.

4.1.3 The Disaster Plan

The pre-earthquake preparedness plan establishes the relationships between the various groups, how they will cooperate and the demarcation of activity areas. Perhaps most importantly, the preparedness plan identifies information needs, information flows and methods of rapid information exchange between agencies.

No disaster plan is likely to predict the exact circumstances to be dealt with – the location, severity and characteristics of future emergencies may be quite different from what is expected – but the methods of working, the areas of responsibility and decision-making, and the flows of information necessary to deal with a disaster can all be planned beforehand.

In practice few disaster plans are ever implemented in the form they are drawn up, but they have a considerable value in focusing the activities of the participants on disaster issues before the event.

[3] Davis and Wilches-Chaux (1989).

4.1.4 Testing the Disaster Plan

The disaster plan needs careful design and testing. The design of the plan should involve all the expected participants. Each agency can be asked to submit its own proposed participation within a master plan drawn up by the central coordinating agency. Testing the plan involves simulation exercises, which can be carried out in limited gaming exercises or full-scale dry-run practices (Figure 4.2). In these tests a scenario for a fictional earthquake occurring at a specified location is normally played out, with incoming *incident reports* (damage and casualty reports), cross-communication of *activity reports* (statements of what each agency is involved in and its anticipated needs) and outgoing *sit-reps* (situation reports on resources and needs). Computer simulation can be an effective way of visualising these scenario exercises. What is tested in the simulation exercises is the information flows, responsibilities and coordination of the agencies involved. If possible a number of widely different scenarios should be used to make sure the disaster plan is adaptable.

4.1.5 Multi-hazard Preparedness Plans

In most countries an earthquake is only one hazard of many that might have to be planned for in a disaster preparedness plan. Earthquakes differ from floods, hurricanes, industrial disasters and other hazards in a number of ways, in the level and type of destruction caused, the geographical extent and distribution of damage, and the degree of warning that can be expected. But the methods of response, the agencies involved, the information flows and other parts of the

Figure 4.2 Disaster plans need testing through simulation exercises and public participation. Simulation exercise in Hospital Balbuena, Mexico City, of an evacuation in an earthquake, while continuing to treat patients and receive incoming casualties

emergency response will have distinct similarities. It is generally accepted that generic emergency response plans are more useful than specific plans to deal with an earthquake or any other individual hazard. An emergency response capability to provide civil protection and containment of any low-probability, high-impact event is more useful than one dedicated to a single scenario. Earthquake preparedness plans should be one specific example of a general emergency preparedness capability for the country or region as a whole.

4.1.6 Communication Systems

Rapid interchange of large volumes of information is essential for the coordination of activities of the very many agencies involved in the emergency response. A general way of providing all the participating agencies with information is through the public media, particularly through radio, which is an instantaneous medium. Increasing use is being made of the internet, posting information on web sites for access by the broad community who still have communications. Radio receivers are common and portable and likely to be used by those affected by the earthquake. The media should be included in emergency planning and play a central role in broadcasting information as soon as it is available. Care should be taken to ensure that reports used by them are accurate and representative. Public media can often be misleading and unreliable – the selective and often exaggerated reporting of the more newsworthy disaster items can often give the impression to outsiders that the earthquake is more severe than it actually is, or focused in a particular geographical locality, omitting other important areas.

Public confidence is boosted by frank and complete media coverage and can act as a communicating medium for the many organisations involved in the emergency response. There is rarely any information which can justifiably be censored or deliberately withheld from the public domain. It is sometimes argued that warnings of follow-on secondary disasters (tsunami etc.) may cause widespread panic or that ongoing rescue reports attract unwelcome sightseers, but there are few reported cases of public misbehaviour and the benefits outweigh dangers. Communication systems are critical for effective disaster response and a special communication system may need to be established as part of the preparedness measures taken against a major earthquake. In a large-scale earthquake, line-based telecommunications within the affected area are likely to be damaged and may be unusable. Such telephone lines as remain operational are likely to be swamped by the general public, either reporting damage or trying to contact friends and family. Satellite cellular phone networks are today the favoured communications systems,[4] but radio networks are also used by emergency teams. Radio-based

[4] Slow response of the government agencies after the 2001 Gujarat, India earthquake was attributed to failure to maintain the cellular phone network installed for such emergency use (India Today, 12 February 2001).

systems require the setting aside of specific broadcasting frequency bands for emergency use. Several bands may be needed in addition to those normally used by the police, fire and ambulance services.

Officials whose responsibilities include emergency decision-making are normally issued with portable radio transceivers so that they can be contacted within seconds if a disaster occurs. A major bottleneck on emergency communication systems during a major crisis is likely to be the volume of incoming incident reports. Telephone switchboards jam and airwaves may become inaccessible as the radio operators at headquarters receive more radio reports than they are capable of processing at the time. Peak traffic loads in communications and particularly in incident reports are important to estimate and if possible simulate before setting up the emergency management communication system. Emergency managers can help by reporting succinctly and may be trained in coded or abbreviated reporting techniques to minimise air-time.

Information about the emergency faced can also be obtained by pre-instrumentation of key sensors. The most important of these for earthquake emergencies are seismometers. A rapid determination of the magnitude and approximate location of the epicentre of the earthquake is essential information in estimating the scale of the emergency being faced. Remotely monitored seismometer networks are routinely used by seismologists to provide location, depth and magnitude of an event within minutes of it occurring. Good communications are needed between seismological observatories and emergency management centres. Other remote sensors may relay important civil protection information from key industry, dams or other facilities whose failure could cause a major threat to public safety.

4.1.7 Information Management

Much of the information that has to be coordinated in an emergency is spatial: the location of incidents, building collapses and transportation routes. A map-room is central to most incident control centres or disaster management headquarters. Computer mapping is increasingly used for emergency management (Figure 4.3), with geographical information systems (GIS) being used to link maps with databases and other information sources.[5] It can be used to estimate earthquake damage in urban areas. The damage estimation methodology implemented in such systems requires a detailed classification of the geology and building stock. Earthquake response spectra are calculated by earthquake parameters and attenuation functions.[6] To estimate the building damage these

[5] An example is EQSIM developed at the University of Karlsruhe in conjunction with INCERC in Bucharest, Romania. EQSIM is a software tool based on the popular geoinformation system ArcView (Baur *et al.* 2001).

[6] The United States Geological Survey has developed TRINET, a system for rapidly mapping spectral values of peak ground motion immediately after an earthquake has occurred.

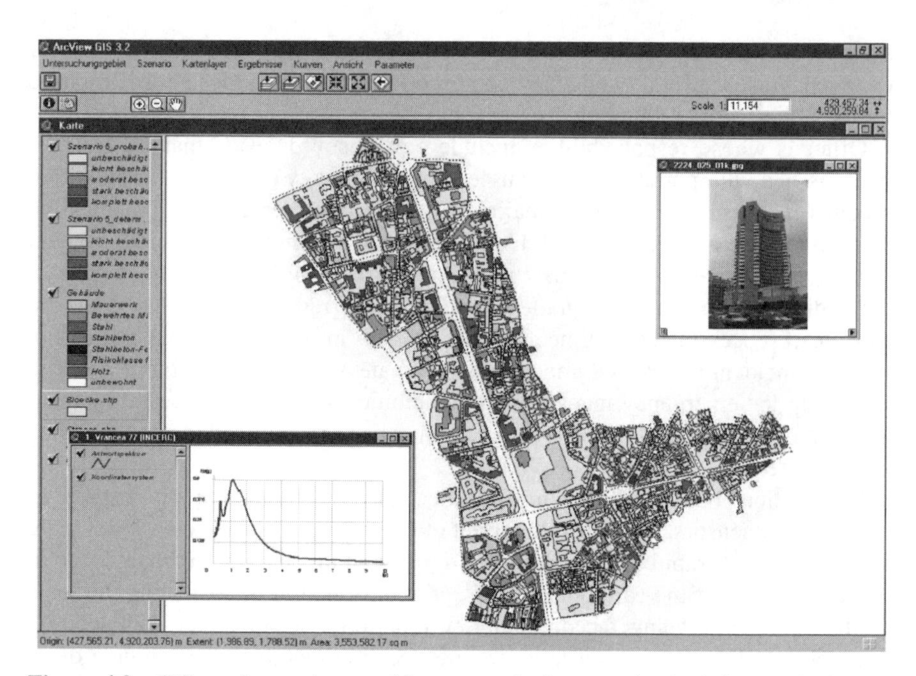

Figure 4.3 GIS can be used to combine maps, databases and calculation methods. The figure shows a damage scenario for an urban area calculated by the earthquake damage estimation tool EQSIM (Reproduced by permission of Michael Marcus from Baur *et al.* 2001)

response spectra are combined with capacity and fragility curves for the different building types. The resulting damage states of the buildings are stored in a database and can be visualised in thematic maps via a GIS interface.

Databases that become useful in emergency situations include resource lists (inventories of stockpiles and government-owned supplies), supply source centres (availability of emergency medical supplies, rescue equipment, tents, etc.), personnel and contacts lists and so on.

Keeping track of the deployment of resources, reserves, requests for assistance and responses to requests is administratively complex, but essential for effective management.

4.1.8 Regional Reconnaissance

In addition to receiving incoming incident reports, it is vital to instigate rapid searches to discover the extent and severity of the impact. It is possible that the worst-hit areas may be unable to report their own damage and reliance on incident reports alone may mean that some worst-hit areas are not reported for some time – the most common cause of high death tolls.

Areas affected by a large-magnitude earthquake may cover thousands of square kilometres. In order to plan an effective emergency response, it is essential to carry out a rapid survey of the extent of earthquake impact. This can be divided into two operations:

(1) regional reconnaissance across the whole affected area;
(2) urban reconnaissance in any large town affected.

The best method for carrying out regional reconnaissance over a large area is by air. Aerial surveys have the additional advantage of being able to cover any mountainous areas in which many of the less accessible villages and towns may be located. Aerial surveys can also report roads blocked by landslides, rivers dammed or other geographical effects of the earthquake. Helicopters are useful and can fly low enough and slowly enough to ascertain damage levels, but are slow and have a short range (Figure 4.4). Light aircraft may be best suited for rapid and wide-scale aerial survey reconnaissance.

If aircraft are not available, a systematic reconnaissance by road will take longer but should be undertaken by as many vehicles as possible, to minimise the time taken.

The regional reconnaissance should be as systematic as possible) with the following aims:

1. To determine the severity of earthquake impact.
2. To determine the geographical extent and spatial distribution of impact.
3. To identify the towns and villages most in need of aid.

Figure 4.4 Aerial reconnaissance can rapidly establish the extent of damage over a region. Helicopter operating from a temporary base in the epicentre of the 1980 Campania earthquake, Italy

Towns and villages should be categorised by their degree of damage, e.g. approximate percentage of collapsed buildings.

It is also important that the reconnaissance is seen as a vitally important operation in the life-saving coordination of the whole region. The reconnaissance pilot has to resist requests to land to help individual stricken communities, before covering the entire region of potential damage.

Damage patterns in earthquakes are not uniform or predictable. In a very general way damage is most severe at the epicentre and attenuates with distance, but from place to place the degree of damage is extremely variable. One village may be almost undamaged while the next, perhaps only a few kilometres away, is destroyed. Towns a long way away from the epicentre can be badly damaged in the middle of an area of generally low disturbance. Regional reconnaissance has to take this into account and be as exhaustive as possible, checking settlements individually rather than making assumptions about the damage attenuation.

4.1.9 Urban Reconnaissance

Urban reconnaissance may be significantly different. In severe cases, large areas of a town may have been devastated, particularly areas of weaker buildings and on poor ground conditions. These areas are likely to need massive emergency resources, and fortunately are likely to be very obvious where they have occurred, so that response can be quickly mobilised.

In some cases of urban disaster over the past few decades, damage has occurred to a small number of buildings across the town, but with a few severe collapses occurring amongst thousands of buildings. These buildings may, however, be much larger structures, containing hundreds of occupants, and have resulted in disastrous death tolls. It is possible for the scale of such a disaster to be unrecognised for quite some time – if, for example, 50 or 100 highly occupied buildings have collapsed, but the remaining 10 000 buildings in the city remain standing. In these cases, the critical need is to identify very rapidly the scale of the rescue operation needed, and find all the collapsed structures amongst the neighbourhoods of perhaps only lightly damaged structures.

An aerial survey of a large town may be able to identify very rapidly which quarters have been most badly damaged, and can spot large fires or blocked streets, but it is not always possible to identify each individual collapsed building from the air. Instead urban reconnaissance relies much more on reports from the public. Many different groups are likely to be helping with the earthquake emergency. It is important that each group carry out a full damage reconnaissance in their own locality. Urban reconnaissance tries to ensure that all cases of building collapse and other incidences of serious damage are reported back to a central authority even if they are already being attended by volunteer groups. Patrols on foot or in well-marked cars, in radio communication with a central control room, take major roads and then minor streets in a systematic pattern.

Again the job of these patrols is to report incidents so that help can be coordinated, rather than to become involved in the management of any one incident, until the reconnaissance is complete.

In areas where there is an array of strong ground motion sensors, computers can be programmed to provide instant maps of probable damage, which can be used to direct the reconnaissance for rescue teams and other emergency management operations.

4.2 Search and Rescue

Rapid rescue of people from collapsed buildings after the impact of a destructive earthquake can save considerable numbers of lives. The principal factors determining the number of people killed and seriously injured after a building collapses are the proportion of people who are unable to escape from the collapse (those trapped by collapse), their injuries and the length of time they are able to survive with those injuries, and how quickly they are able to be rescued and receive medical attention.

4.2.1 Number of People Trapped

When buildings collapse, not all the occupants of that structure are trapped inside. Many are likely to escape just before collapse and some are able to free themselves shortly afterwards. The number of people trapped in a collapsed building depends on the size and type of building, the extensiveness of collapse, how long it took to collapse and how easy it was to escape from the building. In a high-rise building, escape from upper floors is unlikely before collapse, and if it collapses completely, perhaps 70% of the building's occupants are likely to be trapped inside (Figure 4.5). In a low-rise building that takes perhaps 20 or 30 seconds to collapse, more than three-quarters of the building occupants may be able to escape before collapse. Rescuers are therefore looking for a proportion of the occupants in each building. Some knowledge of how many people were likely to have been inside the building at the time of collapse is useful. There are likely to be reports of missing people at the site of a collapsed building, which can be used to try to make some assessment of the number of people trapped.

4.2.2 Survival and 'Fade-away' Time of Trapped Victims

How long someone can survive inside a collapsed building depends on the type and degree of injury inflicted by the collapse. People occupying the building when it collapses may survive if they are not crushed or suffocated. A large number of people killed in earthquakes die of suffocation from dust thrown up by the collapse. Others more fortunate may survive inside the voids that are created

Figure 4.5 Search for victims buried in collapsed structures can continue for many days. Search and rescue activities in the collapse of a reinforced concrete framed apartment building, 1986 Kalamata earthquake, Greece. (Reproduced by permission of A. Pomonis)

within the collapsed building. How long they survive (their fade-away time) when trapped will depend on their air supply and level of injury. Medical evidence for fade-away times for different types of injury is compiled in Figure 4.6.

Different construction types, e.g. masonry and reinforced concrete, have different collapse mechanisms and cavitation characteristics. The void-to-volume ratio of the collapsed structure and the most likely location of those voids are of importance in locating and rescuing trapped people quickly.

Volume-to-void ratios are most favourable for reinforced concrete buildings.[7] The total collapse of masonry buildings provides fewer and much smaller cavities within the rubble than the collapse of frame structures. In the worst case, a victim may be completely buried by building rubble.

Anyone trapped under a heavy layer of earth or dust is likely to suffocate quickly. Estimates of numbers of people being rescued alive after being buried under collapsed earthen building types in Italy, Turkey and China[8] indicate that after six hours less than 50% of those buried are still alive.

Injuries received during the collapse of a structure may also be fatal if not treated quickly. Open wounds and internal haemorrhaging can be fatal within hours without treatment. The recovery of people suffering head injuries is very dependent on rapid medical treatment; data on people suffering closed trauma

[7] Krimgold (1987).

[8] Zhang (1987).

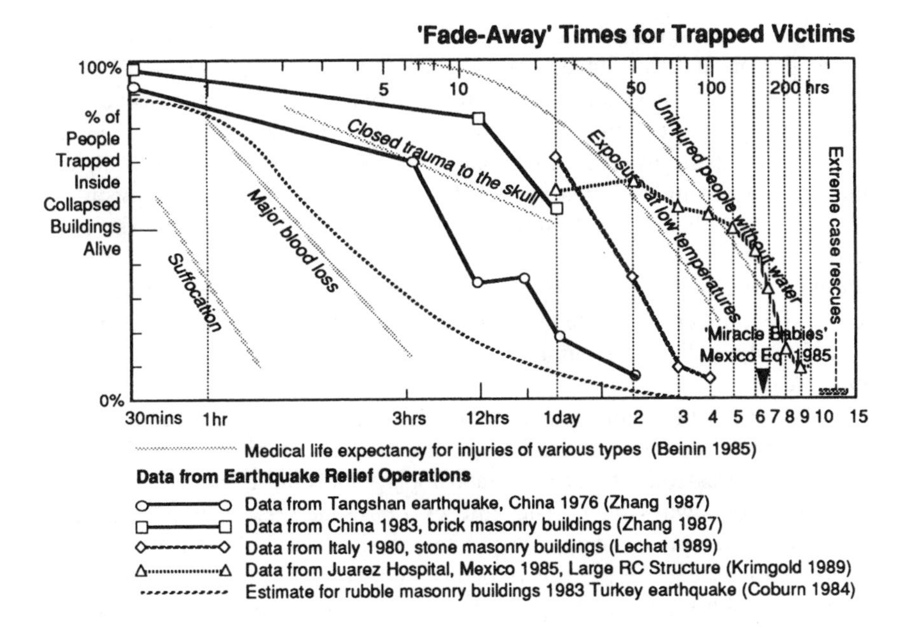

Figure 4.6 Survival rate with time for victims caught in building collapse

to the skull shows that about half die within 24 hours.[9] Other serious injuries similarly need medical treatment and attention within a matter of hours if the victim is to survive.

Longer term threats to people trapped without injury are exposure if the temperatures are low, infection in wounds, dehydration (particularly if temperatures are high), or eventual starvation. There is also the possibility that less serious injuries may become infected or lead to serious complications (see Section 4.3.5 on medical attention at the rescue site).

The length of time that individuals can survive without treatment also depends on their strength and physiology. Older and weaker people are less likely to survive for the same length of time as a healthy, strong person.

Environmental conditions also contribute to fade-away times. In low temperatures, or very high temperatures, survival time is significantly reduced. In warm weather, with perhaps a little rain to provide drinking water in the rubble, a lightly injured person may survive a number of days.

4.2.3 Speed of Rescue

The speed at which people can be rescued from collapsed buildings is largely a question of resources and their distribution. After an emergency situation is

[9] Beinin (1985).

recognised it takes time to assemble and organise rescuers, time to locate a trapped person and time to dig or cut them free from the collapsed building. The most rapid help usually comes from people already at the scene of the collapse, people occupying buildings that have not collapsed or who have escaped being trapped themselves. It is estimated that 90% of people rescued alive are pulled quickly from the ruins by people on the scene.[10] The more people are immediately available to help rescue trapped occupants, the more lives can be saved.

In places where a high percentage of people are trapped, more people will die because the number of people left to rescue them in the immediate vicinity after the earthquake is less; the number of rescuers per trapped person is critical for rapid rescue.

Number of rescuers alone is not the only criterion. Rescuers need to be able to locate trapped people quickly and to free them and possibly to provide medical aid. This requires some basic skills and tools which those immediately on the scene may not always have. The efficacy of rescuers is an important factor; some rescue attempts by inexperienced people may well kill the people they are attempting to rescue. The location of people under rubble and their excavation can be achieved with relatively simple search and rescue techniques and commonly available tools if the expertise is available.[11]

Improving the effectiveness of rescue operations involves the rapid mobilisation of locally available volunteer groups and coordinating them efficiently. They need support, in terms of equipment, transport and medical personnel (see Section 4.3). But most of all local volunteers need to be coordinated by leaders who know what to do. Local emergency services should be capable of providing search and rescue coordination and be trained in specialist rescue techniques.

4.2.4 International Search and Rescue Assistance

The specialist skills employed in the location and rescue of victims trapped in collapsed reinforced concrete structures have become the province of international search and rescue teams. The 1980s and 1990s saw a proliferation in international search and rescue (SAR) teams sent to the scene of a major earthquake by other governments to help with rescue efforts in a gesture of solidarity between the two countries. Non-governmental aid organizations (NGOs) also have SAR capability as a part of their post-disaster relief operations and there are NGOs in several countries specialising in post-disaster SAR.

To a large extent this has reflected the increasing incidence of urban earthquakes causing the collapse of high-rise, high-building-occupancy, reinforced concrete framed buildings. These collapses have evoked a new urban nightmare:

---

[10] Krimgold (1989).

[11] Coburn *et al.* (1987).

the horror of being entombed alive in unyielding concrete for many days on end. International media coverage of mass-collapse disasters has focused on the rescue operations for individual survivors, bringing updates live on TV into homes across the world as the extrication proceeds. To some extent this media focus reflects the time sequence of disaster response. In the first few hours and days many thousands of people may be rescued unreported by media concentrating on initial reports of the scale of the destruction itself. The more difficult rescues and those requiring specialist techniques and equipment remain as observable on-site activities in the following days when international news crews are covering events in depth. The despatch of specialist teams with the capability to help with such difficult rescues is an important public gesture of sympathy and aid by a friendly country.

International SAR teams can make limited but positive contributions to these types of disaster. They bring specialist SAR equipment such as electronic surveillance equipment and experienced techniques to supplement local capability in the more difficult types of rescue situation. The best preparation for fast and effective international assistance is bilateral agreements between nations concerning the procedures to request and to deploy assistance, involving the provision of fast visa procedures, transportation to operation areas, interpreters and connection to the local authorities. Standards for coordination of international rescue teams and for training, composition and equipment are provided by the International Search and Rescue Advisory Group (INSARAG)[12] to advance the quality of international help.

The number of countries developing their own SAR specialist teams for rescue from collapsed buildings is increasing.[13] The coordination of many different groups of semi-autonomous SAR teams at a disaster site is now a key role of disaster management.

Nevertheless, the impact that international SAR teams can have on reducing the overall casualty rates in an earthquake is very limited – the scale of the problem and the need for very rapid assistance mean that international assistance is only useful in occasional cases of difficult rescues of long-surviving victims. Logistical and practical problems of mobilising across national borders mean that international SAR teams usually cannot be in position for two or three days. The information on survival rates shows that the number of people alive inside a collapse after two, three and four days is extremely small, and also that their chances of recovery, even with treatment, are slim. The job of the SAR team by that time is extremely difficult. Their targets are victims who

[12] Further information and the INSARAG guidelines are available at www.reliefweb.int/insarag/ and from INSARAG (2001).

[13] After the Italian earthquake of 1980, four other countries sent specialist SAR teams to assist with the emergency. In the Armenia earthquake eight years later, 19 SAR teams arrived to help the Soviet authorities. Many of these were specialist SAR teams offering their services internationally for the first time.

may well have been capable of calling out a day previously, are now semi-conscious or unconscious, passive and extremely difficult to locate. They are close to death and may not survive even after hospitalisation. The same victims would have been easier to find and rescue would have been more viable had the teams arrived earlier. A significant improvement in the live recovery rate of international SAR teams could be achieved by speeding up their time of arrival at the disaster site.[14]

4.2.5 Strengthening Local SAR Capability

The most effective and realistic method of ensuring rapid arrival times of rescue assistance is for each disaster-prone country to have its own specialist SAR capability. Assistance from foreign countries to help with SAR in a disaster is needed where specialist equipment, experience and organisational techniques are insufficient. If these were to be built up within the country, then local capability would be preferable to foreign assistance. Local teams will always be closer to the event, faster to arrive, conversant with local administrative procedure, customs and language, and responsible directly to the affected community. Significant improvements in life saving after a mass-collapse disaster can only be achieved by strengthening the local capability to deal with it effectively.[15] These measures might include public awareness of simple post-disaster actions, essential principles of rescue from collapsed buildings for local police and fire services, and training in the more complex techniques of location of passive victims and extrication from modern building types for specialist national squads, as outlined in the next section.

4.3 Search and Rescue Techniques

For rescue work after building collapse the five-phase strategy shown in Table 4.1 is used successfully by many rescue teams.[16] The strategy aims to rescue the maximum number of victims with an increasing level of effort of SAR activities and an increasing risk for rescue personnel and victims through the five phases.

[14] In the Gujarat, India earthquake, the first international SAR team on site, the 50-strong Swiss team, arrived three days after the earthquake. Over the next 48 hours, their efforts saved eight live victims.

[15] The military often have a vital role to play in SAR. In recent earthquakes in both India and Turkey, the army played a key role in the entire emergency phase, though with better training and specialised equipment they could have been even more effective.

[16] M. Markus, personal communication.

Table 4.1 A five-phase strategy for SAR operations[a].

Phase	Search	Rescue
1. Reconnaissance and immediate rescue/ evacuation	• Assessment of the collapse area, building type and damage patterns • Hazard analysis: potential further collapses, falling debris, gas, fire, etc. • Interview witnesses and victims for location of further victims	• Immediate rescue • Pick up hurt and easily accessible victims • Mark hazardous areas, No rescue work in unsafe areas
2. Search and light rescue	• Search slightly damaged areas • Search accessible areas • Extended interviews • Objective: assessment of the entire site, knowledge about possible victims and their positions	• Rescue easily liberated victims • Initiate observation of indicators for building stability: gaps, tilted walls, etc. • Prepare for heavy rescue: cranes, heavy tools and equipment, shoring, stabilisation
3. Search areas with expected victims, heavy rescue	• Search all voids and accessible areas with probable locations of victims • Extended use of floor plans and other info • Only trained personnel, canine search and with listening devices • Objective: search of all voids except under rubble	• Penetrate and advance to entrapped victims • Open voids with the possibility of finding survived victims inside • Remove obstacles but mind the weakened structure • Stabilisation • Only trained rescue personnel in hazardous areas
4. Search in rubble and voids below, heavy rescue, specific clearing	• Specific search based on information from witnesses, observations, floor plans, etc., in rubble and small-sized debris	• Penetrate/advance to voids with expected victims in rubble and unstable debris, tunnelling • Mind danger to victims by rearrangement of debris (crush) and fine material trickling (asphyxiation)
5. Final clearing	• Regular stops of clearing work for search in the debris and recently uncovered voids	• Carefully clearing • Help of cranes and excavators directed by rescue personnel • Avoidance of debris destabilisation

[a]From M. Marcus, personal communication.

4.3.1 Likely Locations for Survivors

Most victims who can be rescued tend to be located within voids in the collapsed building. Victims within voids are more likely to survive because not only are they saved from the pressure of building material above, they also have some degree of air supply. In masonry buildings (Figure 4.7), voids and cavitation tend to occur in partial collapse of structural supports; the voids are created underneath fallen floor joists or floor slabs. Protruding joists from rubble may indicate that voids have been created below, which may contain survivors, and standing walls may also have voids at their base.

Doors, tables and other larger pieces of furniture may also have created voids and air pockets for survivors. Staircases, particularly timber staircases, can also create voids in their collapse, and occupants trying to flee from a collapsing building may well be found close to the staircase. Exits (e.g. trying to open doors), escape routes and circulation areas are common locations for buried victims. Other zones of domestic buildings which have strong furniture and may similarly be promising places to search for survivors include the kitchen and utility areas of a house, where refrigerators, washing machines and other steel appliances may provide some protection and where occupants spend a lot of their time.

In reinforced concrete buildings (Figure 4.8), the pattern of collapse of slabs may create cavitation characteristics that could contain survivors. Partially collapsed structures will generally have more survival space inside, and be more accessible than totally collapsed structures. Buildings with some slightly stronger structural elements, like structural cores, may provide some vertical support that will create cavities in the rest of the collapsed structure. Some types of structure, such as deep-beam design, provide greater cavitation potential on collapse than others. From the point of view of locating survivors, the worst type of collapse encountered in reinforced concrete buildings is the *pancake collapse* of all floor slabs, tightly packed one on top of another. Cavitation potential is limited and extremely localised, routes for sound out of the building are minimal and rapid penetration by rescuers into the collapse is impossible. The complete dismantling

Masonry Collapse

Protruding or slanted floor joists may indicate voids that could contain survivors

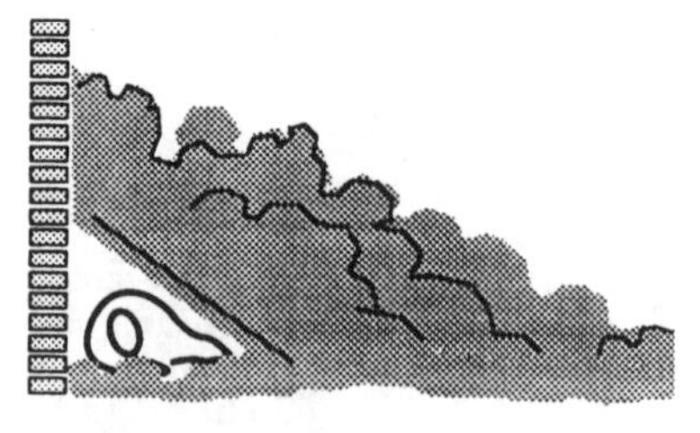

Standing walls may indicate the possibility of a void at their base

Figure 4.7 Likely locations for survivors in collapsed masonry buildings

Reinforced Concrete Collapse

Slab and deep-beam construction may create voids for survival

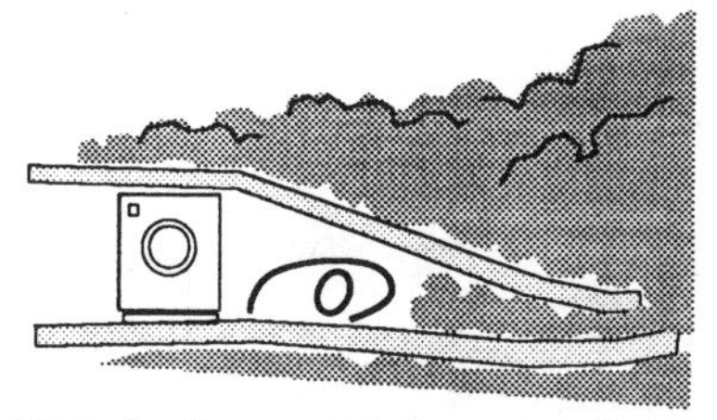

Strong furniture, e.g. steel-cased appliances may resist building collapse pressures sufficiently to create local cavities

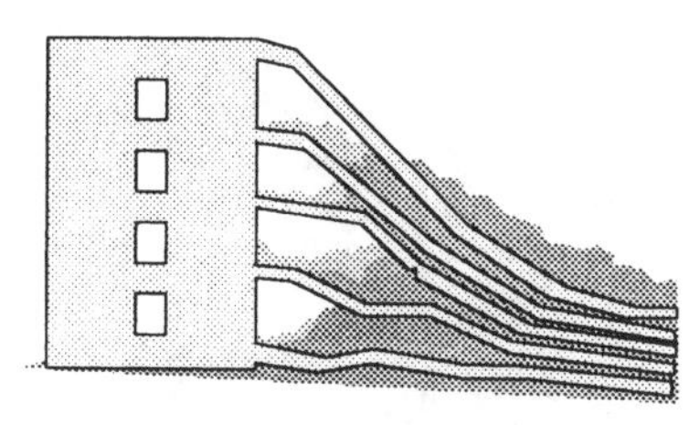

Stronger structural elements, service cores, shear walls etc., may also support collapsed elements to create voids

Structural resilience in failed members may also provide sufficient support to maintain thin survival spaces

Figure 4.8 Likely locations for survivors in collapsed reinforced concrete buildings

of a reinforced concrete building in the search for victims can take many weeks. Concrete is a very hard material to break up and steel reinforcement takes time to cut through. To get people out alive, it is important to locate where they are as accurately as possible to minimise the amount of concrete that has to be cut through.

4.3.2 Finding Survivors

SAR teams need the capability to conduct three primary types of search operations: physical, canine and electronic.[17] Finding people who have been buried by building collapse is heavily reliant on the victims' ability to attract attention to themselves. It becomes increasingly difficult with time to find unconscious victims. It can prove surprisingly difficult to locate even people who are shouting inside a collapsed building. The acoustics of hearing somebody through rubble is illustrated in Figure 4.9. Rescuers are unlikely to hear someone shouting the

[17] Guidelines for SAR teams have been prepared by the International Search and Rescue Advisory Group (INSARAG 2001).

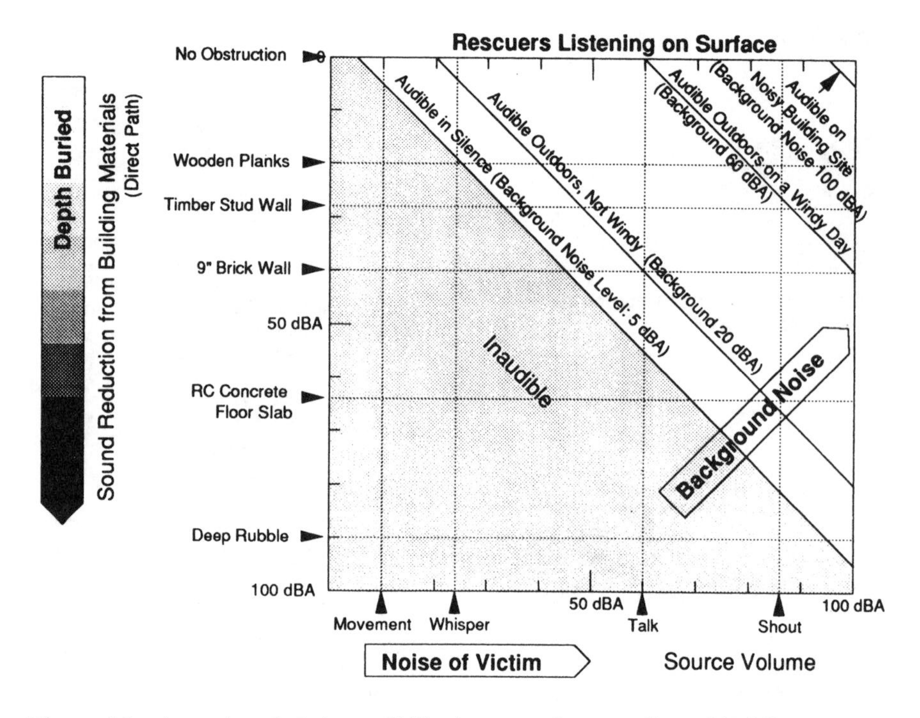

Figure 4.9 Acoustics of victim audibility in rescue from a collapsed building

other side of a reinforced concrete floor slab (floor slabs are designed to provide acoustic privacy between apartments). The best chance of hearing survivors is by means of indirect sound paths reflecting through gaps in the rubble, and a building collapse where the structural elements have broken up gives a better chance of hearing survivors than floor slabs fallen intact. In a pancake collapse of concrete floor slabs, the only sound path is sideways in between the slabs and rescuers should concentrate on listening at the sides and work inwards laterally underneath the slabs. Indirect sound paths do mean, however, that the source of the sound is much more difficult to locate and a person heard calling may be metres away in a different direction from the apparent source of the sound.

Rescuers have to follow the source of the sound in their excavation, and to follow the changes in direction. The victim has to continue to call out during the rescue. Communication back to the victim, through megaphones or other amplified noise sources, is important to keep the victim responding for as long as possible. Someone knocking on a metallic pipe or other piece of reverberant structure is also more likely to be heard because it may give a more direct sound path between source and listeners, losing less energy in transmission.

Of great importance to the acoustics of hearing survivors is the background noise level. A noisy rescue site, with excavation machinery and large numbers of people and maybe helicopters overhead, may make any sounds from survivors inaudible. Some rescue teams demand regular periods of silence during the digging to listen for survivors. These periods may have to be quite extended to catch somebody's occasional cry, and have the disadvantage of slowing up progress on excavation work, but if people are located in this way, the excavation work can be better directed.

Other methods of reducing background noise include working at night, removing any unnecessary sources of noise (keep generators and compressors as far away as possible and behind sound mufflers if they have to be used), and keeping the number of vehicles and people attending close to the site to an absolute minimum. Command and treatment centres should therefore be established well away from the site.

Directional microphones and amplified listening equipment may also contribute to picking up sounds otherwise inaudible. Ultrasonic listening equipment is able to pick up very high-frequency ranges beyond the human ear, less masked by background noise, but only a small portion of sound emitted by a survivor is likely to be in these very high-frequency ranges.

Other methods of locating survivors include canine search, thermal imaging cameras, visual probes and radar. Specially trained dogs can identify human scent and have been used in building collapses to locate unconscious victims. Canine search can be a reliable search method if the dog and its handler are trained for work in the disaster environment. There are, however, some potential problems in using dogs, particularly their handling, transportation and (if sent internationally) quarantine restrictions and possible cultural acceptance problems.

Thermal and other imaging cameras can identify temperature differences in surfaces, and in the case of dust-covered survivors lying exposed but unrecognised, thermal cameras may identify them, but buried survivors rarely give off sufficient heat to be detected through the rubble against background temperature levels. Visual probes consist of thin optical fibres or remotely controlled micro video cameras that can be inserted into the rubble and through small holes to inspect the interior. They have been used to locate victims in a number of building collapses and their wider use may contribute significantly to victim location in future building collapses. The field of view obtained by micro cameras within confined spaces is limited and the interpretation of relayed images requires considerable experience. Radar systems can detect motion from breathing or heartbeat in the range of an antenna even under metres of debris. But many possibilities for incorrect results, e.g. rescue personnel in the range of the antenna, moving curtains or parts of the collapsed structure, rainfall, etc., have to be taken into account when relying on this technology. To date they have been used by only a small number of search teams.

A very basic but effective tool for information exchange to avoid repeated operations at the same buildings by different teams is the marking of buildings where SAR activities took place. The marking contains information about the rescue team, the period of the activities, hazards detected, number and location of victims and of already extricated victims.

4.3.3 Excavation

The knowledge that someone lies injured below a mass of rubble gives an urgency to the rescue. The time available to dig through the layers that trap them may be limited. The safest procedure is to dismantle the building taking away the material from the top down until the lower levels are uncovered. Unfortunately this is also the slowest procedure. Rapid rescue generally requires localised excavation, sinking a shaft into the rubble or cutting holes through material to reach a best-guess position. Localised excavation may require shoring up of dangerous or overhanging structure, building retaining walls for rubble or building structural supports for tunnels.

Wherever possible these structures should be constructed under the direction of an experienced structural engineer. The danger that the rescue team expose themselves to in these situations may be considerable, and risks taken may result in additional and unnecessary casualties. It is vital to plan any excavation in such a way as to minimise the risk of triggering a further collapse, which might be fatal to either the trapped victim or the rescuers. The danger of a strong aftershock occurring in the days immediately after a large-magnitude earthquake is considerable.

In Mexico City after the 1985 earthquake, more than 100 rescuers were killed by further collapses while they were working.[18] To decrease the risks, specially trained rescue team members responsible for the safety during rescue work can interpret damage to the building structure, identify dangerous areas and suggest methods to reach the trapped victims, and to enhance safety by shoring up and observation of critical structural members (see e.g. Figure 4.10). Structural engineers are best suited for this task.[19]

4.3.4 Tools and Equipment

Removing masonry rubble and other broken-up building material can mainly be carried out by hand. If a large pool of volunteer labour is available, rubble clearance can be carried out relatively quickly, forming human chains. If available,

[18] Tiedemann (1989).

[19] The US organisations Federal Emergency Management Agency (FEMA), Washington, DC, the US Army Corps of Engineers (USACE), San Francisco (USACE 1999), and the Applied Technology Council (ATC), Redwood City (ATC 20-1 1989, Gallagher *et al.* 1999), provide guidelines to evaluate the structural condition of building structures to be entered.

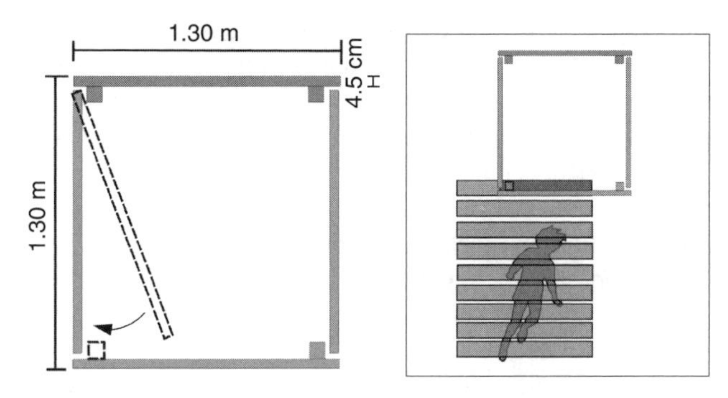

Figure 4.10 Methods of excavation to reach trapped victims in building rubble (after Michael Markus, redrawn with permission.)

lorries may be needed to take this rubble away, to keep the site as clear as possible. Workers need gloves and may need to improvise masks against inhaling the dust on the site. For work to continue into the night, illumination is needed, preferably from construction floodlighting powered by generators, but could be improvised from car headlamps if enough vehicles are available.

For shoring, large numbers of strong timber beams are required, with hammers, large nails and saws to fix in position. Scaffolding poles and extensible props are also useful.

For larger pieces of structure, crowbars and levers may be needed for a number of rescuers to be able to manoeuvre them out of position. Car jacks and lorry jacks may be used to prise blocks of a few tonnes by tens of centimetres. Specialist equipment has also been designed for jacking moderate-sized structured elements apart using air bags that are placed in position and then inflated. Spread over a large surface area, these can move elements of many tonnes. For larger blocks, more specialised and powerful equipment is needed. Construction and excavation machinery may be used to provide the power to move the more massive structural elements. If required, these machines need to be used sparingly. Although powerful, they are imprecise in their control, and may cause unexpected movements of rubble that can kill the trapped victim. If possible, it is preferable to use hand tools to break up larger elements and to reserve the heavy plant machines for dragging away material that is well away from known victims.

Breaking up excavation requires cutters, power tools or pneumatic drills. Cutting through steel reinforcing bars is the slowest part of concrete demolition requiring elaborate steel saws or flame cutters. For this reason, some consideration should be given to where the cut is made through the concrete element to meet minimum reinforcement. In a concrete floor slab, holes should be cut in the centres of areas likely to have been only lightly reinforced, e.g. mid-span and away from edge beams or local stress points that may have additional reinforcement. Where possible alternative routes to cutting through concrete should be considered. For example, instead of cutting down through the roof slabs, it may be possible to dig down underneath the building and to come up inside the structure, or to find existing holes and stairwells and to use these to pass between collapsed slabs.

Very large-scale lifting and jacking equipment, like cranes and winches, can be valuable in rescue operations if very carefully controlled. They may take some time to transport and erect on site. Their use is more suited to the later stages of excavation of a major collapse, where the emphasis has passed from immediate freeing of known survivors to the systematic dismantling of the building remains to retrieve bodies and to check the small possibility of someone remaining alive.

4.3.5 Medical Attention at the Rescue Site

At least one member of the rescue team should be an emergency physician, to advise rescue personnel on medical aspects of retrieving victims, to provide immediate medical attention to victims when located and to act as triage officer, prioritising victims for transportation to hospital (see Section 4.4.2). Some medical treatment can be provided as soon as buried victims are accessible. It may sometimes need a considerable amount of time to free a victim from a collapsed building. Victims may require rehydration, drug treatment and intravenous transfusions *in situ*. In severe cases, amputations may need to be performed. One of the most critical medical complications for trapped victims is *crush syndrome*. A person trapped for more than a few hours with prolonged pressure on a limb or other part of the body builds up toxins in the muscle tissue with reduced blood supply. When the person is finally released, the blood returns to the tissue and the toxins enter the blood supply, which can be rapidly fatal. There are many recorded cases of trapped patients with only light injuries being freed, and appearing initially well, only to die an hour or two later from sudden cardiac arrest. Where crush syndrome is suspected, it is best to treat the patient *in situ*, before releasing the confined limb. Treatment includes intravenous infusions to stabilise the patient long enough to receive dialysis treatment. This involves rescuers clearing sufficient access to the victim before releasing the victim to allow the physician to insert intravenous lines, and may involve the physician operating in a severely confined space.

Extraction of a severely injured victim is a delicate operation, and manoeuvring without causing further injury may be difficult. Stretchers to carry the injured are needed, and it may be necessary to strap patients to them if the rescue route is steep. Where stretchers are not available they may be improvised from planks, doors taken off their hinges or other firm supports. Some SAR teams have specially designed stretcher sledges – aluminium bucket-like scoops for dragging patients over rubble and through tunnels for example.

4.3.6 Transportation of the Injured

One of the greatest needs that rescue and medical treatment teams have is for ways of transporting injured victims to hospital or treatment centres. This need is immediate, and greatest in the first few hours after the earthquake. With good medical care, seriously injured victims can be stabilised at the rescue site, but without early hospitalisation and surgical medical treatment in a suitably equipped operating theatre, their chances of survival are remote. In many large-scale disasters, a shortage of means of transport for the injured has been a critical bottleneck in the victim care process. This is especially true for disasters in rural areas.[20] In some cases of earthquake occurrence in remote regions, only patients capable of walking or being carried by friends make it to hospital. Swift establishment of field hospitals in remote regions may help, but they need to be highly publicised on the radio and placed alongside the main road en route to the major town, for instance, for local people to find them. In remote regions, the transportation of seriously injured over poor roads may also allow their condition to deteriorate. In such a situation, the military and civilians may be mobilised to ferry the injured, or special ambulance convoys could be sent by the authorities into the worst affected areas.

4.3.7 Ending the Search

The decision to stop searching for survivors is always a very difficult one. People have been rescued alive five,[21] ten[22] and even fourteen[23] days after an earthquake (see Figure 4.6). These are often the result of exceptional circumstances; for example, someone with very light injuries and trapped in a void deep in the rubble, perhaps with a water supply or food. The probability of finding live victims diminishes very rapidly with time but there may continue to be a very small chance for many days.

In areas where low-rise masonry buildings have collapsed, all the potentially life-saving voids can be investigated relatively rapidly and a decision made in a few days about the probability of making further live recoveries. But in the collapse of high-rise, reinforced concrete structures, all the voids that may contain live victims cannot easily be explored, and the search operation could continue for many days without any degree of certainty that everyone alive has been located.

Another consideration is the survivability of people who are rescued. Many victims who are dug out alive after many days being trapped are too weak and

---

[20] In urban disasters, by contrast, the limited capacity of local hospitals is likely to be of much more significance for survival rates than the speed of transportation (Fawcett and Oliveira 2000).

[21] Girl found alive under a table in collapsed masonry building, Turkey 1984.

[22] Newly born babies discovered alive in collapsed multi-storey, concrete-framed maternity hospital, Mexico 1985.

[23] Couple found trapped in a cellar underneath collapsed masonry building, Italy 1980.

sick to respond to treatment. Despite even high-quality medical treatment, many lengthily buried patients die in the few days after their rescue. Patients who are unconscious or too weak to attract rescuers' attention may already be too far gone to save. Injury statistics show that a patient without a vocalisation response has less than 25% chance of responding to medical treatment.[24] In situations where resources are limited it is more effective to search widely for all victims capable of making a noise than to make concentrated searches for unconscious people.

There may be no need to declare a formal end to the search for survivors. It is often assumed that at some stage the search should be called off, medical units withdrawn, and public attention shifted towards recovery and reconstruction. This can often seem harsh to those who have not yet given up hope, however unrealistic that may be. Instead the transition can be made gradually, with an increasing emphasis on body retrieval and systematic dismantling of collapsed structures so that should anyone remain alive they will be located. A balance needs to be struck between the benefits of using heavy lifting equipment to dismantle large collapses and the threat these pose to anyone who might remain alive in the rubble.

4.3.8 Dealing with the Dead

It is also important to retrieve as many dead bodies as possible. Relatives need to grieve and to be certain of the fate of those that are unaccounted for. Identifying the dead can be a harrowing and logistically difficult procedure, but a very necessary one for the society affected by the earthquake. In a mass-casualty disaster, the number of bodies greatly exceeds the capacity of mortuaries and conventional funeral facilities. Bodies need to be stored and preserved until they can be identified, documented and buried or cremated. Makeshift mortuaries and identification centres have been set up in sports stadiums, large warehouses and other cool, large, well-ventilated storehouses. In hot weather, decomposition poses a problem and in some cases in the past, authorities unable to provide chilling facilities or chemical preservation have opted to photograph the bodies for identification later, and to dispose of the dead relatively rapidly.

In mass-collapse disasters, many people may remain missing after the SAR. A certain proportion of corpses will be left unidentified and a larger proportion will be unidentifiable. In the wreckage of a building collapse, bodies are not always recognisable or complete. There have been many cases where the number of retrieved bodies is less than the number of people missing. Demolition and wreckage clearance may occur without recognising body parts unless it is carried out very carefully. In some cases rapid demolition may be desirable, but where possible the dismantling of buildings and some degree of rubble sifting is preferable to a blind bulldozing of a disaster site.

---

[24] Noji (1989).

A common fear by the authorities in charge, sometimes argued in favour of bulldozing sites rapidly, is that human and animal corpses remaining in the rubble will become a source of epidemic contagious diseases for the general population or will pollute the water supply. The evidence suggests that this is extremely unlikely.

4.4 Medical Aspects of Earthquake Disaster

A wide range of types and severity of injury are caused by earthquakes. A significant percentage of injuries are not directly caused by building collapse and may be the result of many different earthquake-induced accidents. Some injuries are caused by non-structural building damage, such as broken glass or the fall of ornaments or collapse of parapet walls. But the majority of injuries in a major earthquake are caused by building damage.

Different types of buildings inflict injuries in different ways and to different degrees of severity when they are damaged.[25] Huge amounts of dust are generated when a building is damaged or collapses and asphyxia from dust lining and obstructing the air passages of the lungs is a primary cause of death in many building collapse victims.[26] In earthquakes affecting weak masonry buildings, the earth used as walling or roof material buries and suffocates the victim when collapse occurs.[27] There is also evidence that suffocation can occur from extreme pressures of materials on the chest preventing breathing (traumatic asphyxia). Many victims trapped inside a collapsed structure also suffer traumatic injuries from the impact of building materials or other hard objects, and of these the most common appear to be skull or thorax injuries.[28]

In some earthquakes, head injuries are by far the most common cause of death[29] but may constitute only a small proportion of the injuries requiring treatment in the survivors. Multiple fractures of the spinal column are commonly reported in many victims of some types of collapsed structures, who were either standing or lying down when the collapse occurred.[30] Extensive spinal injuries of this sort appear to be less common in buildings with timber floors and associated more with 'harder' building types with more rigid floors and roof slabs.

---

[25] Beinin (1985).

[26] See reports of dust adhering to lungs in autopsies from Mexico earthquake 1985, and causes of death in Veterans Medical Administration Building, 1971 San Fernando earthquake, California, in Krimgold (1987).

[27] Data from Dhamar Dutch Hospital, after the 1982 Yemen Arab Republic earthquake, and interviews with Army Medical Corps in Erzurum earthquake, Eastern Turkey, 1984.

[28] Data from Ashkhabad earthquake, USSR, 1948, reported in Beinin (1985), and data from Italian earthquake 1980, in Alexander (1984).

[29] Analysis of casualties in Papayan earthquake 1983, Colombia, in Gueri and Alzate (1984).

[30] Beinin (1985).

Another condition reported mainly in the collapse of large, concrete frame buildings is severe crushing of the thorax and abdomen or the amputation of limbs by extreme pressure.[31] Extreme pressures such as these come from large masses bearing down or structural members still connected to the large masses. But the most common types of injury caused in an earthquake are traumas and contusions caused by falling elements like pieces of masonry, roof tiles and timber beams.

More people tend to be injured in an earthquake than are killed. A ratio of three people requiring medical treatment attention to every one person killed is an accepted ratio in mainly rural disasters,[32] but this can vary very significantly with different types of construction affected and with the size of the earthquake.[33] Similarly light injuries requiring outpatient-level treatment tend to be much more common than severe injuries requiring hospitalisation – typically there may be between 10 and 30 people requiring outpatient treatment for every person hospitalised.[34]

The breakdown of types of injury needing treatment may typically be that shown in Table 4.2.

Up to two-thirds of the patients are likely to have more than one type of injury. Most of the injuries are likely to be minor cuts and bruises, with a smaller group suffering simple fractures and a few people with serious multiple fractures or internal injuries requiring surgery and other intensive treatment.[35]

Most demand for medical services occurs within the first 24 hours (Figure 4.11), which is typically before international medical teams will be able to arrive.

4.4.1 Calculation of Medical Resource Needs

In a severe case, e.g. a great earthquake striking a region of predominantly weak masonry buildings, 90% of buildings could be destroyed. If the earthquake

Table 4.2 Types of injury requiring treatment after an earthquake (after Alexander 1984).

Soft-tissue injuries (wounds and contusions)	30–70%
Limb fractures	10–50%
Head injuries	3–10%
Others	5%

[31] Mexico City News, 21 September 1986.

[32] Ville de Goyet (1976), Alexander (1984).

[33] In recent urban disasters, the numbers of seriously injured have been many fewer than the numbers killed. Recent data was reported at the 12th World Congress on Disaster Medicine, Lyons, May 2001 (http://pdm.medicine.wisc.edu).

[34] Alexander (1985).

[35] PAHO (1981).

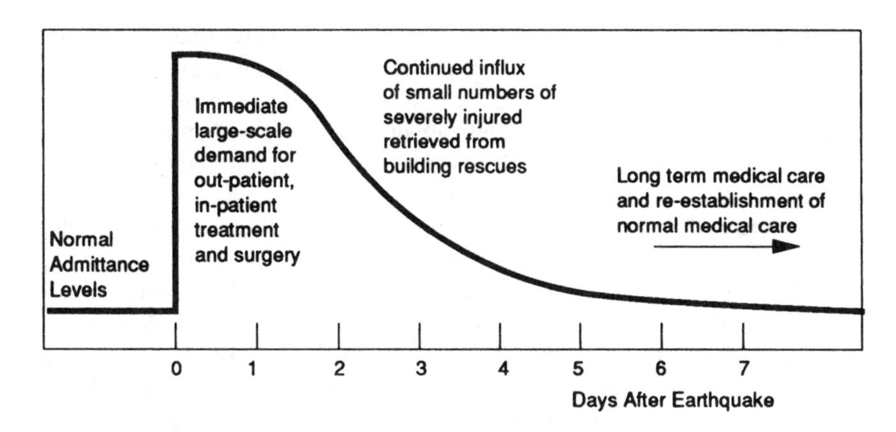

Figure 4.11 Demand for medical services after an earthquake (after PAHO 1981)

occurred at night, catching most people asleep in their homes, the *mortality rate* – the percentage of the population killed – in the towns and villages of the epicentral area could be as high as 30%. The *morbidity rate* – the percentage of the population injured and requiring some level of medical treatment – could be 60–80%. A possible range of severity levels and treatment needed across the population of the epicentral area is shown in Table 4.3, but the limited data available suggests wide variations between different earthquakes and different countries.

Epicentral areas of large-magnitude earthquakes may extend over hundreds of square kilometres and many envelop a number of towns and tens if not hundreds of villages, depending on the population density and settlement patterns of the area. A population of hundreds of thousands or even millions could easily be caught within the zone most strongly affected, leading to a death toll as high as 20 000, somewhere in the region of 50 000 injuries requiring outpatient treatment, 5000 or more people requiring hospital beds and 1000 or more needing major surgery within 24 hours. These medical loads may well be compounded by significant damage inflicted by the earthquake on medical facilities, hospitals, clinics and supply stores, within the affected area.[36]

Table 4.3 Breakdown of typical injury ratios for a population affected by a severe-case earthquake scenario.

Fatalities	20–30%
Injuries requiring first aid/outpatient treatment	50–70%
Injuries requiring hospitalisation	5–10%
Injuries requiring major surgery	1–2%

[36] In the worst urban disaster of the 1990s, the 1995 Great Hanshin (Kobe) earthquake, statistics collected by WHO from 107 major hospitals in the Hyogo Prefecture showed that 717 seriously

A disaster on such a scale would be rare (Table 1.2 shows that only 15 or so earthquakes this century have had death tolls as high as this), but by no means a worst-case scenario. Where the epicentral area enveloped a major city death tolls and numbers of people requiring treatment could be far higher. A secondary follow-on disaster, such as major landslides, dam collapse or urban fire, could push death tolls and medical loads an order of magnitude higher.

The majority of destructive earthquakes, however, will cause lower levels of injury rates, but will still put severe loads on medical treatment facilities. Medical preparedness plans can be built around similar scenario studies and calculations based on the building types likely to be affected, the population densities and settlement patterns, the size and characteristic of earthquakes expected in the region and the medical facilities available in any study area. Guidelines for risk analysis and scenario calculations for human casualty assessment are given in Chapter 9.

4.4.2 Triage

The swamping of medical facilities by such large-scale casualties means that normal standards of medical care cannot be maintained. In a mass-casualty situation, with finite medical resources, medical care provision switches to *triage*: the prioritisation of medical care to those most likely to benefit from medical treatment. The incoming injured are assigned degrees of urgency to decide the order of their treatment. Those with light injuries who are likely to recover whether they are treated or not are assigned a low priority. They may be given initial first aid and given medical attention later when the more serious injuries have been dealt with. Those with severe injuries whose chances of recovery even with treatment are judged to be minimal are also assigned a low priority. Medical resources are concentrated on those with life-threatening injuries who are likely to recover with treatment but who would die without it.[37]

In regions where mass-casualty earthquakes are a possibility, even remotely, the medical personnel should at least be acquainted with triage procedure, if not fully trained in emergency techniques. Non-medical or volunteer paramedical personnel can also contribute greatly to emergency medical care. If they are trained in first aid, particularly management of tissue injury and fractures, they can

injured, 2658 moderately injured and 47 280 slightly injured patients were admitted in the first seven days after the event (Tanake and Baxter 2001).

[37] A disaster response model proposed for the United States (Schultz *et al.* 1996) identifies three phases of the emergency period: a first phase (first hour) during which individual physicians skilled in emergency medicine and equipped with medical backpacks would attend victims nearby; a second phase (1–12 hours) during which patients would be moved to better equipped disaster medical aid centres rapidly established across the affected region; and a third phase (12–72 hours) during which victims requiring further treatment would be moved to collection points for triage, treatment and transportation by ambulance or helicopter to newly established field hospitals or still functioning hospitals elsewhere.

relieve the pressure on the professional staff by initial management of the large volume of moderate injuries. Community volunteer groups can help in earthquake preparedness by maintaining an active membership of volunteers trained in first aid to help in any mass-casualty event. Ideally these volunteers should be trained by and keep a relationship with a local hospital. Simulation exercises can be carried out jointly between hospitals and volunteer groups (Figure 4.2).

Triage classification and referral of more complex injuries require skilled medical judgement. Injury reception areas are usually established at the entrance to or outside of hospitals closest to the damaged area. In the worst-case scenario, a hospital building may itself be damaged by the earthquake and the hospital staff may have to continue emergency treatment without using the buildings. Or staff may be injured or unable to get to work immediately. Hospital emergency plans in earthquake areas have to provide for the contingency of evacuating numbers of patients from wards and critical apparatus from operating theatres, X-ray departments, etc., re-establishing facilities in the hospital grounds at the same time as receiving a massive influx of patients from the earthquake. Hospital emergency plans should include areas set aside for injury reception, first aid and tents to house emergency operating rooms.

4.4.3 Hospital Capacities, Medical Supplies and Resources

Pre-earthquake planning in hospitals and regional health administrations involves studying normal and peak hospital occupancy rates, estimation of spare capacity and likely numbers of beds that could be made available in the event of a disaster. Regional health administrations have a day-to-day responsibility to provide efficient health services, which favours reducing spare, unused capacity of hospitals to a minimum. Possible future mass-casualty occurrences are an argument for maintaining certain levels of spare capacity in medical facilities above the normal operational minimum and studies of likely scenarios will help structure the medical needs of a region.

An emergency plan for the region[38] assesses for each hospital a treatment capacity, defined operationally as the number of casualties that can be treated to normal medical standards in one hour. Treatment capacity depends on several factors including the total number of physicians, nurses, operating rooms, etc. In the United States an average, empirical estimation of hospital treatment capacity is taken as 3% of the total number of beds.[39] Military experience also gives empirical estimates of a hospital's surgical capacity, the number of seriously injured that can be operated on within a 12-hour period. In the United States again,

[38] See for example guidelines for United States Joint Commission on the Accreditation of Healthcare Organizations.

[39] As suggested by the United States Joint Commission on the Accreditation of Healthcare Organizations.

this is approximately equal to 1.75 of the total number of operating theatres.[40] This rate of treatment cannot be maintained over a long period; staff exhaustion, instrument supplies and most critically limitations on medical supplies are likely to reduce treatment rates within 24 to 36 hours of sustained activity.

Medical supplies that are most in demand after a mass-casualty earthquake are wound dressings, fracture settings, intravenous fluids and surgical supplies. Hospital stores can maintain certain levels of supplies, and preparedness plans can help ascertain appropriate stock levels to cope with possible sudden demands for the length of time it is likely to take for emergency supplies to be delivered. Preparedness plans generally rely on delivery of emergency medical supplies into an afflicted region within hours. It is impossible for hospitals to maintain supplies sufficient for a possible disaster, owing to the perishable nature of medical supplies. Most perishable of all are blood banks, and stocks are rarely kept at a high level. Rapid mobilisation of blood supplies and other medical stores into the affected area is a priority.

Blood transfusion centres to obtain donations from the public may have to be set up both in the affected areas and in other regions to replenish depleted supplies and replace blood bank stocks nationally. Fortunately volunteers willing to give blood after a disaster are generally abundant.

4.4.4 Other Aspects of Medical Plans

Other aspects of mass-casualty preparedness plans include changes of organisational structures in hospitals. (*Command team* and more military styles of organisation may need to be adopted.) Simplification of actual medical techniques may be advocated (e.g. the use of splints instead of circular casts for fractures), administrative simplifications (such as tagging patients with standardised triage tags) and rapid redistribution of patients to other hospitals outside the affected area. Plans may even consider scenarios where the medical capability of a very large region or the entire country is exceeded. These plans may envisage the rapid expansion of permanent facilities and staff in the region or the use of mobile emergency hospitals from the military, Red Cross or private sources, or even as a last resort, packaged disaster hospitals from other countries (taking in preference offers from neighbouring countries with the same language, culture and technological level).[41]

4.4.5 Public Health after Major Earthquakes

The loss of sanitation, water supplies, housing and the disruption of normal public health services for a large number of people in an earthquake, coupled with the

[40] United States Joint Commission on the Accreditation of Healthcare Organizations.
[41] PAHO (1981).

presence of numbers of dead bodies in the ruins, often lead to fears that there could be an outbreak of epidemic contagious diseases. The evidence from past events suggests that this is unlikely. The establishment of temporary relief camps may contribute to the potential and the risk of epidemic may be diminished by ensuring the following measures:[42]

- Establish a number of smaller relief camps rather than one large one to restrict concentrations and minimise contagion (sanitation services are better provided in smaller camps).
- Restrict the density of relief camps, spread each camp out if possible (closer human contact increases potential spread of airborne diseases).
- Avoid moving or encouraging large-scale migrations into another region which may lead to the introduction of communicable diseases from one population to another.
- Re-establish public utilities as rapidly as possible, particularly water supply and sewage disposal – insufficient water for washing hands and bathing also promotes spread of contact diseases.
- Re-establish basic public health care services as soon as possible.[43]

It may also be appropriate to set up a disease surveillance system to monitor communicable diseases.

Mass vaccination programmes are generally considered unnecessary and counterproductive by relief agencies. There may nevertheless be considerable pressure to implement vaccination by public and politicians fearful of outbreak rumours. Vaccine may be offered from abroad, and there may be pressure on authorities to be seen to be acting. Vaccination programmes have their own inherent risks, including reuse of inadequately sterilised needles, quality of mass vaccines, lack of cold storage and careful handling, and the generation of relaxed attitudes to health risks by the vaccinated population.[44] Vaccination policy should only be decided at a national level, and preferably as part of a pre-disaster plan. Voluntary agencies should not instigate vaccination programmes on their own initiative.

4.5 Follow-on Disasters

Past experience has shown that death tolls after earthquakes can be multiplied as the result of follow-on disasters, or secondary disasters triggered by the

[42] PAHO (1982).

[43] Ville de Goyet (2000) argues that the prompt resumption of routine epidemic prevention and control measures in use locally before the earthquake is the most effective means of reducing the risk of epidemics.

[44] Mass vaccines sent by an American NGO to help victims of the Kobe earthquake could not be used because they were labelled in English, not Japanese, which contravened local drug distribution regulations.

earthquake and escalating into a catastrophe in their own right. The most important of these are fires, landslides, tsunamis and industrial failures. If they can be foreseen, actions taken during the emergency period may be able to stop them developing into a serious situation.

4.5.1 Fire Following Earthquakes

One of the most severe follow-on or secondary disasters that can follow earthquakes is fire. Severe shaking causes overturning of stoves, heating appliances, lights and other items that can ignite materials. In addition, strong vibration may sever fuel lines or gas connection points causing spills of volatile or explosive mixtures. Large numbers of ignitions of small fires severely tax firefighters. If there is sufficient combustible material in the vicinity of the ignition point, a small fire can grow into a self-sustaining blaze that may trap any occupants still in the building, overcome them with smoke and deadly fumes and finally consume the entire building. Fire is a particular threat in timber-framed buildings and modern apartment buildings, but may also be a significant hazard for masonry with modern furnishings and in temporary or shanty construction.

Where buildings are closely grouped, fire can spread from one building to the next. Multiple ignition points, densely packed combustible housing, prevailing winds and insufficient fire suppression may give rise to the worst urban nightmare – conflagration. Dense urban districts of timber frame housing in Japanese cities and less dense but equally combustible timber frame suburbs of Californian cities are notorious for their conflagration potential in the past. In the Great Kanto earthquake of 1923, thousands of simultaneous fires were ignited, which quickly caught hold, spreading from building to building until whole districts were ablaze. Escape routes for the population were blocked and tens of thousands of people with nowhere to run were consumed in the flames. The city burned uncontrollably for many days, reaching temperatures capable of melting steel, until it finally burnt itself out. In 1906, large parts of San Francisco were burnt in a conflagration that followed a major earthquake. The earthquake was less lethal than the Tokyo event, but caused massive financial losses to the townspeople and the city authorities.

Protection of urban areas against potential conflagrations has been a primary focus of Japanese and Californian earthquake protection policy ever since these events. Most well-planned cities now have regulations governing spread of fire, including building materials of construction and proximity of buildings. Longer term protection methods to reduce fire risk include building code requirements for fireproof construction and urban planning measures to change densities and street layout and ensure frequent hose connection points.[45]

[45] Fires were a significant cause of follow-on damage in the 1989 Loma Prieta earthquake in California, and the 1995 Kobe earthquake, but in each case effective firefighting contained the blaze.

There are older quarters of cities, however, that do remain vulnerable, and large numbers of cities where planning controls are ineffectual. Perhaps the most vulnerable of all are informal housing sectors on the periphery of many rapidly growing cities which might provide the potential for conflagration following an earthquake. An emergency plan for how to tackle such an eventuality, including access routes for fire tenders and evacuation of the population, may save thousands of lives.

Immediately after an earthquake, steps can be taken to minimise fire outbreak and contain the potential escalation of established fires. The professional firefighting forces are the front line of defence. Their staffing levels, equipment quality and resources are critical at this time. Pre-built infrastructure, the water hydrant distribution network and emergency systems may be tested to capacity. The earthquake itself may well have caused damage to the firefighting force's capability – water supply pipes may well have fractured, pumping stations been damaged and in past earthquakes even the buildings of fire stations have collapsed destroying fire tenders and equipment. It is possible that fire brigade personnel are among those injured by the earthquake.[46] The fire brigade's duties may well also include the first-arrival rescue operations in the case of building collapse. If there are a number of building collapses in addition to multiple fire outbreaks, then it is clear that normal fire brigade capabilities will quickly be exceeded. Emergency plans should include mobilisation of reserves and part-time firefighters, call-up networks and reinforcement patterns to bring in fire brigades from outside the affected region, reinforcement from the military or other sources, and incorporation of volunteers and community groups in the firefighting process.

The actions of the general public can be instrumental in minimising fire outbreak if they are suitably prepared. Actions include shutting down all potential ignition sources immediately after an earthquake, carrying out systematic checks of rooms as they evacuate a building, checking neighbouring buildings, extinguishing small fires at source and notifying the fire brigade early of any established fire. Community groups can help by practising fire drills, assembling and checking equipment like extinguishers, buckets and fire shovels, and establishing organisational and warning procedures. These groups should be established in collaboration with local fire brigades and may be part of a more general community or action group incorporating medical volunteers and those concerned with longer term earthquake protection and awareness issues.

If all these measures fail and conflagration takes hold, the scale of the threat to the community is on a scale unlikely to be encountered in normal firefighting operations. Large-scale measures may be needed, such as rapid evacuation of

---

[46] In the 1906 San Francisco earthquake, Fire Chief Dennis Sullivan was critically injured in a building collapse; this loss is reported to have been one of the critical factors reducing the effectiveness of the fire brigade in combating the blaze which followed (Bronson 1986).

populations, demolishing areas of housing to create fire-breaks and fighting the fire from the air.

4.5.2 Industrial Hazards

Earthquakes damage machinery, structures and industrial processing plants. There are many industries in seismic areas, some located relatively close to population centres and employment catchment areas, whose failure could pose additional hazards to the population. These include processes using or refining hazardous chemicals, or involving bulk fuel storage or combustible or explosive materials. Some processes involve fuels or materials which are not themselves dangerous, but which would give off noxious fumes in the event of a major industrial fire. Industrial facilities are generally designed to much higher engineering standards than most other structures, but earthquakes are extreme events and test such engineering to its limits.[47] Any unseen weakness in a system is likely to fail and even small failures can cause catastrophic results. Major industrial accidents occur even without earthquakes, and disasters such as the poisonous chemical gas release at Bhopal in India in 1984 have shown that such hazards can affect a large number of people.

Dams may also fail, threatening communities downstream. A standard procedure after any sizeable earthquake should be an immediate damage inspection of all dams in the vicinity, and the rapid reduction of water levels in reservoirs behind any dam suspected of having suffered structural damage.

The worst scenario for an emergency planner is damage to a nuclear power station in an earthquake. The Chernobyl disaster in 1986 was not caused by an earthquake, but demonstrated the catastrophic impact of failure in a nuclear facility, with the enormous resources required to stabilise the situation, and the public hazard of release of radioactive gases into the atmosphere. Facilities such as nuclear power stations are generally designed to very high standards of earthquake resistance and the chances of their failing are very small, but earthquakes are extreme and unpredictable events and failure can never be ruled out. These low-probability, high-consequence scenarios have to be considered in emergency plans.

4.5.3 Landslides Triggered by Earthquakes

Landslides, debris flows and rockfalls triggered by earthquakes are also a major cause of risk to the population. In the earthquake in 1970 in Ancash, Peru, a giant debris flow was triggered in the mountains that washed down into the

[47] The Izmit Refinery of TUPRAS, one of Turkey's four major refineries, was severely damaged in the 1999 Kocaeli earthquake, the blaze lasting many days and causing a serious hazard to victims and rescuers (EEFIT 2002b).

valleys below, burying two towns and their 40 000 inhabitants under 20 metres of mud and boulders. Most of the dead in Guatemala City in the 1976 Guatemala earthquake, and again in San Salvador in the 2000 El Salvador earthquake, were the inhabitants of houses sited on the steep slopes on the outskirts of the cities when large-scale slope failures took the ground from beneath them. There are many recorded instances of mountainsides disintegrating in earthquakes, sending cascades of boulders down into the towns at their base.

These hazards may not easily be preventable in the emergency phase of dealing with the earthquake; prevention is mainly a matter of identifying potential slope instabilities preventing development near them (see Chapters 6 and 7) or possibly carrying out geotechnical engineering to stabilise threats if appropriate.

In the emergency phase, awareness of this possibility may help populations maintain a vigilance and possibly evacuate areas if minor rockfalls, slope failures or debris flows suggest that a more severe failure is imminent. In some cases the major land failure is triggered by an aftershock, having been primed by the main shock. Some major debris flows start slowly with a minor trickle and then are triggered in waves. In these cases there may be sufficient warning for action by a population that is aware of the possibility.

Other consequences of major rockfalls and debris flows include damming of rivers and blocking roads. Debris flows damming rivers cause land upstream to flood and may suddenly breach, sending waves of water downstream; both of these consequences may pose additional hazards to human settlements. In areas where roads needed for relief activity cut through mountainous regions or run along steep slopes, the emergency plan should include rapid deployment of road clearance and repair gangs to ensure that rescue teams and emergency supplies can get through.

4.5.4 Tsunamis

A tsunami or sea wave may also follow an earthquake and cause damage to coastal installations and settlements. A large-magnitude, shallow-depth earthquake with its epicentre in the ocean causes vibration waves on the surface of the water above. These waves are markedly different from the usual, wind-driven waves in having a very long wavelength, an extremely rapid speed of travel and a low attenuation. Their amplitude at the source of the earthquake is small – a few tens of centimetres – and the ripples spread outwards with speeds around 1000 km/h depending on the depth of the sea over the epicentre. These waves can travel thousands of kilometres, from one side of the ocean to another, weakening only very gradually as they travel. As they reach the coastal shelf approaching land, the diminishing depth of water slows up the speed of the wave causing it to increase in amplitude. The wave becomes slower and builds up in height. The wave can become metres high and breaks onto the shore violently, washing inland and damaging coastal installations. Tsunamis are also exacerbated by bays and inlets along the coast that constrict the wave as it travels inland, forcing it even higher: the Japanese word *tsunami* means literally 'bay wave', as this is where the smaller tsunamis are most commonly observed. Tsunamis tens of metres high have been recorded and there are historical reports of massive walls of water crashing inland higher than the tallest trees, washing away houses, pounding docks and carrying ships far inland.

Large-magnitude earthquakes in deep water just beyond the continental shelf have been recorded close enough to land to damage structures and then to inundate them with their tsunami shortly afterwards. This is, however, comparatively rare and most tsunamis are caused by earthquakes in deep water a considerable distance away from the coast – earthquakes which may be too far away for the coastal communities to feel.

Some protection from tsunamis can be achieved through the construction of sea walls, beach defences, shoreline tree plantations and other physical planning and protection measures. These need designing carefully, perhaps also as a defence against cyclone-driven sea surges, and are part of the range of long-term measures that need to be carried out well before the occurrence of any event.

The only civil protection measure against a large tsunami is to evacuate the population close to the coast further inland and to high ground. To do so requires considerable preparation and logistical resources. Tsunami warning stations are now located at many points in the Pacific Ocean and can detect the sea wave when it is first created. They can predict the scale of impact of a tsunami at various coastal locations possibly several hours before it arrives.

Good detection, communication and rapid warnings are useless if there is not a full social infrastructure, ready to act on the warning, already in place. Evacuation measures are discussed in Section 3.5. Evacuations cannot be improvised successfully, and require the population to recognise the alarm, know what to do and to undertake it without panic. Resources, such as transport and facilities for the population at their refuge areas points, need to be pre-planned and possibly rehearsed beforehand.

4.6 Shelter, Food and Essential Services

In the day or so immediately following the earthquake the priorities are undoubtedly medical and rescue needs. Saving the lives of those injured or trapped far outweighs most other needs. However, the other needs of the population suddenly deprived of homes, contents and possessions, urban services and other essentials cannot be ignored and will assume greater significance as soon as the life-threatening situation stabilises.

There is an urgent need for shelter for the population made homeless by building damage, possibly also needing food if large areas of buildings are destroyed along with their contents. There will be needs for drinking water, clothing, sanitation and basic comfort provision. Most of all there will be a need to restore public confidence, and to impose demonstrably some sort of order on the chaos.

The first few days of the earthquake emergency, and how it is dealt with, will also pave the way for the earthquake recovery, described in the next chapter. Decisions made about immediate shelter provision or short-term expediencies to overcome other needs have significant implications on the longer term reconstruction.

The provision of basic shelter and living needs for the dispossessed in the immediate first few days of the earthquake emergency will depend a great deal on the scale of the earthquake impact, the wealth and surviving spare capacity in the community not destroyed by the earthquake and the weather conditions, resilience and expectations of the affected society.

Decisions on whether to build temporary houses, or to stay in tents or to build core houses, or go for accelerated reconstruction all influence the timescale and strategy of reconstruction. Decisions on where to locate temporary camps will affect the spatial planning on new settlements and long-term reconstruction. In Chapter 5, the issues of housing, the decisions on providing shelter during reconstruction operations and the pros and cons of temporary housing are discussed. In the earthquake emergency, shelter for the homeless is one of the urgent needs for which some solution is needed in the first few days.

4.6.1 Improvising Shelter for the First Day or Two

To a large extent the solution of immediate shelter and material needs has to be met by improvisation locally. If the weather is not too bad, people may sleep outdoors for the first one or two nights – particularly with aftershocks threatening to cause further damage (see Section 4.7.3). In bad weather immediate shelter needs can be improvised in undamaged buildings, particularly public buildings like schools, town halls or other undamaged community buildings that might be pressed into service, or people can sleep in cars. As far as possible, other families in the local community whose houses are not damaged should be encouraged to take in the homeless for a day or two until longer term arrangements can be made. Experience shows that this is likely to happen without official encouragement. Many of the homeless are likely to find temporary accommodation with nearby family and friends, if their houses have not been as badly damaged. This is commonly seen in rural communities, where kinship ties are stronger and geographically closer than the more dispersed social communities found in towns. But some official appeals and encouragement (like promising a guest allowance) may be helpful to unaffected households taking in strangers.

4.6.2 Problems with Temporary Evacuation

In some earthquakes in the past, faced with fairly severe levels of destruction, fears of epidemics, apparent shortages of spare accommodation and imminent bad weather, the decision has been taken by the disaster authorities to evacuate the entire population from the worst damaged areas. Although apparently logical in the face of all the difficulties, this has almost always been detrimental to the recovery of the region and bad for the affected families.

The effective abandonment of the badly damaged region for a number of weeks or months causes deterioration of the buildings, property, livestock and cultivation, and the economic recovery of the community. It severs the population from its usual environment and makes it difficult to return to begin the process of physical and economic reconstruction. Members of an evacuated family are psychologically separated from their place of work, familiar surroundings, their possessions, animals, gardens and fields, and the effort required after weeks away to return to a damaged and deteriorated home to start rebuilding is far more demanding than if they had stayed. Many families may never return and may choose to make a new life elsewhere. Evacuated shop owners and traders may be unable to reopen a successful shop in their temporary refuge and may be unable to continue trading when they return. The impact on the agricultural, commercial and economic activities of a region caused by even a short-term evacuation of the population may be severe.

Worst of all have been decisions to evacuate the women and children, leaving the men to participate in the reconstruction. This causes emotional stress in breaking up family units at the time when family coherence and mutual support are most needed to survive the personal and economic disasters that they have each suffered.

For these reasons it is usually better not to evacuate a population unless there is a real and imminent danger of a secondary hazard. Logistical resources needed to evacuate a population and to service its needs in another area can be better used in bringing those needs to a population remaining in position. Even the temporary hardships of a winter in minimal shelter are preferable to the long-term hardships of economic and social collapse of the area. The population remaining in position will direct its energies towards clearing up the damage and re-establishing some semblance of normality and order out of the chaos in ways that would be impossible if it had been evacuated.

4.6.3 Tents

The most useful form of immediate shelter for very large numbers of homeless people is undoubtedly tents. Tents are relatively easily stockpiled and transported, rapidly erected and can provide adequate climatic protection against quite extreme conditions. They are also safe against aftershocks or another strong earthquake. Tents are difficult to erect in hard urban landscapes, or on steep gradients or in strong winds, but in most other situations can be pitched close to the damaged house (important to householders wanting to protect possessions or tend the gardens) or on adjacent public open space.

Tents are stockpiled by humanitarian organisations, such as the United Nations High Commissioner for Refugees (UNHCR) and the Red Cross, but stocks are limited as warehousing is expensive and tents degrade. Lead-times for manufacturing and delivering large numbers of tents are months, rather than weeks. Major crises usually, therefore, result in the use of a number of different types of tents from a variety of sources. Problems often occur in identifying how each type should be erected and used. The distribution of plastic sheeting is also a useful temporary shelter options, where sufficient structural materials can be found to support it, and where the climate is not too severe (Figure 4.12).

Tents have to act as surrogate houses for families for a number of days or weeks. They have to provide climatic comfort, protection from rain and ground water, visual privacy and storage space. Families will tend to protect valuable possessions and electronic goods (radios, TVs) inside their tents (Figure 4.13). Tent specifications vary widely, but larger units with space to stand, made from durable waterproof fabrics and with an integral ground sheet, are minimal needs. The tent needs to be relatively easy to erect and supplied with erection instructions in the language of the affected population. Tents used as standard equipment by humanitarian agencies should be used, or the agencies referred to for specifications, and great care should be taken not to purchase inappropriate tents.

The plan should include stockpiles of suitable tents and plans for their transportation and distribution. It ought to be possible to provide each homeless family with a tent within a couple of days after the earthquake has occurred. Where no tents are available shelter can be improvised using a combination of locally available materials and plastic sheeting.[48]

Figure 4.12 Plastic sheeting supported on a timber framework provides a sufficient temporary shelter to re-establish primary school classes in the immediate aftermath of an earthquake: the scene in Sukhpur village, Gujarat, India in February 2001

[48] Davis and Lambert (1995).

Figure 4.13 Tents have to act as surrogate houses for families for a number of weeks, keeping valuables and salvaged house contents out of the rain. Temporary camp in Muratbagi village, 1983 Erzurum earthquake, Turkey

In cold weather, great care is needed in assembling a 'package' of shelter, bedding, clothing, calorific intake and heating appropriate to climate and culture. 'Winterised' or 'arctic' tents are usually extremely expensive, having been developed initially for cold-weather expeditions, and require the intensive use of space heating. Low-cost insulated liners have been developed for winterising the standard tents of humanitarian agencies. It is important to consider both a suitable insulated flooring and fire protection measures. Tents can be heated with diesel oil, gas or even solid-fuel space heaters, provided that they are sealed stove-type heaters and that each heater comes with chimney pipes to allow the combustion gases to be vented outside the tent. Flues must be isolated from the tent using a manifold, to prevent the tent catching fire, and must allow for the movement of the tent in windy conditions without either leaking or overturning the heater. Again, tent heaters were often developed for military or specialist expedition use, and are often expensive. However, suitable and safe tent heaters have been developed and provide adequate heating if they are available.

Attention needs to be given to the cost and logistical implications of maintaining a fuel supply, as salvaged building timber is an expensive fuel source in the longer term.

4.6.4 Food and Water

Food and drinking water are also unlikely to be immediately available to the dispossessed in the first day or so following the earthquake. In mass collapses food stores are likely to be buried (although some may be salvageable later), gas supply, electrical power supply and other fuels may be disrupted and cooking facilities could be out of action. Piped water supplies are likely to be lost with underground pipe ruptures, damage to pumping stations and possible destruction of water tanks and wells. The influx into the disaster area of many rescue workers, emergency personnel and volunteers may also increase the need for food provision.

Improvised mass-cooking facilities and any undamaged school, factory and private canteens can be pressed into service. Mobile kitchens from the military and many charitable organisations such as the Red Cross/Red Crescent can usually be deployed within a few hours. Water bowsers, special tankers and as many tanks, buckets and water containers as possible need to be employed for distribution and storage.

Food and water provision centres may need to be established in each badly damaged locality. In addition to establishing the facilities for mass catering, including food and water storage, cooking facilities, eating utensils, washing-up facilities and so on, there also needs to be a regular distribution system established to deliver food and water to each centre.

For water, WHO guidelines suggest the minimum needs for drinking, cooking and basic cleanliness in temporary camps are 15–20 litres per person per day. Water needs are slightly higher in mass feeding centres: 20–30 litres per person per day; and highest in field hospitals and first aid stations: 40–60 litres per person per day.[49]

In urban areas, emergency water systems can be rapidly established using plastic water pipes rolled out along the side of roads to standpipes or communal water distribution centres supplied from water tanks, filled by tanker or from a surface or ground water source. Such supplies require treatment such as sedimentation using alum sulphate, and batch chlorination, or in longer term situations, slow sand filtration. Repair to the water supply system is an obvious priority for emergency recovery. In re-establishing a water system damaged after an earthquake it is recommended to raise the water pressure and increase the chlorine concentration to protect against any polluted water that may seep in through damaged pipes.

[49] Assar (1971).

4.6.5 Sanitation and Field Camps

Loss of water supply also means that washing facilities may be lost and the sanitation drainage system inoperable. It is also likely that underground sewerage systems will be damaged even if water pipes are not (large-diameter, concrete-cased or masonry-lined main sewers are more vulnerable to earthquake ground motion than smaller bore, water supply pipes). There may be a need to establish temporary public sanitation systems, such as field latrines and communal washing and bathing facilities. The establishment of such facilities, particularly if they are part of large field camps, needs care and experience.

Field camps should be avoided, unless they can be sited safely near the homes of those affected. Camps need siting to avoid becoming waterlogged, being exposed to extreme weather or other hazards. Guidelines on the size of camp (e.g. 10 000 people), their density (8 m between tents) and layout of facilities (100 m maximum walking distance to water supply points, for example) are important and fairly well established in temporary camp guidelines.[50,51] Such guidelines are not manuals but offer best practice to inform decision-making in particular circumstances. Facilities such as latrines, ablution blocks, canteens, laundry facilities, garbage collection and disposal points all need experienced design. Organisations such as the UNHCR and some humanitarian organisations have experience in setting up field camps and the establishment of such emergency facilities should primarily be their responsibility, wherever possible.

4.7 Re-establishing Public Confidence

There will be an urgent need to restore public confidence after the earthquake. Earthquake damage is chaotic and ugly. The sight of shattered buildings is pitiful and demoralising. The evidence of the disaster is all around and inescapable. There is a need for all the groups involved in the emergency, the voluntary NGOs, the government officials and the affected communities themselves, to restore a sense of control and to impose some sort of order on the chaos.

4.7.1 Rubble Clearance

An immediate operation to clear up, make streets safe, and stabilise damaged buildings raises confidence, boosts morale and demonstrates a collective will to fight back from the disaster. The authorities need to take the initiative in this by having a strong immediate presence in the areas of damage, with police or other officials on the streets, putting up barricading, establishing signs, taping off areas

[50] INTERTECT (1971).
[51] UNHCR (1999).

of continued risk to the general public and other visual evidence of establishing control. Affected populations are often concerned about leaving their homes, as they might be looted of possessions and building materials. The removal of human and animal corpses is a high priority, for morale as much as public health.

The general public, community groups and NGOs can greatly assist by instigating a clean-up campaign to clear rubble, remove broken glass and bringing areas of moderate damage back to normality. The labour supply is generally available and although the equipment may be limited, the operation should be initiated immediately. Some administrations have chosen to seal off areas of damage and to prohibit people from entering, because of the fear of further injury from the damaged buildings. This may be justified in the case of badly damaged and unstable structures, and obviously in cases of doubt it is essential to err on the side of safety, but demarcation of stay-away areas should be limited as far as possible to individual buildings and sensible rubble fall-out zones. The abandonment of entire areas of towns to earthquake damage should be a measure of last resort.

4.7.2 Making the Streets Safe

Damaged buildings that still pose a threat to passers-by have to be made safe. This should be carried out under the supervision of experienced engineers and the disaster management authorities should deploy emergency public safety engineering gangs to make streets safe. Priorities for attention are buildings on main thoroughways where damage has caused partial wall failures (Figure 4.14), building

Figure 4.14 Use of timber to prop unstable masonry. Aftershocks may continue for several weeks, and propping of the façades of damaged masonry buildings is essential to maintain safety for the public. Propping in Bagnoli Irpino, after the 1980 Irpinia earthquake in southern Italy

elements poised in precarious positions or where there are other visual clues to instabilities, such as bulging masonry. Not all cracked walls are unstable but detailed examination is needed to determine the stability, for which there is not enough time in the emergency period. Where there is doubt, it is best to prop the suspect element to restrict future movement.

Where possible, threatening elements, such as overhanging slabs of masonry, dislodged roof tiles or damaged architectural ornamentation, should be dismantled. Smaller elements can be picked off by hand from a hydraulic maintenance platform.

Large overhanging or tilting slabs of masonry or walls about to fall should be demolished if they threaten public space. This requires heavy machinery, particularly to pull or knock down dangerous elements from a distance without personnel having to get too close.

There may be legal limitations to carrying out extensive demolition of private property by disaster authorities without the owner's permission; however, most public authorities have powers to take emergency action for public safety, and a judgement will have to be made in each situation. The emergency public safety engineers should be briefed on their legal powers before setting out on their assignments and may be advised to make photographic records of damaged buildings where they make interventions.

Where demolition is not an option, buildings can be stabilised by propping, tying or providing temporary shoring, as indicated in Figure 4.15. The best material for shoring is large-section timber beams, and large quantities will be needed for stabilising any sizeable area of damage.

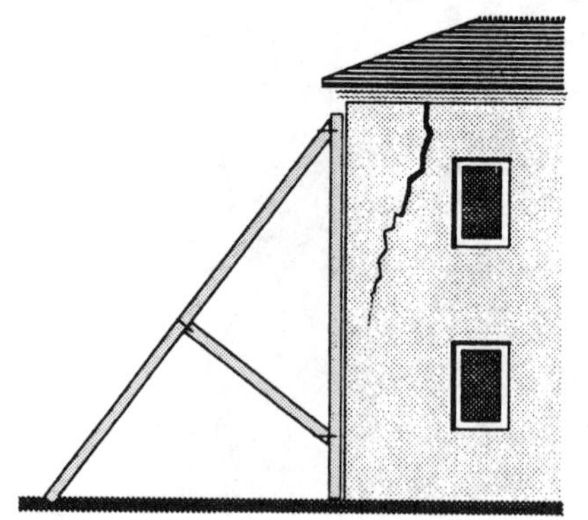

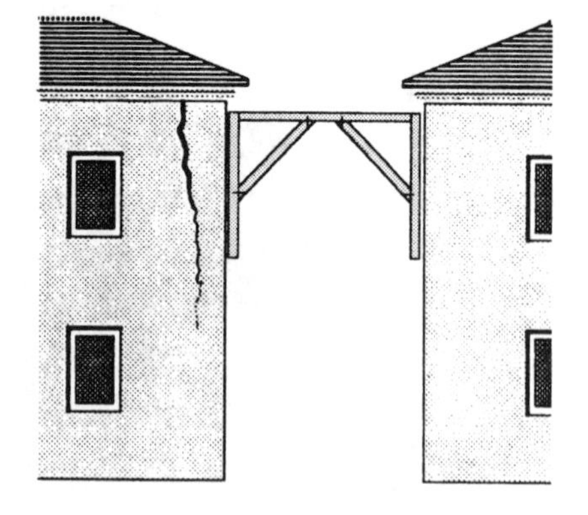

Figure 4.15 Making damaged buildings safe is important in areas of public access

Scaffolding and extensible steel props can also be used, and may be preferable in cases of providing lateral supports for large façades.

Design of propping is a skilled operation, and should be undertaken under the direction of a qualified engineer. Propping is designed to provide restraining forces in the event of any further movement by the damaged element. The timber beams or propping member will provide compressional support if both ends are securely anchored. The support to damaged elements, and to masonry in particular, should be provided by a spreader-plate – a timber beam or beams held flat against the wall to distribute the point load over an area of masonry.

Another important part of making safe and stabilising damaged buildings, particularly buildings of special importance, is making them weatherproof. Rain and water penetration into damaged structures, particularly masonry buildings, and into cracks in the masonry will not only ruin building contents, but also weaken the structure and is likely to cause further collapse. Weatherproofing can be carried out using tarpaulins or plastic sheeting or by building more durable protection, such as temporary corrugated steel sheeting, iron roofing or even temporary concrete block walls to replace lost masonry. Cracks and dislodged masonry elements can be grouted with a mortar filler to stop water penetration, which can be removed later when the crack is repaired.

4.7.3 Aftershocks

After any major earthquake there are likely to be a series of smaller shocks as the crust realigns itself around the major fault movement. These will occur most frequently in the first few days after the event, becoming gradually fewer and fewer in time, but some shocks may be felt months and even years after the main shock.

The magnitudes of the shocks will be related to the size of the main shock. There is great variation in the number and sizes of aftershocks that follow different earthquakes, and some seismic regions are more prone to aftershocks than others, but as a rough guide, at least one earthquake of one degree of magnitude smaller than the main shock can be expected, around 10 aftershocks of two degrees of magnitude smaller and a large number of smaller events three or more degrees of magnitude smaller. Thus for a magnitude 7.0 event, it is very probable that an aftershock of around 6.0 will occur during the emergency phase, and perhaps 10 of 5.0.

These aftershocks will cause further damage, particularly to damaged structures and buildings weakened by the main event, but it is rare that an aftershock causes severe damage or has an impact on the same scale as the primary shock. It does, however, pose a distinct threat to any of the emergency personnel working near damaged buildings. SAR teams, firefighters, damage surveyors and others working in the damaged zone should be aware of this possibility at all times. Entry into damaged buildings should be minimised and one member of the team should always remain outside. The main danger is probably falling roof tiles, pieces of brick and smaller building elements. The emergency teams have to keep one eye always on the building skyline above them (and the other on rubble, holes and emergency water pipes in their path on the ground), should wear hard hats and other protective clothing.

The main effect of a strong aftershock is, however, largely psychological. The population that has recently been through the horror of a major quake feels the start of it again. This can cause panic and is severely demoralising. People who may have moved indoors may move out again and spend longer camped outdoors; some of the population may give up and decide to move away from the area at this point. The emergency administration and voluntary groups will have to combat this loss of public morale in the way they work. The best way to minimise loss of morale after aftershocks is to publicise their likelihood well in advance. Immediately after the earthquake, information disseminated to the general public should include warnings that strong aftershocks are probable. When aftershocks do occur they should be well publicised and attention drawn to the diminishing rate of aftershock occurrence as time goes by.

With a moderate-magnitude earthquake there is also a small possibility that it could be followed by a similar-sized or even larger magnitude event.[52] This should be borne in mind by the emergency management authorities, but the chances are low–less than one major earthquake ($M \geqslant 6.0$) in 1000 is a double event or followed by a larger magnitude earthquake. Taking specific steps or issuing public warnings against this eventuality is likely to freeze the entire emergency and recovery operation. If sensible precautions are taken against a large aftershock this will also give protection should a larger earthquake occur.

Further Reading

Davis, J. and Lambert, R., 1995. *Engineering in Emergencies: a Practical Guide for Relief Workers*, Intermediate Technology Publications, London.

Davis, I., Wilches-Chaux, G., 1989. *The Effective Management of Disaster Situations*, Disaster Management Centre Guidelines No. 1, Oxford Polytechnic, Headington, Oxford OX3 0BP, UK.

IFRC, 1996. *World Disaster Report, 1996*, Oxford University Press, Oxford.

IFRC, 2001. *World Disasters report 2001: Focus on Recovery*. IFRC, Geneva (www.ifrc.org).

PAHO, 1981. *A Guide to Emergency Health Management After Natural Disaster*, Scientific Publication No. 407, Pan American Health Organization, Regional Office of the World Health Organization, 525 Twenty-third Street, NW, Washington, DC 20037, USA.

[52] In the 1997 Umbria–Marche earthquake the main shock ($M = 5.8$) was preceded by a smaller ($M = 5.1$) shock nine hours earlier, which ensured that many people had evacuated their homes, but also led to casualties among those surveying the damage.

PAHO, 1982. *Epidemiological Surveillance After Natural Disaster*, Scientific Publication No. 420, Pan American Health Organization, Regional Office of the World Health Organization, 525 Twenty-third Street, NW, Washington, DC 20037, USA.
Stephens, L.H. and Green, S.J., 1979. *Disaster Assistance: Appraisal, Reform and New Approaches*, New York University Press, New York.
UNHCR, 1999. *Handbook for Emergencies*, United Nations High Commissioner for Refugees, Geneva.

5 Recovering from Earthquakes

5.1 Opportunities and Challenges

The physical destruction wrought by an earthquake has to be repaired. Reconstruction can be a daunting prospect. The scale of destruction can be extensive and it is often difficult to know where to start. The reconstruction also provides opportunities: the opportunity for a new start and to make improvements on the situation that existed before the earthquake.

There is, however, more to the reconstruction than replacing the physical fabric of the buildings and structures damaged by the earthquake. In a destructive earthquake, the factories, shops and commercial buildings that house the economic activities of the region may be incapacitated. Without the economic engines to drive it, the recovery of the region cannot really progress.

The recovery process can be broadly classified into:[1]

- The *immediate relief period*, generally lasting a few days.
- The *rehabilitation period*, from the end of the relief period for a number of months.
- The *reconstruction period*, which may last a number of years, even tens of years in some cases.

The previous chapter was concerned with the immediate relief period and the activities needed to deal with the earthquake emergency. Inevitably even during the first days after the earthquake, some consideration will be given to the reconstruction task that lies ahead and preparations have to be made for the long-term recovery. Many of the decisions made in the first few days have a significant influence on the long-term recovery and future prosperity of the

[1] UNDRO (1982).

earthquake-affected region. This chapter is concerned with planning a successful recovery during the rehabilitation and reconstruction periods.

5.2 Sectoral Recovery Plan

Damage from an earthquake is likely to be suffered by all types of physical fabric, and to have an impact on many sectors. Reconstruction is commonly planned sectorally. The programme for reconstructing schools buildings, for example, will be planned and costed separately from that for repairing damaged roads and rebuilding bridges.

A sectoral approach is useful because groups of facilities are the responsibilities of different agencies and require different skills to understand the reconstruction needs. A typical breakdown of sectors for a region is given in Table 5.1. The sectors considered important are likely to vary slightly from one region to another.

5.2.1 Coordination of Sectoral Plans

There is also a need to coordinate and integrate the reconstruction plans of the various sectors and to balance budgetary requirements between different sectors. There will also be logistical and practical coordination to be reconciled: a finite resource of building materials and construction equipment may need to be allocated between competing sectors, and priorities assigned. There are also obvious overlaps and benefits from combining some operations, such as for example building schools and housing in the same construction operation or using similar administrative procedures to disburse reconstruction grants to a number of sectors.

All sectors are to some degree interdependent: food processing factories need rebuilding at the same time as agricultural production is revitalised; commerce, industry and the service sector need to be helped to recover at the same time as rebuilding housing. This interdependency of sectors is discussed in more detail below. The reconstruction operation is essentially the recovery of a community from multiple losses and is much broader than the physical rebuilding of damaged buildings and cracked pipes. It is the revitalisation of economic production, of regenerating jobs and income, re-establishing lifestyles and repairing the social linkages of a community.

5.2.2 Loss Estimation

An important prerequisite for planning the recovery is accurate information on the losses of each sector. Assessment of losses should be initiated as soon as possible using a structure such as that of Table 5.1. Detailed loss inventories may take weeks to compile, however, and there will be a need for an approximate estimate of damage in the first few days after the earthquake.

Table 5.1 Earthquake damage surveyed by sector.

Sector	Loss inventory
Housing	Number of dwellings unrepairable. Number of dwellings in need of repair. Number of people homeless
Health	Damage inventory of hospitals, clinics, primary health care centres. Loss of beds and operating units. Inventory of lost or damaged equipment including vehicles, machinery, operating supplies
Education	Damage to schools, universities, kindergartens. Numbers of lost classroom places. Loss of school equipment
Urban services	Electricity: damage to power stations, transformers, transmission lines, substations
	Water: damage to pump stations, pipeline networks, water tanks, water towers, wells, reservoirs, water processing plants, pollution into water supply
	Gas: damage to gas pumping stations, pipeline networks, leaks, gas storage
	Sewage: damage to sewage treatment plants, sewer pipes and underground chambers, surface water drainage networks, flood drainage networks
Communications	Telephones; cellular networks and land line systems
Transportation	Roads
	Bridges
	Railways
Special facilities	Ports, airports
	Rail terminals
	Dams
Government and local services	Admin offices, fire, police stations, town hall, village hall, prisons; loss of administrative records
Manufacturing industry	Damage to factories; damage to manufacturing machinery; loss of stored products; disruption to supply of fuel, raw materials, electricity, water, waste disposal; workforce laid off, earnings lost, downtime in production
Retail and service industry	Shops, business premises, warehouses, fuel stations, food distribution, transport and supplies, destruction of stock, lay off of employees
Commerce, financial and professional services	Loss of premises; commercial offices; small businesses; professionals (doctors, lawyers, dentists, etc.); clinics; studios; disruption of communications; disconnection with established public/client base; extent of needs to re-establish phone, mail, business communications, internet systems and servers
Agriculture	Damage to agricultural building stock; loss of livestock; damage to equipment, vehicles; market gardening, greenhouses; damage to food processing plants, food and produce storage
Tourism and leisure	Damage to hotels, guest houses; tourist facilities; restaurants; negative publicity dissuading tourists
Other sectors	Cultural: museums, monuments, statues; townscape
	Leisure, sports, clubs
	Community/theatres/cinemas
	Religious: churches, mosques, temples

The surveying of damage item by item takes considerable time and labour resources, and estimation of building repair costs is notoriously difficult (see Chapter 2). One method of gathering loss assessments is to arrange for the associations of each industry, the professional societies representing different commercial sectors, the trade unions and other bodies to report on their own sectors. Information in the aftermath of a disaster is likely to be confused and perhaps partisan, so where possible information could be requested from more than one source as a cross-check.

Evaluation should include:

1. Loss estimates (in value) and numbers of units.
2. The same as a proportion of the total existing.
3. An estimate of the proportion of the losses likely to be covered by insurance and the total shortfall of losses to be borne by the owners themselves.
4. The effect of the losses on the economic production of the sector (the percentage of productive capacity lost, and estimated downtime).
5. The replacement resources needed to resume pre-disaster production levels particularly identifying time-dependent resources, such as person-days needed to resume pre-disaster production levels, and likely bottlenecks.
6. The effects of lost production on employees. Numbers affected by the lost production, particularly any on reduced pay, suspended or laid off as a result of lost production. The extent of lost production on any casual labour normally involved.
7. The implications of damage occurring in other economic sectors (e.g. supplying raw materials) and the effects of damage in this sector on other economic sectors (e.g. retail outlets).

5.3 Repairing Economic Damage

An earthquake can have a major impact on the economy of a region. Damage may have been inflicted on industry, damaging factories and destroying machinery. In addition to the replacement cost of factory buildings and contents, the production of the factory may be halted, and the manufacture of its products stopped, causing the workforce to be unemployed. Downstream activities, like shops selling their products or other factories using the products in their own manufacturing, are likely to suffer as a result of the factory's lost production. In a competitive market, a factory or company temporarily prevented from trading by an earthquake is likely to have its business taken by competitors. Even if the damaged business is not in open competition or is protected in some way while it recovers, a prolonged suspension of activity will harm its economic viability. If the factory is closed for a long time, the businesses it supplies will suffer. If the factory is not trading, it will find it difficult to pay its workforce and, without assistance, it may have to lay off its employees and the local retail economy will suffer as a result. If the housing that the employees live in is also damaged, employees with reduced incomes will find it harder to recover from earthquake losses themselves, and to repair their houses.

The economic recovery of a region will depend very critically on the nature of the different economic sectors that were active in the earthquake-affected area, on the damage that has been caused to each and on the interdependency between them.

Not all earthquakes are harmful experiences to the long-term economy of a region. Some regions have used an earthquake reconstruction to accelerate economic growth and the earthquake can be seen in retrospect to have had a beneficial effect on the economic development of the area. On other occasions, the occurrence of an earthquake has been a terminal blow for the economy of a region, leaving it crippled, depopulated and in decline. It has been argued that earthquake impacts accelerate the economic trends that were already in progress before the earthquake occurred.[2] An area of growth and strong economic development is better placed and more likely to use reconstruction resources for positive expansion whereas an area already suffering from population migration and economic decline is liable to find emigration accelerated by the earthquake and outside investors less willing to put capital into reconstruction. A good economic basis for the reconstruction plan is essential to underpin the physical rebuilding.

Unless there is a coordinated revitalisation programme that ensures that the economic production of the area is re-established and jobs are restored, families may find that they have assistance in rebuilding their house but no means of restoring their income. A good earthquake reconstruction programme recognises the economic structure underlying the region and assists the economic recovery in parallel with the physical rebuilding. Some of the issues to be considered in planning the economic recovery of agricultural, industrial and service economies are discussed below.

5.3.1 Recovery of Agriculture

Agricultural and particularly horticultural economies are, on the whole, not as vulnerable to earthquakes as industrial economies, but even moderate losses may have wide repercussions in food supply for a region and be difficult to bear for farming communities with a low income.

Fields and growing crops sustain little earthquake damage, but other facilities, e.g. irrigation systems, barns and outbuildings and agricultural equipment, may suffer. Damage to irrigation systems, dams and water supply pipelines can cause serious damage to growing crops. Collapsing animal sheds may kill livestock and additional losses may be suffered by animals being destroyed during the

[2] D'Souza (1984).

emergency period – in cases where strong earthquakes have hit animal-rearing communities in the colder months, losses of over 50% of the livestock they depended on have been recorded.[3] Collapsing sheds and barns can destroy stored crops, animal feed, tools and farming machinery. Market gardening harvests have been destroyed by earthquake damage to glasshouses during frost.

The sudden distraction of the population by an earthquake during harvest activities could lead to the loss of a crop, but damage to other sectors is likely to contribute more to the economic impact on the agricultural sector. Difficulties of transportation to get food to market during the emergency or reconstruction period, and damage to food processing factories, may make it difficult for farmers to sell their produce. And there have been a number of recorded occasions where farms undamaged by the earthquake have suffered because nearby towns were badly damaged and inundated with food aid sent from outside the region, and no longer bought produce from them.[4]

Recovery of agriculture can be assisted, for example, by:

- Helping subsistence farmers to rebuild their reserves and re-establish their household economy.
- Targeting assistance selectively on the poorest and most economically vulnerable, with support for livestock replacement and repair of damaged infrastructure such as water storage facilities, farm roads and bridges.
- Taking the opportunity to upgrade inefficient pre-earthquake farming practices.

5.3.2 Recovery of Industry

Industrial sectors tend to be complex with many interdependencies, a variety of ownership types and a major role in the prosperity of a regional economy. Earthquake damage in a region with many factories and processing plants is likely to have consequences for the economy well beyond the earthquake-affected area, possibly affecting the national economy and international export markets. Losses to industry are likely to be heavier from the lost production while the facilities are closed for repair than from the damage itself.

The interdependency of many factories and industrial facilities on each other and on the transportation system and physical infrastructure means that earthquake damage to any link in the chain can have consequences for many undamaged facilities: damage to a quarry producing raw materials or damage to a factory making parts for assembly in other plants may mean that production is halted in all the places they supply which may, in turn, affect other factories dependent on them. This interdependency means that economic production

[3] Aysan (1983).

[4] Farmers claimed that the food aid ruined their market after the Kalamata earthquake in Greece, 1986.

in industrialised areas is highly vulnerable to earthquake disruption. Damage to roads and rail communications, ports and freight-handling facilities means that factories cannot be supplied or distribute their output. Loss of electrical power supply, gas supply or water supply will halt production in factories and sudden disruptions of utilities may damage cooling systems, kilns and boilers. Industrial economies also tend to provide mass employment concentrated in a small number of facilities: damage to a single factory can have an impact on the whole local economy.

Many larger companies with a portfolio of investments, owning other factories in other regions unaffected by the earthquake and fully covered by insurance, may have the resources to re-establish their earthquake-damaged operations. They may on the other hand also take the opportunity of the earthquake not to rebuild or reopen a factory that was unprofitable or whose repair cost appears uneconomic against low profit margins. If a number of companies simultaneously take similar decisions on a number of marginal plants, the earthquake can result in high unemployment in the area and a depressed local economy.

Similarly, many smaller private companies, owning only one or two factories, with few assets and possibly underinsured, may find themselves simply unable to raise the full funding needed for repairs even if the factory is profitable. The true extent of this exposure to economic collapse of small and medium-sized industrial operations is often underestimated and only realised after the earthquake has happened.

Measures which can be taken to support the recovery of industry include:

- Converting earthquake-damaged areas into new special economic development zones (EDZs) with special privileges in import/export tariffs, and removal of some planning restrictions.
- Ensuring that insurance claims are settled as soon as possible.
- Providing other economic incentives for investment in new plant and equipment.
- Ensuring that the resources and infrastructure needed to support the recovery of industry are given a high priority in the emergency and reconstruction operations.
- Supporting damaged industrial enterprises to maintain their workforce during reconstruction.
- Encouraging improvisation to return to some degree of production at the earliest possible opportunity.

5.3.3 Recovery of Small Business and Retail Sector

In every type of economy, agricultural or industrial or commercial, there is an infrastructure of shops, small businesses and trades that supply the population with its day-to-day needs and supply industry and other commercial companies with goods and services. These range from food stores, high-street shops, supermarkets, department stores and other retail outlets to hairdressers, dentists, banks, and other professionals providing services. Service industries may make up a large proportion of an urban economy and typically provide incomes for a lot of people.

Service industries can be badly hit by earthquake damage to building stock. Retail industries cannot function without premises. Stock may be lost or ruined by building damage. Disruption of normal transportation systems may disrupt supplies. The greatest long-term damage to industry, however, is likely to result from prolonged disruption of customers having access to its products or from normal business.

Service industries and retail businesses tend to be much more closely integrated into residential areas than manufacturing industry. The fortunes of service industries are closely linked to those of the residential areas: where residential building stock has suffered high levels of damage, the service industry will also be badly affected.

Shortages of food, clothing and everyday necessities among those affected by the earthquake may be caused as much by the closure of food shops, clothes shops and local stores as by damage of homes. A considerable proportion of the affected population – and the proportion will vary considerably from one community to another – earn their livings from retail and service industry employment. It is important to them and the community as a whole that trade continues.

The recovery of the small business sector can be assisted by:

- Allocating enough temporary accommodation to the small business sector and service industries.
- Compiling and distributing a directory of temporary new trading locations, and using the media to publicise new locations to assist in re-establishing trading contacts.

5.3.4 Commercial Offices and Central Business Districts

The larger scale office, clerical and managerial activities of companies may also be disrupted. The central business districts (CBD) of a city or commercial business parks tend to be concentrations of office buildings containing the headquarters, management or operational centres of many larger companies. This may include financial institutions and commercial sectors vitally important to regional, national or international trade. Larger commercial office buildings tend to be engineered to higher structural specifications than much of the domestic and service industry building stock and it is likely that structural damage levels in the commercial sector will be less severe. Damage may still be heavy, and there have been instances of high-rise commercial buildings being affected by low-frequency ground motion, for example, in cases where smaller, domestic construction was less damaged.[5]

Priorities to support recovery in these areas are:

- The re-establishment of urban services, power, telecommunications and transport networks.
- Re-establishing business confidence by demonstrating commitment to full reconstruction.

5.3.5 Tourism and Leisure

Confidence is also important to regenerate another service industry: tourism. Tourism – income generated from people visiting and passing through – often plays a large role in the local economy. An earthquake is likely to cause severe damage to the hotel and tourist trade. News of the earthquake discourages visitors from elsewhere from visiting the region.[6] Undamaged hotels are likely to be used as temporary living accommodation during the emergency and rehabilitation phases so that additional visitors cannot be accommodated. The reputation of the area as a nice place to visit may be ruined and the name of the region may remain synonymous with earthquake destruction for a long time. If damage has been caused to the historical buildings or physical features for which the area is renowned, the tourist trade may take a long time to recover.

This damage to the tourist trade can be combated to some degree by turning the reconstruction itself into a tourist attraction. The reconstruction of the historical centres of Gemona and Venzone after they were destroyed by the 1976 earthquake in Friuli, Italy, was widely publicised (see Figure 5.1) and is now well worth visiting. The publicity generated by the earthquake to the outside world can, with the right media approach, be used positively to project an image of strong recovery, confidence, the availability of new, and better, facilities for the visitor and the indomitable spirit of the local population. Priorities to support the recovery of the tourist industry include:

- Re-establishing confidence by widely publicising the reopening of closed facilities and the availability of accommodation.
- Providing price discounts and other deals to attract visitors back.

---

[5] For example, Mexico City 1985, Bucharest 1977.

[6] The global media reporting of the damage to the Basilica of St Francis in Assisi in the 1997 Umbria–Marche earthquake in Italy resulted in a catastrophic downturn in the tourist industry in the years immediately following, even though most of the tourist infrastructure was undamaged.

Figure 5.1 Postcard of earthquake damage to the historic centre of Gemona, Italy, in 1976, on sale to tourists after the painstaking renovation of the town

5.4 Physical Reconstruction

In planning the physical reconstruction after the earthquake, severely damaged areas need to be treated differently to the larger, less damaged areas surrounding them. Urban areas may need a reconstruction policy different from that required for the rural areas. The type of society that has been affected, its degree of self-reliance and expectation of living standards will also greatly affect the way that reconstruction is carried out and the decisions made. One thing that all the areas are likely to have in common, however, is the urgent need to resolve uncertainties and to begin reconstruction.

Delays in reconstruction create uncertainties in the population and in potential investors. They reduce the psychological momentum for the reconstruction that often builds up towards the end of the emergency period, and they delay the all-important resumption of normal lifestyle and economic production for everyone. The most damaging delays are those before starting the reconstruction, time taken to plan or make strategic decisions: time when everyone affected is wondering what, if anything, is going to happen. An early and highly publicised launch of reconstruction work and visible signs of activity are needed to sustain public morale and encourage everyone to participate in recovery.

An early start is more important than a fast reconstruction programme. A fast construction programme tends to need a lot of planning, so is usually slower to start. A lengthy programme with construction being done well and jobs carried out thoroughly is preferable to a rushed job. Providing people can see progress being made, an 'early and steady' (early start, steady progress) philosophy to the reconstruction is preferable to a 'late and fast' (late start, fast progress) one.

Often political pressures, promises of construction miracles or over-ambitious targets lead to projects designed around speed and quota targets rather than quality and achievement. Over-hasty programmes result in poor-quality construction standards, storing up problems for the future, and reduce the possibility of using the programme to benefit the local construction industry and local economy, as described in Section 5.6.

In order to make an early start the strategic decisions have to made quickly. The detailed decisions can be worked out while the reconstruction is underway, but procedural decisions and broad locational priorities for rebuilding do need to be clearly defined at an early stage. There also needs to be an early decision whether to go for temporary housing or accelerated reconstruction (see below), but much of the detail can be worked out later.

Delays in making these key initial decisions often occur from waiting for a full evaluation of the damage and in the political process while seeking consensus among the various groups involved. In some cases, too, delays occur because decision-makers do not know what the best strategy is in a given situation. Experiences from past earthquake reconstructions can help to indicate what has been successful in certain conditions and what has failed in others.

5.4.1 Reconstruction of Severely Damaged Settlements

Damage from an earthquake tends to be most serious near the epicentre and reduces in severity with distance from this point. As a result there are likely to be many more settlements with moderate and light damage than have suffered heavy damage. A large proportion of the houses needing to be rebuilt are near the epicentre and the main reconstruction activity and media attention naturally focus on those places with high-percentage damage. In a severe earthquake the levels of destruction around the epicentre can be near total.

5.4.2 Reconstruction of 'Destroyed' Settlements

The reconstruction in those 'destroyed' settlements needs to be treated as a special case, unlike the other settlements damaged by the earthquake. Problems have commonly arisen from trying to extend the same policy, financing, housing eligibility and other procedures for rebuilding destroyed areas as for those with lower levels of damage. The distinction between 'destroyed' and 'damaged' is somewhat arbitrary as there are likely to be settlements with every level of damage, but in settlements where over three-quarters of the buildings need to be rebuilt,[7] the situation is qualitatively different from settlements where, say, less than half of the buildings need to be rebuilt. The areas defined as 'destroyed' should be kept to a minimum and defined by an explicit structural damage scale, e.g. neighbourhoods or villages with 75% of buildings with unrepairable damage greater than or equal to damage level D3.[8]

In these destroyed areas, the damage is probably too extensive to repair piecemeal. There may be no recognisable features remaining to identify land boundaries or the old layout of the streets. The collapse of many buildings may mean that the community has suffered many casualties, the social structure of the community is likely to be in turmoil, leadership could be gone and communications between members will be disrupted. Planning for the reconstruction of these areas needs to incorporate the recovery and reconstitution of the social order of the community, helping it to re-establish a sense of identity and place. This is helped by drawing the community groups in to the decision-making – it is important that it is not assumed that the disrupted community is powerless – helping the community to choose its representatives, re-establish its structure and discuss its own future.

Anything that hinders that process of recovery has to be removed temporarily. It may be worth considering whether any existing restraints like land development laws, employment regulations, foreign investment restrictions or building codes could be temporarily lifted within the destroyed areas to enhance the initial recovery period, reimposing them when the area is strong enough to take them back on board. Encouraging investment through creating economic development zones (EDZs) (see Section 5.3) or offering land free to private builders may be initiatives that will save the community.

As with other damaged areas, it is not necessary or desirable for outside help to build everything for the community or to hand over a gleaming new settlement; the main objective is to get the community back into its own construction activity, participating in the control and creation of the reconstruction as part of its revived viability.

5.4.3 Urban Reconstruction

Towns and cities represent high investments in infrastructure and services. Towns contain a complex linkage of economic activity, industrial production, commerce, trade and activities reliant on the urban infrastructure that service a wide region

[7] 75% of the buildings needing to be rebuilt does not mean that 75% of the buildings collapsed. Buildings are generally unrepairable, i.e. cost more to repair than to demolish and rebuild, when they are heavily damaged and have multiple cracks or structural distortions. In the damage scales given in Chapter 9, this is usually equivalent to heavy damage $D \geqslant 3$. Common damage distributions compiled in Chapter 9 show that when 75% of buildings are damaged $D \geqslant 3$, the percentage of buildings collapsed is likely to be around 25%.

[8] See Table 9.4 for damage definitions.

around the town. The physical rebuilding of towns needs to focus on repair and restitution of urban services as a first priority. Underground utilities – pipes, cables and sewerage systems – are generally less vulnerable than buildings but it is harder to locate failures and to repair them. Roads, bridges, power supply lines, transformer stations and many of the other civil engineering works that make up the physical infrastructure are similarly part of the first priority for reconstruction resources.

Urban Design

The physical planning of the reconstruction of a town, particularly one with high levels of damage, represents an opportunity for change: a chance to design a better town or rerationalise old anomalies. Kenzo Tange's modernist boulevards for Skopje, Yugoslavia, restructured from the old traditional street patterns in the reconstruction after the 1963 earthquake,[9] the replanning of Lisbon after the earthquake of 1755[10] and, more recently, the 1988 Leninakan earthquake in Soviet Armenia[11] are all examples of a major new urban design being introduced across the rubble of a destroyed city. Not all such urban redesigns are implemented or are successful if they are.[12] The complexities of land ownership, the emotional ties of the community to the old places and the existing urban infrastructure still in place mean that it is easier, cheaper and more popular to rebuild using the old structure of the settlement. For a community struggling to return to normality after the shock of the earthquake, it is psychologically important to recreate familiar localities, street plans and meeting places.

Relocation

The most radical method of reconstructing afresh is to abandon the old site of a settlement and build a new town on a different site. Relocation of urban and rural communities suffering high levels of damage has been a common policy in some countries (see e.g. Figure 5.2). This is sometimes advocated because the ruined settlement contains such horrors or the task of demolishing everything seems so complex that abandonment and starting again on a new site appears more attractive. This is more of an option with smaller towns and villages than with large cities – only in the most unusual circumstances would it be worth considering the abandonment of the vast infrastructural investment that is represented in a major city, even if it is heavily damaged.[13]

[9] UN (1970) and Ladinski (1989).

[10] Kendrick (1956).

[11] NCEER (1989).

[12] Davis (1978).

[13] The Tangshan earthquake of 1976 in China, in which 250 000 died, was one such case.

Figure 5.2 Relocation of damaged villages is rarely popular with occupants or successful in the longer term. This Turkish village, badly damaged in the Bingöl earthquake of 1971, was relocated and government houses built on the flat plain above the slope, visible top left. Prevailing winds and poor water supply on the plain meant that the site was soon abandoned and the villagers rebuilt their own houses back in the sheltered and irrigated site of the original village

The appeal of a greenfield site, where everything can be started from scratch and construction operations are more straightforward, is clear for the urban planners and construction agencies. It is sometimes understandably but incorrectly assumed that because the settlement was badly damaged, this must mean that the present site is unsafe for earthquakes. Siting considerations for earthquake risk are discussed in Chapter 7, and ways of improving earthquake safety in reconstruction urban planning is considered further in Section 5.7.

Relocating a settlement for these reasons is rarely popular with the occupants of the settlement, or successful in the longer term. Most settlements are located where they are for a reason which may be water supply, trading routes, fertility of land or a combination of these and other reasons. The larger the settlement is, the more successful that site has been.

The decision on whether to relocate a damaged settlement is a difficult one and should be very carefully considered – the advantages of a new start and logistical expediency may be greatly outweighed by the cultural severance and economic implications of a different site. Resiting can have a severe impact on the local economy – studies of resettlement in rural economies demonstrate the changes in agricultural productivity, fuel consumption and domestic economy caused by the relocation of communities to sites of different microclimate, fertility and water supply.[14]

Attachment to a place is a common reaction to relocation; however, the reasons behind the willingness to stay in the original place can provide insights into more acceptable and successful relocation when it is the only alternative.

Research[15,16] has highlighted the reluctance of communities to move elsewhere. Typically a core group of people, usually the older members of the society, are highly resistant to moving while the younger generations may be more receptive if the new location provides good public services and job opportunities.

The capital investment to turn any relocation into success can be very high. For the rural areas, small towns and underprivileged urban areas the cost may be sometimes justified if compared with the risks of living in hazardous areas, but in well-established urban areas, politically and economically it is unlikely ever to be viable.

In areas where the population is increasing rapidly, relocation of communities to safer areas or attempts to reduce population densities in risk zones, in the long run, may prove to be an unsuccessful exercise. Examples illustrate[17] how difficult it is to depopulate even a badly damaged old town centre in favour of a newly located reconstruction site.

The relocation of a settlement should not be carried out without fully understanding the reasons why it is sited where it is and demonstrating that any proposed new site is significantly more advantageous.

Emergency Building Codes

In urban areas there is often a conflict between the need for a speedy start to the reconstruction and a desire to study in detail the options for reconstruction, to investigate the structural damage fully to ascertain whether revisions are needed to the building design code and to look at the microzoning of the town to determine earthquake effects in the ground conditions for future development planning.

In many cases the authorities put a freeze on reconstruction, prohibiting private owners from rebuilding or developing new sites and delaying public reconstruction projects until the studies are completed. To carry out these studies properly requires many months or years – often the authorities underestimate how long it is likely to take – and a delay of this type has severe penalties in creating uncertainty, reducing the psychological momentum for the reconstruction and slowing down the all-important economic recovery.

[14] Coburn *et al.* (1984a).

[15] Aysan and Oliver (1987).

[16] Aysan *et al.* (1989).

[17] D'Souza (1986).

A better approach is to promulgate an emergency building code very rapidly so as to allow reconstruction to start immediately, while the studies continue in parallel. The emergency code is an interim temporary code requiring better design standards which operates until the code revision is finalised. It is conservative and could include a 'better-safe-than-sorry' microzoning of the town requiring stronger construction in any of the areas suffering most severe damage: as the microzoning studies allow more detailed identification of any poor ground conditions, the requirements can be relaxed.

Use Existing Master Plan

Where a city or damaged region already has an existing plan for growth or long-term strategic objectives, the earthquake reconstruction should follow those plans and use them as the guiding principles for its own plan. A reconstruction is often a good opportunity to implement elements of an existing plan. The reconstruction then builds on existing studies and plans developed under less demanding circumstances. A reconstruction master plan that begins from first principles is likely to be carried out insufficiently or will delay the reconstruction. The pattern of earthquake damage will naturally affect the master plan to some degree, but should not affect the overall strategic objectives. Its main effect will be to determine the locations and facilities where the implementation can take place during the recovery period.

5.5 Housing and Shelter Policy

One of the major concerns for most earthquake reconstruction is how to rebuild the housing damaged by the earthquake. Large numbers of families may be homeless and exposed to the elements. This crisis of shelter lends an urgency to the problem: humanitarian concerns, political pressures and practical considerations mean that housing reconstruction is likely to be given a high priority in the reconstruction schedule.

Four alternative models of the post-earthquake housing recovery process have been distinguished:[18]

1. The redevelopment model: complete redevelopment of the devastated area by the national government.
2. The capital infusion model: infusion of outside aid targeted to low-income housing provided by government, international aid and NGOs.
3. The limited intervention model: assumes private insurance will cover some losses, property prices will adjust, and government will assist only the poorest.

[18] Comerio (1998).

4. The market model: complete reliance on market forces to adjust, adapt and reconstruct after the disaster.

Each of these models has been adopted following some of the earthquakes of the last 30 years. The redevelopment model was adopted following the 1976 Tangshan disaster in China and the 1988 Leninakan earthquake in Soviet Armenia; it could only be considered in a socialist economy of the type which hardly exists anywhere in the world today. The capital infusion model is the one most frequently adopted for major disasters in both developing and industrialised countries, e.g. the 1985 Mexico City earthquake and the 1993 Latur earthquake in India. It was also followed for reconstruction following the 1999 earthquakes in Turkey and the 2001 Gujarat earthquake in India. The limited-intervention model is currently the basis of stated US disaster-recovery policy, although the recovery following the 1994 earthquake was (largely for political reasons) exceptional in that it contained many elements of the capital infusion model.[19] By contrast, the market model was essentially the basis of the recovery from the 1995 Kobe earthquake in Japan.

5.5.1 Emergency Shelters and Temporary Housing

During the emergency phase in the days following the earthquake, the many families whose houses are damaged are housed elsewhere, in empty public buildings or hotels, staying with relatives, living in tents or some other emergency arrangement. Planning for getting the homeless more satisfactorily housed has to begin at once so that the emergency situation is not prolonged.

In severe earthquakes it can be expected that reconstruction of all the damaged and destroyed settlements will take some time to plan and to execute, and it is often assumed that some form of intermediate stage involving emergency shelter or temporary housing needs to be provided while this is happening. However, the effort, resources and enthusiasm that are put into the intermediate stage of temporary housing are all too often at the expense of the final reconstruction. Strategies for reconstruction that have opted for an intermediate stage of temporary housing generally take far longer to achieve final reconstruction and often run into problems raising the resources for the later phase. In many cases the intermediate 'temporary' stage has become permanent and the final reconstruction stage is reduced in scale or fails to materialise altogether. Thus, where it is possible to achieve a programme of accelerated reconstruction, the use of temporary housing should be avoided. The argument for accelerated housing reconstruction rather than using temporary housing is expanded in the next two sections.

The perceived urgency and scale of the need often means that radical solutions are turned to in order to solve the logistical problems. There is sometimes

[19] Comerio (1998).

a distinction made between 'emergency shelter' (durable shelters which are designed to be distributed in the first few days) and 'temporary housing' (prefabricated or system housing which requires some preparation of the site). There have been very many attempts to design emergency shelters – structures that can be supplied rapidly, erected quickly and produced en masse as dwellings for the homeless victims. Some of the solutions have been ingenious, others ridiculous, ranging from polyurethane domes and plastic igloos to frame kits and many types of prefabricated housing units (Figure 5.3).

Temporary housing imposes its own characteristics on an emergency situation. The urgency and practicalities of erecting units rapidly mean that flat sites are preferred to sloping ones and rectilinear, dense layouts of houses are the most convenient site plan. This may mean building the temporary field camps on sites some distance away from the damaged settlement or building camps on agricultural land.

In many cases, by the time the emergency housing has been provided, the communities have already improvised shelter for themselves or, in some cases, completely rebuilt or repaired their own houses. Even if the emergency shelters are provided rapidly enough to be useful, they are often abandoned as soon as the family has any practical alternative, so the wastage rate of these expensive units is high: donors often find that they have wasted large sums of money that could have been better used.

Figure 5.3 Temporary housing. Temporary settlements are often needed in the worst-damaged areas, and they are likely to remain for years, so they need to be planned so that they can be incorporated into long-term resettlement plans. Containers adapted for temporary housing after the 1980 Irpinia earthquake in Italy remained in use for many years

Detailed studies of the communities provided with temporary housing have shown that it has an overall negative effect on their recovery from the earthquake.[20] Disadvantages of temporary housing include the practical difficulties of upgrading, expanding or moving back from the shelter into normal permanent housing,[21] the creation of new 'temporary' settlements away from the damaged settlements which are insufficient for full normal resumption of life yet draw resources and the focus of activity away from the old settlement,[22] and perhaps most importantly, the alien nature of the houses being culturally unfamiliar at the exact time when, psychologically, the community is trying to reassemble normality and needs familiarity.[23]

The more experienced disaster relief agencies now recognise that the treatment of such communities as 'victims' – that is, as passive unfortunates, incapacitated by the disaster and likely to freeze to death unless someone provides shelter for them – is wrong. Communities affected by disasters have been found to be resourceful, ambitious and hardworking in their efforts to recover from the setback of the earthquake damage. What they need is support for their efforts rather than someone else to do it for them.

Policies for disaster assistance that focus on assisting the *process* of the community housing itself are more successful than outside agencies *providing* shelter. Assistance to the process of reconstruction means supporting the building activities that the community initiates and allowing the community to control the way that reconstruction happens. This may be difficult for some donors to accept; and planners and technical specialists may also find it frustrating that they are not in charge as they often feel they could control it better. Technical specialists are likely to be able to visualise major opportunities in the rebuilding to rerationalise the layout of a neighbourhood or to build greatly improved types of houses with modern facilities. The ambitions of the communities affected are likely to be more modest, usually to rebuild their familiar neighbourhoods and villages, obviously improved if possible, but to re-establish the lifestyles, streets and houses they had before the earthquake (Figure 5.4). Their visualisation of the reconstruction is based largely on what was there before, with a limited horizon of improvements, perhaps bringing more houses up to the standard of the most prestigious house in the neighbourhood. If this control can be retained by the community and this is accepted by the disaster assistance agency, a partnership will be developed to build a foundation for a strong recovery.

However, in Section 5.4 we argued that the reconstruction of the destroyed settlements should be treated as a special case, and it may be appropriate to provide temporary housing in these circumstances, to provide for the large number

---

[20] Davis (1978).

[21] Aysan and Oliver (1987).

[22] Coburn *et al.* (1984a).

[23] Oliver (1981).

Figure 5.4 Reconstruction plans often underestimate the capability of the local community to participate in its own recovery. His village in ruins, a builder begins reconstruction after the 1990 Manjil earthquake, Iran

of people made homeless. This will be especially important in colder climates. Location and design of this temporary housing will need careful consideration to minimise the potential negative impacts discussed above, and where it can be provided in the form of core houses suitable for incremental reconstruction (see below), this will be a benefit to long-term recovery.

5.5.2 Assistance for Accelerated Recovery

If accelerated reconstruction is to be the policy adopted, assistance from outside is likely to be needed in providing building materials (or assisting with the gathering or manufacture of building materials), construction equipment, building tools and other scarce resources. Logistical support, transportation management and helping expand the existing construction capacity of the area, supplementing any shortfalls and providing other assistance, like organisation of additional labour, are also important inputs from disaster assistance organisations at this time.

External assistance to help a family rebuild a damaged house is probably most effective as a financial contribution. This is often administered as a grant or loan on favourable terms. The finance is used by a householder to commission the repair or reconstruction of their house from a builder of their choice.

Sometimes there may be restrictions on the type of house built or a requirement to build a standard house type. Builders participating in the scheme may also be controlled either by government registration or through training, or as a government employee, financed through reconstruction credit schemes. In less free-market situations, housing developments and apartment buildings are built as part of a public housing scheme or government construction project, with the completed dwellings being allocated to householders, possibly with loan repayment conditions attached. Whichever method of housing procurement is used, the more familiar it is and closer to the pre-earthquake method of obtaining living accommodation, the better the reconstruction will develop into recovery.

Where loans are offered rather than grants they are usually optional but are generally preferred by agencies as take-up of the loans indicates higher commitment by householder, and repayment of loans can be used to finance future projects in a revolving fund. The repayment terms do, however, need careful gearing to the economic capabilities of the affected community – there have been examples of schemes with poor take-up because of higher repayment requirements than general income levels[24] and schemes with very high rates of householders defaulting on their loans.[25] Loans also preclude the very poor, those without an income or those without collateral. If it is possible to integrate low-income groups into the standard reconstruction programme using special-case financial arrangements this is preferable to creating ghettos of special low-income housing schemes.

5.5.3 Core Houses and Incremental Reconstruction

A method of achieving accelerated reconstruction in rural communities and in those that are more self-reliant is to assist each homeless family to build at least a 'core house' – a small, solid structure sufficient for immediate needs and which can be added to later to form a larger, permanent home, or useful as an animal shed or outhouse. The householders can, of course, build more than that depending on their own resources, but a minimum structure should be available to everyone whose house is uninhabitable through earthquake damage.

Building materials, tools and technical support need to be provided for each house. The most important need is likely to be building materials – materials that are familiar to the householder, speedy to erect and rapidly deliverable. Examples may include concrete blocks for walls, timber, roofing sheets and glass for windows. In many cases building materials from the damaged houses may be retrievable and reusable in the reconstruction. Joinery elements like doors and windows may be retrievable, but are often too damaged to be reused, so a supply of doors and windows may be an important part of any core house programme.

---

[24] Thompson *et al.* (1986).

[25] Coburn (1987).

Core houses need to be watertight, thermally adequate to survive cold seasons and as well-built as can be managed. Proper foundations should be dug. Mobile technical support units visiting the affected areas to distribute materials should also provide advice and guidance on robustness of construction. It may be that such programmes can be linked with builder training programmes, as described in Chapter 6, Section 6.6.

If the construction of a core house entails the demolition of the damaged building and clearing the site some limited technical assistance in phase planning may be needed.

5.6 Reconstruction and the Construction Industry

The availability of building materials and a strong local construction industry are essential for successful reconstruction. The reconstruction operation itself requires a supply of building materials and a major input from the construction industry. For any sizeable earthquake the quantity of materials and the output capacity of the construction industry needed will be considerably greater than the normal availability of materials and normal output of the construction industry within the affected region itself. It is often assumed that outside contractors should be used to rebuild the stricken areas and that supplies of materials will have to be brought into the region from outside.

Often in rural areas or in provincial towns, the reconstruction is carried out by contractors brought in from other large cities, or by government construction agencies brought in for the job, or even international contracting companies from other countries. International aid or international financing for earthquake reconstruction often is tied to international contractors. Bilateral aid for example may entail the use of contracting companies from the donor country or for international aid, standard contracting procurement procedures may require international tendering which favours external companies being brought into the damaged region.

Although understandable, there are significant disadvantages to this approach, the most important being that the normal building industry of the region can suffer long-term damage.

Any reconstruction operation that leads even inadvertently to a reduction in involvement by the local building industry is storing up problems for the future, when the outside contractors withdraw and the local building industry is once again expected to resume operations, usually with expectations that all new construction will be earthquake resistant and to higher specifications. A reduced involvement of the local building industry in reconstruction means a loss of local skills – cases of unemployed local builders migrating to other towns are not uncommon – a suppression of local capability in terms of equipment and practical expertise, a reduced identification of the local community with its own reconstruction, and most importantly, a loss to the local community of income that the labourers and builders generate.

Earthquake reconstruction is therefore one of the best ways to inject capital into the local economy, through the building activity needed to replace damaged structures. The opportunity is wasted if the investment is taken directly out of the area by outside labour, totally imported building materials and construction companies based elsewhere who will take their profits away from the area that needs it most.

The logic for using external contractors is based on the arguments that:

(1) the scale of the building work needed is far beyond the capacity of the local building industry;
(2) a rapid construction programme will mean a quick return to normality; and
(3) a higher quality of construction will be provided by bigger and better construction companies.

However, to achieve the long-term goal of full recovery, the reconstruction programme should give preference to local construction companies as the major vehicle for construction activity (see e.g. Figure 5.5). Wherever possible, these local companies should be assisted to grow, increase their capability and retain control of the construction programme. Government and other agencies can

Figure 5.5 The construction activity undertaken during the reconstruction can be the focus of a community's economic recovery if planned carefully to maximise the involvement of the local building industry. Villages in southern Italy under reconstruction after the 1980 Campania earthquake

assist this operation by establishing a procurement and tendering procedure that gives preference to local companies. Larger companies bidding for construction contracts can be required to work in partnership with local companies to provide management and training for their employees and to employ local labour.

Subdividing reconstruction contracts into smaller parcels by district and sector has also proved successful in past earthquakes, allowing certain sectors such as housing, schools, etc., to be reserved for the local contractor companies, or even one- and two-person builder operations, while other sectors such as bridges, hospitals, etc., are opened to bids from large-scale contracting companies. The principle of using and assisting the local building industry to carry out as much of the work as possible should be strongly promoted.

5.6.1 Building Materials Supply: an Integrated Plan

Similarly the use of locally produced building materials for the reconstruction will also provide economic returns within the earthquake-affected region and maximise the usefulness of capital investment in revitalising the local economy. The demand for building materials for any sizeable reconstruction is likely to be much larger than normal production levels in the immediate vicinity. A better response for the longer term prosperity of the area is to try to increase the building materials production output of local manufacturing capability as far as possible and to supply as much as possible of the materials needed from local sources.

This will be more feasible in rural and semi-industrialised situations and in regions where the higher percentage of construction is built using local materials anyway – it may be less relevant to fully commercialised building industries, where materials are normally purchased from a wide range of suppliers nationally.

Planning for the reconstruction should include an integrated plan for building materials supply. This should cover all the sectors and incorporate the expected needs for housing, schools, hospitals, industry and all the other sectors. A global assessment of the total materials needed will emerge from the sectoral damage analysis, and the integrated building materials plan should propose what proportion of that need can be met locally, how much can be obtained from suppliers in other parts of the country and what has to be imported from other countries. The preparation for the plan will include an appraisal of current or pre-earthquake building materials usage in the area, what and how much is produced in the region, how much output capacity could be increased and what new plants could be established.

The philosophy of using smaller scale production facilities to provide quicker start-up time, production closer to the point of use, and greater control by the community itself over building material manufacture is being used increasingly in the reconstruction of earthquake damage in semi-industrialised and developing regions.[26] Small-scale production technologies are proving increasingly successful in building material manufacture in developing regions.[27]

Choice of building materials for use in reconstruction should largely reflect the policy of preference for local production. Designs for new construction should be determined by the materials available rather than trying to introduce new forms and materials. Designs incorporating brick infill walls may be inappropriate in areas where stone is normally used and plentifully available, for example. Reconstruction designs should largely respect the existing traditional building forms, materials and architectural style of the region. As far as possible, as argued in Section 5.5, control over the design and construction of buildings should be left to the owners and the member of the community that use them.

5.7 Turning Reconstruction into Future Protection

In the aftermath of the earthquake, the replacement of destroyed buildings and the reconstruction of a damaged community present a significant opportunity to make the new community safer against a possible repetition of the disaster some time in the future. After a major disaster, the replacement of possibly large sections of a town and the rehabilitation of a significant percentage of the townspeople give an opportunity to bring about changes that will reduce the impact of the next earthquake.

Changes are possible after a disaster where they would not be possible beforehand. Funds are available, everyone is aware of the hazard and generally agreed on the need for protection, the political climate is sympathetic and there are political opportunities to push through change where it is needed. But the window of political opportunity and the period of availability of financial assistance are usually short. The key to making maximum use of the opportunity is pre-planning and an awareness of how best to achieve mitigation within reconstruction activities. The post-disaster emergency period is not usually the best time to make crucial decisions concerning the long-term future of a city, and yet experience has shown that many reconstructions are planned rapidly, immediately after the event, with little studied consideration of what contributes most to future safety.

There are at least five important considerations that affect reconstruction planning and the policies that are likely to be most effective in bringing about future safety:[28]

1. The return period of the earthquake.
2. Pre-existing plans for the future development of the city – including seismic risk studies for future protection.

---

[26] As practised in the reconstruction programmes of Iran (1990), Yemen (1982) and Ecuador (1987).

[27] Spence and Cook (1983), UNCHS (1990).

[28] Aysan *et al.* (1989).

3. Profile of the communities affected, including the economic basis and cultural preferences of the various groups affected.
4. The scale of the disaster.
5. The resources available for reconstruction.

Of these only the last two factors are unknown before the earthquake. Preparedness planning should establish the longer term aims of a mitigation programme in general, so that in the event of a major earthquake, the reconstruction can be channelled towards well-established mitigation aims rather than having to improvise a new strategy plan.

The way the reconstruction is carried out can have a major effect on the future safety in addition to what is reconstructed – the process is as important as the end product. Social and economic recovery of the affected communities and the reduction of the overall vulnerability of the city to the impact of future earthquakes require integrated and comprehensive policies covering a wide range of activities. A major reconstruction offers the opportunity to introduce comprehensive mitigation measures into the ongoing processes of planning, administration and construction. It also provides the impetus to channel financial resources where they are needed and prompts a political willingness to implement policies.

Political pressures for rapid recovery should take second place to systematic studies of long-term needs. The emphasis should be placed on creating the economic building blocks, the cultural continuity and the spatial framework for future development rather than on construction showpieces. Institutionalising reforms in the building industry and construction process will be more important in the long run than building an instant earthquake-proof town.

5.7.1 Reconstruction after Earthquakes with Long Return Periods

It is obviously important to capitalise on the opportunities, funds and incentives present after an earthquake to improve the building stock and restore public confidence. However, unless the opportunity is also taken to instigate much longer term protection measures and to carry the lessons beyond the areas immediately affected by the earthquake, stronger reconstruction may not be a very effective way of reducing future earthquake losses. The risk of future earthquake losses varies considerably from place to place. Return periods for earthquakes striking the same place twice are usually considerably longer than the lifetimes of individual buildings, and in areas of long return period, the construction of stronger buildings has a sense of "closing the stable door after the horse has bolted". The return periods of high intensities from shallow-depth, near-field events in most sites across the world can usually be counted in centuries. In eastern Turkey, an area of relatively high seismicity, for example, the area affected twice by a damaging earthquake ($I \geqslant$ VII) within 100 years is only 2% of the high-risk areas. In southern Italy the return period of damaging earthquakes at any particular location within the Apennine region is less than once in 350 years. Most earthquakes occur in areas which have not recently experienced a destructive event.

In most locations in the world, where high intensities have a long return period, using a reconstruction programme to make a place safer against a future earthquake should mean planning long-term strategic developments for the city rather than short-term upgrading. Long-term strategies that should be considered include:

- Using the reconstruction to maximise long-term economic development for the region.
- Controlling future urban land-use patterns in the reconstruction to minimise risk.
- Structuring the morphology, street layout and landownership in the reconstruction to improve urban safety and density of development.

The immediate reaction of most city authorities after a disaster is to rebuild damaged buildings in strong, earthquake-resistant construction. This is a natural reaction, but its effect may be largely symbolic and psychologically reassuring rather than an effective method of reducing the losses from future earthquakes. Major monuments may last hundreds of years, but ordinary residential building stock in a city may have a lifetime of 30 to 100 years depending on pressures of development, housing markets and fashions of housing style.

5.7.2 Historical Reconstruction and Present-day Risk

The end results of the reconstruction-into-protection process are of special importance to the study of mitigation as are the examples of recent reconstruction policy and their intended results. It is possible to look at examples of towns that were rebuilt after historical earthquakes that are now, many years later, facing up to the threat of a repetition of a destructive event. (See the boxes on the following pages.) Studies of urban seismic risk in Noto in western Sicily, destroyed, relocated and rebuilt after a massive earthquake in 1693, and of Bursa in western Turkey, repeatedly damaged by large-magnitude earthquakes and considerably rebuilt after the destructive event in 1855, and also of Quetta, in northern Pakistan, reconstructed after the earthquake of 1935, give insights into the long-term nature of earthquake protection from decisions implemented in the aftermath. The case study of Mexico following the destructive earthquake of 1985, provides a twentieth-century comparison.

Reconstruction case study Noto, Sicily, an eighteenth-century reconstruction

The city of Noto today. The older stone masonry *palazzi* of the historic centre (background) are being abandoned in favour of new reinforced concrete villas seen in the foreground

After the earthquake that destroyed their city and killed an estimated 3000 people in 1693, the citizens raised considerable sums to rebuild their city safely.[29] After extensive public debate, the decision was finally made to relocate the city from its ruined site to a new location over 10 kilometres away where it could be laid out along the latest principles of city planning. The new city layout, along wide streets and punctuated by a series of Baroque architectural monuments, provided an urban framework within which the townspeople could rebuild their family houses. Most rebuilt in the grand style, building large and strong Italianate *palazzi* in dressed stone to replace the vulnerable timber-framed or rubble houses of the ancient town.

Now, 300 years on, the town council is again facing up to the threat of a return of a destructive earthquake, forecast with a return period of between 200 and 1000 years. Apart from the civic and religious monuments, less than 1% of the building stock now at risk was built as part of that eighteenth-century reconstruction.[30] The rest of the buildings were built in subsequent centuries, replacing the older buildings as they deteriorated, infilling vacant blocks and

[29]Tobriner (1982).

[30]Coburn *et al.* (1984b).

expanding onto areas surrounding the city. And the few remaining eighteenth-century *palazzi*, so much more robust and earthquake resistant than the buildings they originally replaced, are now, after years of gradual deterioration, among the most vulnerable of the existing buildings – in 1997, part of the dome of Noto's cathedral suddenly collapsed without assistance from any earth tremors.

The policies of the eighteenth-century reconstruction for which today's population have cause to be grateful in making the city safer against the next earthquake are the strategic decisions on relocation, replanning and restructuring the local economy made at the time. For example:

- The relocation of the city away from its ancient defensive site onto a site closer to the rich agricultural plains and a secure water supply ensured the prosperity of the townspeople subsequently leading to a continual upgrading of building quality.
- The choice of site on a firm, travertine hilltop – one of the flattest rock sites in the region – reduced the potential for landslide and slope failure that claimed many lives in the old city in the 1693 earthquake.
- The rationalisation of the city's street layout with wider avenues and lower densities of housing has made the streetscape safer and more accessible to emergency services, than if it had been rebuilt on the old site.

Reconstruction case study Bursa, western Turkey, a nineteenth-century reconstruction

The extending suburbs of Bursa in 1985. The direction of expansion can have a big effect on the city's future earthquake risk

The earthquake of 1855 that damaged the historical city of Bursa, once the capital of the Ottoman Empire, destroyed revered monuments, including several of the main trading bazaars, and caused serious fires that consumed sections of the residential areas in the city. The reconstruction that followed was chiefly funded from Istanbul, the nation's capital, and consisted of wide-scale restoration of the monuments and a resettlement of the population.[31] Many feared another earthquake, even stronger, and it is reported that morale among the townspeople was low. The protection measures included the separation of houses and the use of masonry instead of timber frame for buildings where possible. Tall minarets were demolished as a hazard to the population. Rocks were cleared from the slopes above the city to reduce the risk of future rockfalls.

For the city of Bursa today, again affected recently by the 1999 Kocaeli earthquake, hazard analysis shows that there is a relatively high level of seismic hazard in the area. A 'seismic gap' close to the city has been identified by seismologists, which may indicate the likely location of a large earthquake in the near future.[32] Detailed seismic risk analysis of the modern city shows that the major contribution to the present-day seismic risk has little to do with the earlier reconstruction. There is little evidence in today's city of the changes in the building stock that occurred following the earthquake. The buildings built before 1920 now constitute only 3% of the building stock. However, some of the reconstruction activities of 1855 have had an impact on the subsequent risk of the city. The population reduction after the 1855 earthquake reduced the regional importance of the town, which limited its nineteenth-century growth. This was reversed in the 1950s when a major industrialisation of the Marmara Sea region included car factories and major investment in Bursa, and caused a very rapid growth in the city. The city continues to grow at well above the average rate for Turkish cities and its centre has retained its historical siting on the firm rock hillside of Uludağ Mountain. Losses in future earthquakes will be highly influenced by the direction of expansion of the city suburbs in years to come. Expansion out onto the alluvial plains could mean significantly higher earthquake losses in a future earthquake than if the suburbs continue to expand along the rock mountainside or onto firmer ground nearby.[33] Building quality and engineering design will be important in reducing future losses but the main potential for earthquake losses will be the older twentieth-century buildings. The reconstruction project to make Bursa safer in 1855 had little concept of the massive changes that Bursa would undergo a century later.

[31] Kuran (1986).

[32] Coburn and Kuran (1985).

[33] Akbar (1989).

Reconstruction case study Quetta, Pakistan, an early twentieth-century reconstruction

Reinforced masonry is a resilient and cost-effective way of building in earthquake areas. 'Quetta bond', first developed after the 1935 earthquake in northern Pakistan, is still in use today

Quetta is one of the major cities of Pakistan, with a key military significance. In 1935 it suffered a major earthquake which destroyed almost every building in the city and many surrounding villages and claimed an estimated 20 000 lives.[34]

[34] Jackson (1960).

Because of its strategic position, relocation was considered impossible, and the seismologists' report pointed out that as a result of the energy released in the earthquake of 1935, Quetta could expect to be safe from another such event for some time.[35] The national government ordered instead that the city should be rebuilt on earthquake-resistant principles, and a building code was drawn up, which was in many respects a forerunner of modern codes.[36]

General regulations were specified governing the shape, height and spacing and materials of buildings. For important buildings, a system of steel frames with brick infill panels was specified; brick masonry buildings were to be built according to a new bonding system which incorporated concrete ring beams and vertical reinforcement (later known as Quetta bond).[37] For the poor, various systems using timber frameworks clad in lightweight materials were proposed, and the heavy mud roofs which had caused so many deaths were banned. Reinforced concrete frame construction was not recommended, as it required too high a level of skilled work.

For a time this code was effectively enforced throughout the city. But the following decades brought war, then independence, then a mounting and still critical refugee problem. It was impossible to maintain the tight controls on building which were possible in the years following the earthquake. Over the years since the earthquake the population has grown more than five-fold; pressure on space has made the demand for higher buildings irresistible; the timber required for the cheaper code buildings is now unobtainable; and the municipal engineer is too preoccupied with public health problems to be concerned with control of building standards.[38]

Today the vast majority of the population live in unauthorised buildings of poor masonry materials, extremely vulnerable to earthquakes; even in the city centre buildings of reinforced concrete are constructed with no proper provision for earthquake forces. The recurrence of the 1935 event today would without any doubt be a disaster on a much larger scale than before.

Quetta's experience demonstrates that the introduction of a building code alone will not be sufficient to ensure future standards of protection; a continuing awareness of the earthquake risk, a degree of public control over building, and above all the economic means to pay for protection are all needed if protection is to be effective.

[35] West (1935).

[36] Quetta Municipal Building Code (1940).

[37] Spence and Cook (1983).

[38] Spence (1983).

Reconstruction case study Mexico City, a late twentieth-century reconstruction

Strengthening of an existing reinforced concrete frame building by the addition of steel cross-bracing. One of a large number of public buildings in Mexico City strengthened this way following the 1985 earthquake

Mexico City has suffered three damaging earthquakes since 1957, each with a level of ground motion strong enough to cause structural failure and collapse in some of its weaker buildings. The earthquake in 1985 resulted in the highest level of damage in the city's history: over 600 buildings collapsed and more than 7000 people were killed. The high levels of damage were as much due to the poor quality of building in the 1960s and 1970s as they were to the fact that this was the strongest shaking to hit the city this century. The particular characteristics of the ground conditions in the city – built on a deep and ancient drained lake bed – make it likely to experience strong ground motions much more often than most other cities elsewhere. Any distant earthquake occurring up to 400 km away from the city may cause the saturated weak soils below the city to amplify the shaking. A damaging level of ground motion may be generated in this way every

15 years or so. The effects are, however, highly localised, and earthquake motions repeatedly damage the same area within the city – the area around the historic centre in which about 1.5 million of the 19 million inhabitants of the city live.

In Mexico City, the short return period of the earthquake and the characteristic patterns and repetition of damage in the same locations make mitigation through reconstruction an important priority. This has been well appreciated by the authorities in charge of structuring the reconstruction. Mitigation measures taken after the earthquake included:

- a large-scale programme of reinforcement of several hundred government buildings, schools, hospitals and other structures;
- a massive public housing reconstruction programme which has gone far beyond replacement of earthquake-damaged buildings to upgrade poor-quality and vulnerable housing in the city centre;
- a complete revision of the urban master plan for the city, including a rezoning of the city, proposals for decentralisation and reductions in allowable densities;
- a programme of renovation, strengthening and reuse of historical buildings;
- an urban upgrading programme to revitalise the city centre, to regenerate economic and environmental conditions and reduce vulnerability of the communities most at risk;
- the revision of seismic building codes, enforcing a considerable increase in earthquake resistance of engineered buildings.

5.7.3 Exporting Improvements beyond the Reconstruction Area

It is seismically probable that the areas most likely to be hit by the next earthquake are areas outside those badly damaged in the last earthquake, but probably within the same seismic region. To make a significant impact in the losses from earthquakes in the region as a whole, the reconstruction can be used to promote mitigation activities outside the damaged areas, into zones where the likelihood of an earthquake is equally severe, but perhaps on a shorter timespan. The contradiction here is that, while the actual risk may be higher, the immediacy of earthquake danger is not so obvious to the general public in the areas which have not experienced an earthquake recently, and the incentives and opportunities for the occupants of those areas to carry out disaster mitigation activities may also be much less.

The fringes of earthquake-affected regions are often important and fruitful areas to instigate earthquake protection projects, both because the population tends to be very aware of the recent, nearby earthquake and because the areas are still under threat from future earthquakes.

In general, any reconstruction aiming to instigate mitigation measures against future events should aim to export its lessons to areas with significant future risks.[39]

[39] The 1999 earthquake in Kocaeli Province, Turkey, triggered considerable earthquake mitigation activity in neighbouring Istanbul Province in the years immediately following.

5.7.4 Relocation of a Badly Damaged Community

Severe damage to any settlement almost always gives rise to suspicion that the damage is due to poor ground conditions, localised active faults or some site-specific hazard. A reduction in earthquake risk, it may be argued, could be achieved by resiting the community on a safer site, and the justification of future earthquake protection is often advanced for relocating a town or village. Problems with relocating settlements in a reconstruction were discussed in Section 5.5. The costs of relocating almost any sizeable established settlement will be prohibitive and unlikely to be justifiable in earthquake protection terms. If it is possible to influence locational planning, the opportunity should be taken to introduce land-use modifications within the damaged settlement. Urban planning measures for protection are discussed in Section 6.3. Better protection can usually be provided more cost-effectively by stronger design and construction standards of buildings on the existing site. Upgrading design and construction standards and building stock management are discussed in Chapter 6.

The conclusions that can be drawn from this are that in order to make a significant impact in the losses from earthquakes, mitigation measures have to target the long-term reduction of vulnerability for the wider city, aimed at improving not only the existing and future building stocks but also the general living standard of the communities at high risk and the decentralisation of the vital urban functions.

5.7.5 Deconcentration of Cities and Services

The dispersal of elements at risk over a wider area to make them a more difficult target to hit is a key strategy in mitigation planning. Reconstruction after an earthquake is a good time to instigate deconcentration measures. Deconcentration may involve reduction in densities, dispersal of elements and restructuring road layouts. Practical measures to decentralise urban areas on a regional basis may be accelerated by the impact of an earthquake. The principle of restructuring risk by compartmentalising utility sectors, and spreading elements at risk, such as buildings, industry or services, around the city is a valuable one, and one that can be widely applied. The full range of deconcentration measures possible, to reduce densities, regulate urban form and protect utilities, is described in more detail in Section 6.4.

Further Reading

Aysan, Y. and Oliver, P., 1987. *Culture and Housing after Earthquakes: A Guide for Future Policy Making on Housing in Seismic Areas*, Disaster Management Centre, Oxford Polytechnic, Headington, Oxford OX3 0BP, UK.

Comerio, M., 1998. *Disaster Hits Home: New Policy for Urban Housing Recovery*, University of California Press, Berkeley, CA.

Cuny, F., 1983. *Disasters and Development*, Oxford University Press, Oxford.
Davis, I., 1978. *Shelter after Disaster*, Oxford Polytechnic Press, Oxford.
NCEER, 1989. Proceedings of Conference on *Reconstruction After Urban Earthquakes: An International Agenda to Achieve Safer Settlements in the 1990s*, September 1989, National Centre for Earthquake Engineering (NCEER) at University of Buffalo, Red Jacket Square, Buffalo, New York, USA.

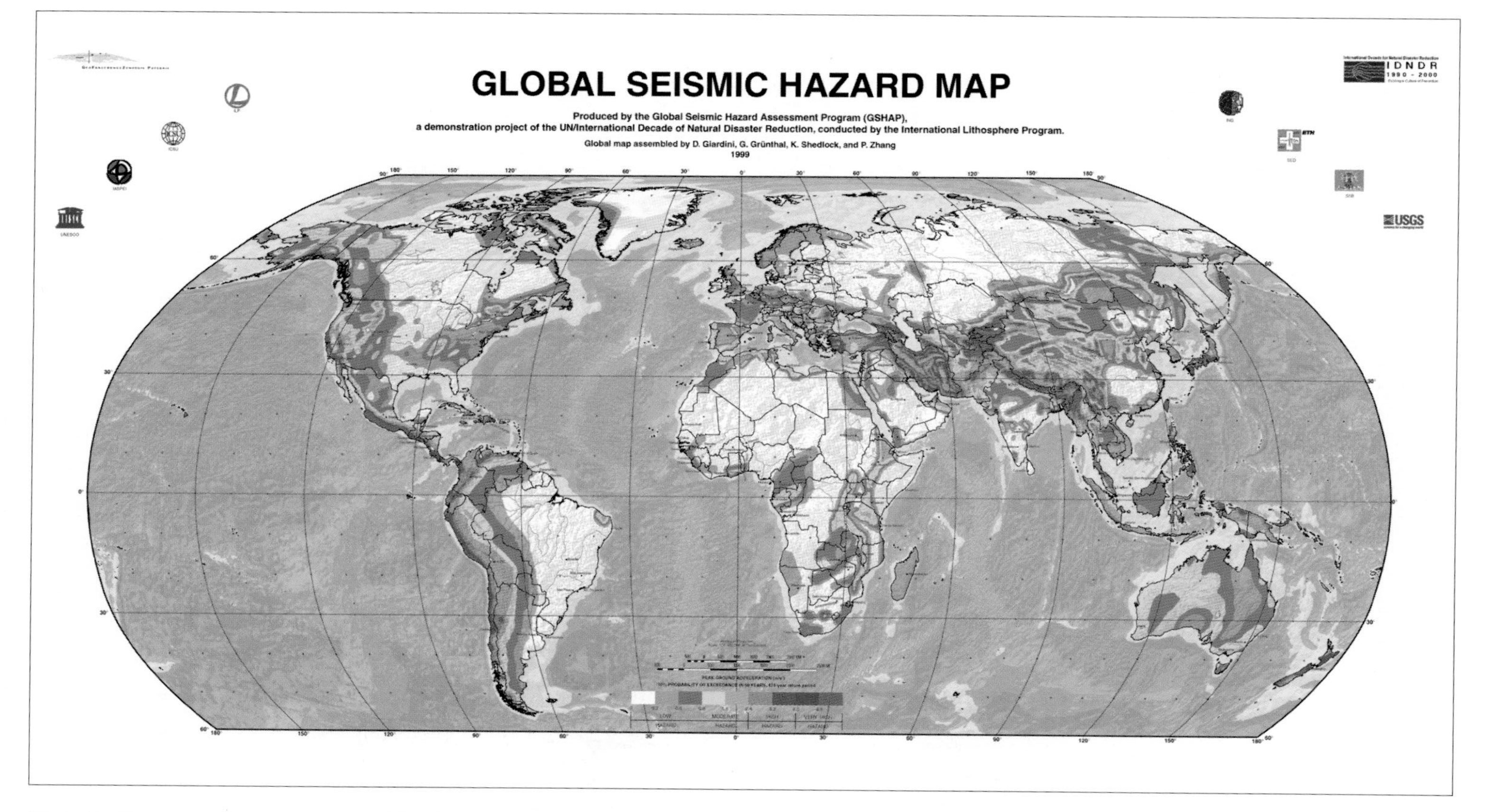

Plate I Earthquakes across the world. Global distribution of seismic hazard as designed by GSHAP. The map shows the peak horizontal ground acceleration with a 10% probability of exceedance in 50 years (Source: GSHAP, http://weismo.ethz.ch/GSHAP/)

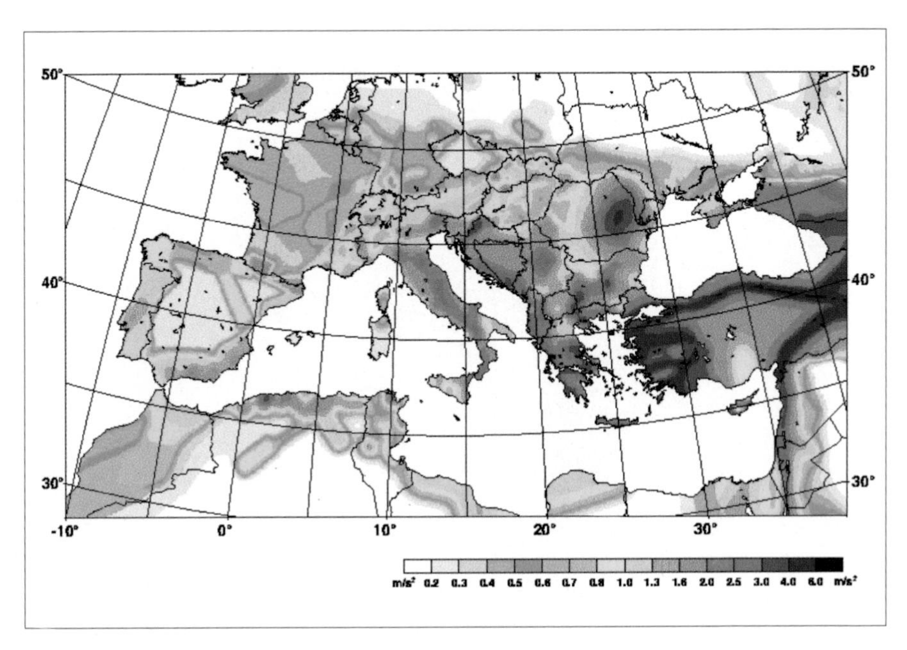

Plate II Seismic hazard map of the European area derived from the GSHAP (Grünthal, 1999). The map shows the peak horizontal ground acceleration (m/sec^2)with a 10% probability of being exceeded in 50 years (Source: GSHAP, http://weismo.ethz.ch/GSHAP/)

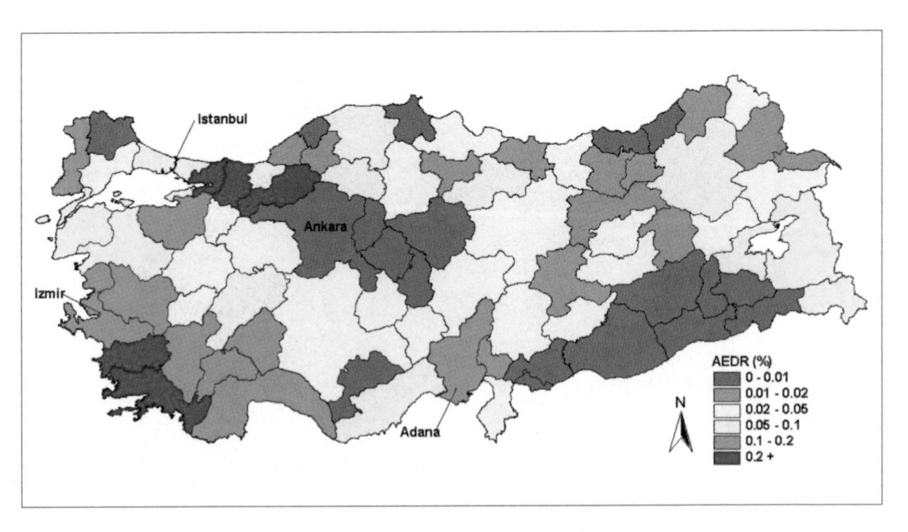

Plate III Insurance risk mapping: annualised earthquake damage ratio for Turkey shown at province level (from Bommer *et al.* 2002; copyright Cordis Consulting Ltd.)

6 Strategies for Earthquake Protection

6.1 Creating a Safe Society

As described in Chapter 2, everyone is a stakeholder in the likely losses from an earthquake and has an interest in earthquake protection, from individuals to companies, professional risk managers, financiers and government agencies. If you live in an earthquake area, your safety depends on the strength of the buildings you spend your time in and the precautions you take in your daily life. Companies can protect their operations and their staff by minimising the vulnerability of buildings, equipment and contents. Risk managers can transfer and manage their risk by buying insurance. National governments take the lead in setting building codes, safety standards and establishing a safety culture.

Previous chapters have been concerned with the event of an earthquake and with the emergency which a damaging earthquake causes. They have discussed how to act in that emergency, and how to prepare for it. The following chapters are principally concerned with strategies for making the community safer in the event of an earthquake and defining the roles that various groups may play in bringing about earthquake protection. These groups include:

- individuals and community groups
- private corporations or organisations
- urban authorities
- national governments
- international aid and development organisations.

The process of instigating protection measures may be initiated by many different groups and the protection of the various elements at risk and the actions needed are likely to be the responsibility of a number of different groups and sectors that may need to act together for protection to be achieved.

6.1.1 The Safety Culture

Everyone can contribute to the safety culture. Different groups play different roles within it. National and local government bodies can establish the framework of legislation and social consensus within which the safety culture can grow. Urban and regional planners can create a safe physical environment and infrastructure. Building designers need to have a constant awareness of the safety issues when producing the building stock. Private companies and organisations need to protect themselves, their employees and their employees' livelihoods through a consciousness of earthquake protection measures. Community groups can participate in their own protection both directly, through taking collective action to reduce their earthquake risk, and indirectly, by lobbying for action to be taken on their behalf by the authorities. And most importantly, individuals must learn to protect themselves.

In the following sections the activities and strategies that make up the roles of these different but complementary groups are outlined.

6.2 Personal Risk Management

If you live in an earthquake area, there is a lot that you can do as an individual to protect yourself and your family.

6.2.1 A Safe Home

Your safety depends first and foremost on the buildings where you spend most of your time. Most important of all is your home. If the building where you live is strongly built and, best of all, designed by an engineer to resist earthquakes, then the chances of your building collapsing or suffering severe damage are small, even in a strong earthquake.

When choosing a new house, or finding somewhere to live, or perhaps building a house yourself or having one built, it is important to consider the earthquake resistance of the house among all the other considerations in making the choice.

An indication of which building types are most earthquake resistant is given in Chapter 8. The most important indicators are the materials of construction of the structural system and the sizing and connections between the structural elements.

Very old buildings or damaged and cracked buildings are likely to be more vulnerable than newer buildings or buildings in good condition. The detailed assessment of how strong a building is to resist an earthquake is complex and really requires a qualified structural engineer to make a definitive assessment. If concerned about your building, consult a qualified engineer and ask whether your house conforms to current building regulations on seismic resistance.

If your house is decidedly vulnerable, it may be possible to carry out structural strengthening to make it safer, and different ways to do this are described in Chapter 8, or it may be that you should be starting to think about moving house, finding a building that still suits you, but that is also a bit more earthquake resistant. Nobody is going to move home just to reduce their earthquake risk, but if you are thinking of moving anyway, this could just encourage you to do so. In houses with extensive maintenance needs, particularly those of unprocessed building materials, 'radical maintenance' or rebuilding of major parts of the structure may be desirable, to rebuild in a stronger, more robust construction.

Simple strengthening improvements to the existing structure like tying together the roof beams to make a rigid roof element, described in Chapter 8, may be able to be carried out without major disruptions. Additional bolts, e.g. in timber-framed housing to strengthen joints and to anchor the timber frame to the foundations, may be cheap and easy to fix yourself and in some earthquakes have saved the house from slipping off its plinth and becoming a total write-off.

Internal posts and beams can be fitted relatively easily into buildings whose load-bearing walls are in a suspect structural condition, a form of 'retrofit' strengthening which supports the roof in the event of wall collapse and has saved many lives in past earthquakes.

6.2.2 Safety Indoors

There are many items in and around the home that, even if the building is undamaged, can cause injury or cause costly damage in the event of a sudden earthquake. Heavy items of tall furniture may overturn in earthquake shaking, valuable items on shelves are likely to be thrown off, and heaters, cookers and other sources of flame can be overturned and cause fires. An occasional check around the home, fixing tall furniture to the wall, reinforcing shelving, keeping items low and safe will make your living environment a safer place. An earthquake can strike at any time, so it is a good idea to get into safe habits of not leaving any naked flames unguarded or precariously balanced valuables.

Outdoors, a common killer is the collapse of garden walls. Check your walls are in good condition and if necessary buttress them at intervals along their length.

6.2.3 A Safe Workplace

Other places where you spend a lot of time should similarly be appraised for their seismic safety. Your workplace is important. If the building is vulnerable or machinery, hazardous materials or naked flames are left for long periods in situations where an earthquake could trigger an accident, it is important that this is brought to the attention of the management. If necessary consult your fellow workers or labour union representatives. Places where your family spends a lot of time – the children's schools, for example – should be examined from the earthquake safety point of view.

6.2.4 Earthquake Insurance

Earthquakes can deal you and your family a severe economic blow. The costs of repairing damage and replacing lost possessions mount up to large sums very rapidly. Often damage is caused in unexpected ways – building contents ruined by dust or water leakage, cars written off by a falling street lamp or other unforeseen accidents. If you are unlucky you could lose your house – often the most valuable asset of a family.

Precautions to recover from this level of loss are not easy. The most straightforward protection is earthquake insurance – in many countries earthquake insurance is a relatively low-cost addition to an insurance policy on house and contents for fire, theft and more common risks. The financial hardships that an earthquake can inflict make insurance an important component of a family's protection.

6.2.5 Participate in Earthquake Protection

If you recognise earthquakes as a problem you will want to protect yourself against them. Equally important you will want your community, your public services and utilities and your environment around you to be safe against earthquakes. Community groups campaigning for higher protection levels need your support. Neighbours who do not understand what earthquakes are or how an earthquake could affect them need your help to explain it to them. Drills and public awareness activities need your participation. Emergency planning and implementation may need your leadership. Politicians fighting for tougher anti-seismic legislation need your support. In earthquake country, everyone's participation in public safety is vital. Only you can make earthquake protection happen.

6.2.6 Assistance from Community Groups

Communities living in earthquake areas can protect themselves against earthquake risk more effectively than anyone else can. Earthquake risk depends on the houses they live in, the building stock in the vicinity and the environment all around them. The ability to withstand the impact of an earthquake physically may be less critical than the ability to withstand the impact of an earthquake economically and socially. The ability of a community to provide support to its members in the recovery period, the independence of a community so that it is not dependent on outside support to survive and to rebuild are all qualities of its people and social structure. Informal groups, social networks and formal community associations can all play an important role in improving earthquake protection. In some communities special organisations have been set up for earthquake preparedness and to promote protection issues.[1] Other already existing community groups in high-risk areas, e.g. tenant associations or neighbourhood schemes, have incorporated earthquake safety into their other activities.[2] Community groups look after the weaker members of their society, the old and the disabled, and ensure that they are not forgotten in the protection activities.

Information Dissemination

Information is a critical resource for earthquake protection and one of the most important roles for community groups is to obtain and disseminate the facts about earthquake risk, explaining the possibility and likely consequences of an earthquake and the actions that should be taken in the event of one occurring. Community groups can encourage people to protect themselves by providing information on which building types are safest to live in, methods of strengthening existing buildings or upgrading existing homes, directories of builders qualified in earthquake-resistant construction, earthquake safety issues in the home and workplace, and so on. Publicity information, newsletters and publications may all form part of the community's dissemination of earthquake protection information.

Emergency Preparedness

Community groups can also promote participation in emergency preparedness activities. A community disaster plan may be drawn up, involving specific roles for individuals including damage assessment, house-to-house checks and contacting each family in the community, firefighting, search and rescue (SAR), first aid, making contact with authorities, supervision of an evacuation, emergency, food, water and power provisions. Community groups can help organise practice drills, school participation and special events to raise public awareness and ensure that everyone knows what to do.

Simple first aid, firefighting and rubble-clearing equipment and perhaps emergency supplies may also be maintained by community groups, regularly checked and practised with.

---

[1] See for example the California earthquake preparedness organisation BAREPP in the early 1990s.

[2] See the description of Tokyo Community Preparedness Groups in Berger (1985).

Community Construction Projects

The role of some community projects groups may extend to community action, with the members of the group contributing finance, labour and materials to help in the protection effort.[3] Communities may build their own public buildings, like schools, meeting houses and religious buildings, through group effort or fundraising. These structures should be demonstrably earthquake resistant and can be used as shelters or emergency headquarters in the event of an earthquake that could damage other structures in the community. Community construction might also be employed for houses or agricultural buildings, with earthquake-resistant house buildings being built through communal effort for older or weaker members of the community unable to build for themselves, or for housing for the whole community.

In settlements at risk from rockfalls or landslides, community construction projects might include slope stabilisation projects, construction of barriers or retaining walls, and planting foliage barriers against rockfall, or tree planting to stabilise loose soil cover on slopes.

Community Lobby Groups

The community is best placed to decide the protection level it needs. Most of the major government protection actions or programmes by others to improve earthquake safety levels have been initiated as a result of political pressures arising from the communities affected. Public safety is an important public issue and community groups can legitimately and effectively campaign for higher protection levels. Issues of revising building codes, passing legislation, allocating public expenditure, taking action on unsafe structures, or implementing planning and building controls, improving protection levels on schools, hospitals and other public services and other protection-related objectives can be promoted through community action. Representation to political leaders, petitions, public meetings and press campaigns can all promote protection measures and bring about change that benefits the community. Experience has shown that a large-scale public campaign, presenting a carefully considered agenda of actions and maintained over a long period, is able to take advantage of any political opportunities, such as a disaster at home or abroad, or a disaster scare or prediction, to have their programme implemented.

Community-initiated Projects

A community can attract resources for its own projects by initiating or formulating ideas and requesting assistance from a development or government organisation.

---

[3] Maskrey (1989).

Many development organisations prefer to support community-initiated projects rather than projects suggested for them by outside agencies. Assistance can be sought from a number of sources. If there is a local development organisation working in the vicinity, the type of organisation that might be helping nearby communities with agricultural projects or water and sanitation works, it could be approached. NGOs working in physical, social or economic development in any particular country often have to be registered on a national charity register, usually held by a government agency, which could be a starting point for finding an organisation that could provide technical or financial assistance. Government agencies may also provide support to community-initiated projects through rural development support, urban development schemes, public works activities or economic assistance programmes.

6.3 Corporate Risk Management

An earthquake can have a major impact on a private company, on an industry or a manufacturing operation, on any college campus, medical complex or any other organisation responsible for a number of buildings and employees. The managers of organisations can take a number of precautions and safety measures to safeguard their continued operation, to protect the well-being of staff and public using their facilities and to minimise economic losses. An organisation should take sensible measures to protect itself and its personnel otherwise it could find itself legally liable for people hurt by earthquakes on its premises.

Earthquake insurance is an important tool used by most commercial companies, to cover property damage, business interruption and third party liabilities for injuries suffered by employers and visitors. The potential financial losses are measured against the available resources and insurance cover is bought to manage the risk of any shortfall. A vital factor determining the safety of the organisation is the earthquake resistance of the buildings and equipment it uses. Building vulnerability assessment is discussed in Chapter 9.

6.3.1 Structural Safety of Buildings

An inventory of the buildings that the organisation owns or leases should be prepared. The earthquake resistance of the buildings should be assessed, preferably by qualified structural engineers. This assessment should include whether the buildings comply with current seismic building codes, ordinances or other legislated standards that apply to existing buildings. The assessment should identify any highly vulnerable structure or major weak spots in otherwise sound structures.

The current condition of structures should be examined and any deterioration of buildings noted, particularly any deterioration (settlement, water penetration, etc.) likely to affect the structural competence of a building. Foundation conditions may also be important in places where siting is suspect or subsidence is evident.

The check should also include the earthquake safety of non-structural parts of the building, including cladding, parapets, signboards and other pieces of the building that could shake free in the event of an earthquake and hurt someone.

Any serious problems or threats to personnel identified in the buildings used by the organisation will obviously have to be acted upon. Problems of non-structural threats may involve putting additional fixings (bolts, steel straps, brackets, etc.) to make sure they are safe if shaken in a strong earthquake. Non-structural items that cannot easily be made safe should be removed or demolished. Buildings suffering from deterioration may need repair or improved maintenance to prevent further deterioration. The causes of deterioration should be identified and preventative action taken: for example, water penetration may need repairs to roofing or copings, settlement may need underpinning of the foundations. A regular maintenance and checking programme for building stock and facilities should enable potential problems to be identified early and preventative measures taken.

Options for a Highly Vulnerable Building

A building that is identified as highly vulnerable (through deterioration, or having a substandard structure, or where a serious defect or design flaw is identified in it) is likely to need more radical action. One option may be to move to another building, selling up, terminating the lease or otherwise disposing of the vulnerable building and finding a more earthquake-resistant building to move the operations into. If this is not possible, the building can be strengthened. Remedial works can be carried out to strengthen the structure of an existing building (see Section 8.8). This may be straightforward and involve limited cost, but it can also be expensive and disruptive. A structural refurbishment of a building is likely to require the temporary evacuation of the building and housing of the operations elsewhere, until the structural work is completed and the operations can be reinstated. If the structural refurbishment required is extensive, it may be worth considering demolition of the building and building a new one to new engineering standards.

Change of Building Use

Financing the structural upgrading of weak structures, or site redevelopment or relocation, may all take some time to bring about, and could be programmed over a number of years. In the short term, it may be possible to reduce risks by rationalising changing building use and relocating some activities from one building to another. An organisation using several buildings where some are more vulnerable may find it possible to reduce occupancy levels in the more vulnerable buildings and to move any important activities and valuable contents to the more earthquake-resistant buildings, using the weaker buildings for less critical or lower staffed activities.

6.3.2 Hazardous Facilities

Organisations operating facilities that in the event of failure could cause a serious threat to the general public need to take particularly stringent safety precautions. In most cases industries using or manufacturing hazardous chemicals, or storing quantities of flammable or particularly polluting materials, are already likely to be bound by statutory requirements and planning restrictions. Earthquake hazards, however, exploit the weakest link in any facility and a plant with only one defect can still fail with catastrophic results. Protection standards for structures, equipment and pipelines may need to be several orders of magnitude higher than those for conventional structures because of the severe consequences that such a failure could entail. The expression of protection objectives in terms of the probability of failure for both conventional structures and hazardous facilities is discussed in the next section. Regular reviews are likely to be needed of both the physical facilities and operational practices. Detailed plans are needed for contingency actions to minimise consequences in the event of failure (see Section 4.5).

6.3.3 Protection Objectives

The decision on what constitutes a threat and what level of structural resistance is required in the organisation's building stock is unfortunately not altogether clear cut. At the very minimum, the buildings should comply with current levels of legislated codes and statutory requirements. Failure to do this could render the organisation legally liable and open to litigation in the event of anyone being hurt in its buildings in an earthquake. The legal requirements represent minimum levels of safety and it may be worthwhile for the organisation to protect itself to higher standards than the minimum. Many countries have no codes for existing buildings, and only apply statutory requirements to new ones being designed. Buildings built before a code was introduced are usually exempt from that code and are probably less earthquake resistant than the code considers desirable.

An organisation may wish to protect itself to higher levels of safety standards than those assumed in the building codes. The objectives of the seismic component of the building codes (see Section 8.6) focus mainly on prevention of building collapse to save lives, and explicitly or implicitly accept a building becoming unusable or suffering damage levels in large earthquakes. A commercial company, or a medical complex or some other organisation, may not accept the loss of building serviceability that is assumed in these minimum levels of code protection. It may be very important to prevent damage completely or to be able to continue to use the buildings to continue manufacturing or to provide the services even in the event of a relatively strong earthquake, so the organisation may demand a higher level of protection than is provided by the statutory codes. If the protection objectives of an organisation can be articulated explicitly, the engineering studies of the earthquake resistance of the buildings can then be made much more clearly. The protection objectives can be formulated by classifying the organisation's buildings and facilities into classes of protection. For example:

- Failsafe structures: structures which should not become unusable in the event of any earthquake that can reasonably be foreseen during their working lifetime.
- Functionally protected structures: structures which should continue in function in all except the most severe earthquakes. Design should prevent structural damage (likely to cause temporary suspension of function for repair work) in all earthquakes that have, for example, a greater than 1 in 100 chance of occurring during the projected duration of the important function being carried out in that structure.
- Occupant-protected structures: structures which should not pose a threat of injury to their occupants through structural failure in any earthquake that can reasonably be foreseen. Design should prevent collapse of any part of the structure occurring in all earthquakes that have, for example, a greater than 1 in 1000 chance of occurring during the occupancy period expected for the building.

But the cost element will always be an important factor in the decision. Setting acceptable levels of risk, and balancing considerations of costs and benefit, is discussed in Chapter 10. Alternative protection objectives could be proposed, including ranking building protection by number of people occupying the building, length of time during the day that occupancy is sustained, and other measures of the importance attached to each structure. Protection could be prioritised by income generated for an organisation in each building or, for an organisation responsible for historical structures or museums, say, by cultural importance of the building and its contents.

The principle of setting protection objectives as explicitly as possible remains the same in each case. It should be possible to define a level of damage that is unacceptable (collapse, structural damage, any visible damage, etc.), the duration of time over which that level of damage is likely to be unacceptable (the lifetime of the structure, or the planned duration of an activity taking place there, or some other period) and a probability – the odds that appear acceptable for that outcome. No building can ever be made entirely earthquake proof, but the probability of failure can be reduced to smaller and smaller levels by increasing the level of resistance designed for. Very critical structures, like container vessels for nuclear power stations, have typical protection levels with a probability of less than 1 in 1000 of structural damage occurring for a 10 000-year earthquake. The decision on the level of protection needs to be made in this way. The directors or managers of an organisation may take professional advice, but apart from statutory minimum requirements, the probability levels and safety factors implemented are directly their decision.

The expression of protection levels is needed in these terms for the designers of the buildings or facilities. If the managers of the organisation do not decide explicitly, the designers of the facilities will make those decisions for them, using their own assumptions.

Each level of protection implies certain levels of cost and the scheduling of buildings or other facilities into categories of protection may need to be budgeted carefully and adjusted according to budgetary constraints.

6.3.4 Non-structural Hazards

Many injuries and much of the cost and disruption from earthquakes come from the contents of buildings, equipment, machinery and other non-structural elements. A review of the earthquake safety of non-structural items in the organisation should be carried out in addition to the review of protection of buildings and other major facilities. The measures to improve the earthquake safety of non-structural items can often be achieved more immediately and at lower costs than structural safety improvement.

A review should be made of the structural stability and robustness to violent shaking of all the contents of a building, the equipment in it and any major pieces of machinery. A room-by-room review is likely to reveal many items that could cause injury to the occupants in the event of violent shaking. The exact severity of the earthquake likely to be experienced will vary considerably. A rough guide might be to imagine all items in the room suddenly pushed sideways, rocked through 45° or given a jolt vertically sufficient to lift them into the air. On the upper floors of tall buildings, lateral movements are likely to be significantly larger than elsewhere.

Large and Heavy Furniture

Anything tall and large, and likely to overturn if tilted through 45°, should be restrained through strong fixings to a wall or bolting to the floor. Shelving, showcases, tall filing cabinets, wardrobes and other large pieces of furniture are particularly likely to pose a threat to occupants. Heavy objects stored on high shelves can be prevented from falling by putting lockable doors on the front of the shelves or holding them in place with other restraints. Warehouse storage racks insufficiently braced and restrained can cost considerable sums in lost stocks.

Glass

Glass cabinets, showcases, glazed screens and other glass items capable of shattering can cause serious injury and must be firmly fixed and, where possible, made from toughened glass. Any fixtures into or hung from walls – shelving brackets, mirrors, wall-lights, pictures – should be tested for firmness and, if necessary, redrilled or given additional fixings.

Fire Sources

Sources of flame or electric filaments in boilers, heaters, space heaters, pilot lights, cookers, etc., need special precautions to prevent violent vibration causing fires. Unrestrained cookers or heaters can overturn. It is possible that boilers can be thrown off their fixings and wall-mounted boilers can become detached from the wall. Gas pipes should be flexible and capable of differential movement between the pipe and the appliance without overturning the appliance. Fixed gas supply piping should be replaced by a flexible, jointed system. Some gas valves can detect strong earthquake vibration and shut down the gas supply automatically, and a system of this sort may be a suitable safety precaution for gas-supplied organisations. Similar automatic shutdown procedures may be worth installing on manufacturing processes, machinery and other operations that could be harmful or costly if allowed to run on unchecked.

Fragile Items

Valuable and delicate items may need special fixings and restraints. Museum displays, presentation stands and wall-hung works of art should be secured. Expensive equipment – desktop computers, TVs, projectors – can be anchored on short leashes that prevent them being thrown off worktops but still allow some latitude in positioning.

Machinery

Manufacturing equipment and machinery is commonly damaged in earthquake vibration. Large pieces of machinery capable of overturning should be bolted down. Supports and mountings for machinery that could slip or collapse should be reinforced. Machinery fitted with vibration mounting (usually fitted to reduce the vibrations from the machine being transmitted into the floor) experiences reduced earthquake vibrations, and it may be worth fitting sensitive machinery with vibration mountings for earthquake protection.

Liquids

Pots and containers of hot liquids, or dangerous reagents or other hazardous materials, should be carefully checked to see that they are never in a position in which they would be unstable if shaken suddenly.

Escape Routes

An important check is the exit routes for personnel in the event of an earthquake. The corridors, stairs and safety doors must remain clear and unthreatened by hazards even in the event of strong earthquake motion. Routes must be operable even in the event of mass panic: doors must open outwards, balustrades and handrails should be able to act as crowd barriers and withstand the crush of a potential crowd. Stairwells must be robust and structurally sound. Checks should be made that glass in roof-lights over stairwells is well secured against fracture in earthquakes and any ceiling lighting or other fixtures in the stairwell itself are extra secure.

These are examples of the types of checks and precautions that will make the workplace or the general environment of the organisation a safer place against the sudden occurrence of an earthquake. The interior fixtures, fittings, furniture and equipment are always changing. The checks and precautions against non-structural seismic hazards therefore have to be taken regularly. The facilities or property management in the organisation should be trained to make regular, thorough checks of earthquake safety in the non-structural elements of the buildings.

6.3.5 Emergency Planning, Employee Training and Earthquake Awareness

A plan should be developed for the organisation detailing what should happen in the case of a major earthquake. This should include what to do during the earthquake shaking; evacuation of the buildings after the earthquake has finished (or during the earthquake if easy and safe to do so); care for young, elderly, sick or infirm people; safe shutdown of any machinery or processes; procedures for extinguishing any potential fire sources and making hazardous situations safe; checking personnel and accounting for missing persons; first aid and dealing with distressed people; checking and reporting damage; damage limitation measures; measures to inform the workforce of whether or when it is safe to return to work or whether to discharge; procedures for orderly return to work or orderly departure home.

The plan should define tasks and designate roles for activities to specific members by name. Employees need to be trained in the emergency plan and their role in it. Training should include the extent and seriousness of the earthquake threat, the policy of the organisation and all the preparedness measures being undertaken by the organisation to increase protection against the earthquake threat. Training should also encompass the threat from earthquake in the home and encourage employees to prepare their families and to improve earthquake protection in the home. Training might also include first aid skills, possible elementary fire suppression and search and rescue techniques.

Employee training is best undertaken as classroom activity, reinforced by simulation exercises and participatory drills. The organisation should also mount an intensive employee public awareness and information campaign through bulletin boards, notices and posters, company newsletters, memos and other regularly published materials or media.

6.3.6 Increased Self-sufficiency

A very common consequence of an earthquake is the temporary disruption of utilities, power supply and communication systems. The chain of delivery through extensive networks of pipes, cables and substations is easily broken by a failure in any of the linkages in the chain. In a large network there can be many failures, and it may take days or weeks for the utility company to find and mend all the failed elements. Water supplies, electrical power, piped gas, sewage disposal, telecommunications and other services dependent on physical networks may all be knocked out for long periods of time. An organisation that is otherwise intact may find itself unable to function because of disrupted services. It is possible to protect against this eventuality by increasing the self-reliance of the organisation to operate without continuous provision of services.

Stand-by Generators, Fuel Reserves and Water Tanks

An obvious back-up for failed power supplies is stand-by generators. Fuel to operate the generators for several days' normal use will enable the organisation to remain unaffected by power blackouts from most moderate events and supplies for a week or two should see it through some of the more serious events. Water supply can be ensured by maintaining reservoirs in water tanks (water tanks on towers need strong engineering construction as they are notoriously vulnerable to earthquakes). Sewage disposal can be maintained for some time if there is a capability to switch to temporary septic tank systems in place of mains drainage. Low levels of gas usage may be able to be maintained by stored gas, but high levels of gas usage may need to be capable of being supplied by an alternative fuel source – oil, or solid fuel. An operation that is capable of running on more than one type of fuel and that has stored fuel reserves capable of lasting for some time is less vulnerable to disrupted supplies in general.

Communications Back-up

Communications are often critical to an organisation's operation – perhaps internally between departments but more importantly to clients, suppliers and other contacts outside. Cable communication systems are not easily protected – in addition to possible earthquake damage, telephone switchboards are usually jammed with emergency calls following the earthquake. Internet servers and networks are notoriously vulnerable to physical disruption. Companies that work in e-commerce or use the internet for essential business communications need disaster plans and back-up servers located outside the region likely to be affected by earthquake. Cellular phone networks have also been found vulnerable to disruption in major earthquakes. Radio systems are less vulnerable to earthquake disruption and may be worth installing as a back-up for communications. Internal communications within an organisation can be maintained through UHF radio systems – these are usually sufficient to cover a large site or campus. City-wide communications can be maintained on VHF radio, within specific wavebands usually requiring a licence. Conversations over a radio system of this sort are less secure – that is, other people can eavesdrop – but could be critical in an emergency. A larger radio communication system can enable contact to be maintained with places far beyond the area likely to be damaged in an earthquake.

Maintaining Transportation Links

More serious disruption to an organisation's operations may be the possible enforced isolation if road and rail linkages are cut either locally or in the region. Inability to receive or make deliveries for any length of time may cripple the operation of an organisation, particularly manufacturing operations unable to receive raw materials or spare parts and unable to get finished products to market. An ability to be flexible in transport mode will help, using road if rail links are cut and vice versa. A storage capability to stockpile several days (or weeks) of output, with freezing or preservative capabilities for perishable goods, may make immediate despatch less critical. Similarly, increasing the margins of stock operations, although perhaps expensive in warehousing capacity, will make the operation less vulnerable to disruptions in delivery of supplies.

The less reliant the organisation can be made on continuous services being provided from outside, the less vulnerable it will be to disruption from a future earthquake.

6.3.7 Information Protection and Business Contacts

Many businesses, particularly small businesses, suffer badly from the loss of information or records in the earthquake damage. Files can be lost in destroyed buildings, ruined by fires or by water leakages caused by the earthquake, wiped from computer memories or simply thrown into disarray by the overturning of filing cabinets. Protection of commercial records from earthquake damage is an important consideration. It is possible to formulate filing and archiving procedures to protect against earthquake-induced information loss. A measure of protection can be ensured by keeping copies of important documents on back-up servers, or physically in separate filing cabinets, preferably steel cased and low level. Archives may be safer if kept in a separate building. Hard copies of important computer files, and back-up disks, should be similarly 'hard filed'.

The chaos ensuing after a major earthquake is also extremely disruptive, again particularly for a small business. Communications may be cut and routines shattered. If the business itself has lost its premises, or is forced to close temporarily, potential customers trying to make contact will be unable to do so. Contact should be re-established as soon as possible by informing customers and clients about the continued delivery of services and goods, and any relocation address, through advertising, mail, telephone or personal contact. Disruption is likely to be minimised if part of the organisation's normal activities involve informing clients and customers, suppliers, subcontractors, staff and other business contacts of emergency plans that would affect them, including information channels likely to be used to confirm continued operations, contingency plans and enquiry contact points. Information about an organisation's emergency plans is unlikely to frighten off customers and may encourage confidence if it is presented in the context of a range of activities being undertaken by the organisation to improve earthquake protection for staff and customers.

6.4 Urban Risk Management

6.4.1 Urban Planning

The layout and development of cities, the location of infrastructure, key buildings and utilities and the physical development of the built environment all affect the consequences of an earthquake. The urban planner, the regional planner, engineers designing the layout of utility networks, transportation routes or key installations, and anyone whose job is to locate facilities within a city or whose decisions affect the use of land, all have a role to play in reducing potential earthquake impact.

Urban planning departments are usually a part of local or regional government, and activities of the management of private building stock, seismic design code enforcement and other local government measures for earthquake protection may well be a central part of the responsibilities of an urban planning department. If not, the linkages between land-use master planning for earthquake protection and other urban planning protection measures and the control of building quality are so interrelated that the development of effective earthquake protection measures needs a strong coordination between the groups with those responsibilities.

As with all urban planning, effective management of the development of a city depends on understanding the processes that are making it the way it is. The trends in land prices, the locational preferences for various industries, activities and communities, the demographic trends of the population and many other factors are all driving forces shaping the city. Urban planning is the attempt to direct those forces using limited means and a small repertoire of legislative and economic powers. The concerns of urban planning are many: to ensure a sanitary, pleasant and safe environment for the population, to provide adequate services to the people and workers in the city, to enable the city's activities to be carried out more easily and to plan ahead for the future. Many of the concerns of earthquake protection also parallel these objectives: limiting the densities of development and concentrations of population, protection of service provision and facilitation of continued economic activities.

By its nature, urban planning is long term. Master plans have to encompass decades of expected growth, and it is evident that earthquake protection is necessarily a long-term process.

Adding Building Stock Management to Land Use

Where earthquake protection may be different to normal urban land-use planning is in the emphasis on building stock management, i.e. the influencing of the process of creation and maintenance of privately owned buildings in addition to land use and location. This process-orientated approach in combination with locational aspects may require a slight reappraisal of planning methodology.

Earthquake protection should be seen as an additional element of normal urban planning. It should not be a separate activity from other planning operations, but rather an integral part of the planning process – another factor to be weighed in the decision-making and balanced against other factors: when siting a new school or planning a new residential suburb, earthquake risk should be weighed against the transportation implications, cost of land, suitability of the local environment, cost of providing services and so on. Where there is a choice of sites with an identifiable difference between them in earthquake susceptibility, this should influence the choice – if all other factors are equal the less susceptible site should be chosen. If not, the cost of building the school to higher standards of earthquake resistance or imposing stricter controls on the residential structures should be balanced against other costs and advantages of the sites. Where a site of higher seismic hazard is chosen, the facilities and building stock built on that location must be built to higher standards of earthquake resistance. Thus the integration of seismic building code enforcement and building stock management with land-use planning becomes critical.

Microzoning and Vulnerability Mapping

From the discussion in the next few sections it will be seen that earthquake protection planning at an urban scale involves both the location of elements in the city and the quality of elements in those locations. Earthquake protection planning at the urban scale requires two additional maps to the urban planner's usual map collection:

(1) the *seismic microzoning map* of the geological earthquake hazards and
(2) the *seismic vulnerability map* of the buildings and facilities of the city.

The addition of a seismic microzoning map in preparing land-use plans or development master plans may be fairly straightforward and comparable to other preparation and study maps that contribute to the planning process. However, the seismic vulnerability map encompasses the physical attributes of the building stock in a more comprehensive way than is usually needed for other planning activities. In addition to the characteristics of function, plot development, density and perhaps number of storeys that are commonly used to map the building stock for land-use planning, earthquake protection needs information on construction materials, structural form, height and size, engineering design quality and age and other broad indicators of seismic vulnerability (see Chapters 8 and 9 for vulnerability classification of building types) with which to classify the earthquake resistance of the building stock.

Building Stock Data

Information is needed across the city, from district to district, about the numbers of different types of building classified by their seismic vulnerability together with their functions and occupancy. This is usually built up from building census data if it already exists or can be obtained by carrying out building surveys on a street-by-street basis, but useful data on the physical characteristics of the building stock can also be gathered from aerial survey interpretation, planning applications or other documentation, or assumed from historical urban development patterns and existing land-use plans or zoned from other information sources.

Seismic vulnerability mapping and building stock inventories can be time consuming if carried out in detail, but may only be needed at an approximate level to give enough information for urban protection plans. The broad identification of the building types most at risk from a future earthquake and the parts of the city which are likely to be worst affected may be relatively easily identified. The policies of upgrading the most vulnerable building stock sector and proposing land-use plans that reduce earthquake risk in the city are likely to be obtainable from relatively simple analyses.

Land-use Planning and Seismic Microzoning

Some types of ground are safer than others in earthquakes. In addition to the numerous ground failures caused by earthquake vibrations, such as landslides, slope failures, liquefaction and rockfalls, it is well known that different types of ground vibrate more severely in earthquakes and so cause higher damage levels to the buildings built on them. Siting considerations for earthquake protection are discussed in Chapter 7.

Seismic microzoning, or the identification of various ground conditions in terms of their earthquake hazard across an area at the scale of a city or conurbation, is an important tool for urban planning to incorporate earthquake protection. Methods of microzoning are described in Section 7.4. The seismic microzoning map, even if fairly coarsely defined, can be used as an additional information resource for urban planners to incorporate earthquake protection considerations into their normal land-use planning decisions. The map may define areas of likely ground motion amplification, potential slope failures, landslides or rockfalls and potential liquefaction.

The delineation of the city and its environs, particularly its potential areas of expansion, into areas of relative severity of ground motion shaking likely to be experienced in a future earthquake can help shape a safer city. It may be possible to avoid building on some areas of potentially higher hazard altogether – a zone of very high hazard might be left as park area or the areas of city expansion might be encouraged out in an opposite direction (through preferential provision of transportation routes, urban services, etc.). By building on areas of potentially lower hazard, future earthquake damage can be reduced. This method of damage reduction has the advantage that if locational planning is possible, there is no direct capital investment required to bring about increased safety. There are a number of indirect costs involved – land prices may be higher in one area than another, or there may be increases in transport costs or needs for additional infrastructure – but in many cases the total costs to the community can be far less than those involved in the construction of stronger building stock. Where choices of location are limited, or the arguments for locating in an area of higher seismic hazard for other reasons are convincing, structures or infrastructure built in that location must be built to a higher standard of earthquake resistance. The matching of engineering code requirements and building stock management with land-use planning therefore becomes critical.

High-intensity Amplification

The potential effectiveness of land-use planning for safety will vary considerably from case to case. Different types of ground affected by the same earthquake waves may vary in their severity of shaking and consequent destructiveness by one or more degrees of intensity. Stiffer soils, or hard rock, may be shaken with ground motion of intensity VIII while softer ground close by, like shallow alluvium, is shaking more severely, closer to intensity IX. From the vulnerability studies outlined in Chapter 9, this would mean that around 75% of weak masonry buildings built on the soft ground could collapse, killing perhaps 14% of their occupants, whereas only 40% of the same building types built on the rock would collapse, killing less than 5% of their occupants. There is generally more difference between the performance of different ground types at higher intensities, so for moderate levels of earthquake shaking locational planning is less effective in reducing losses. But where high intensities are possible, the microzoning of a city or town can play an important part in earthquake protection. An example of using urban land-use planning for earthquake protection is shown in Figure 6.1.

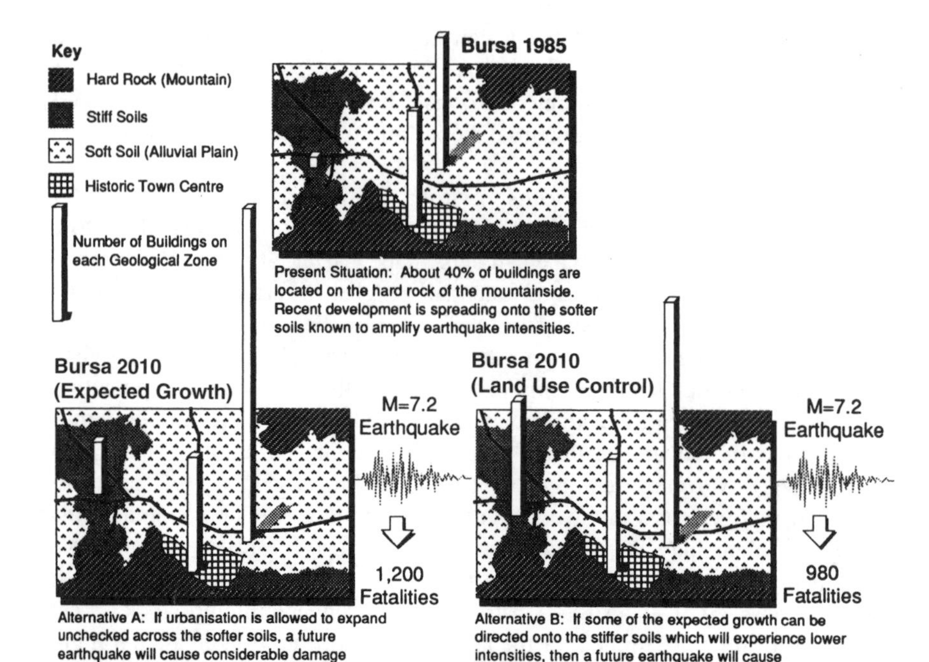

Figure 6.1 Study of earthquake implications for planning of new city suburbs in Bursa, Turkey (after Akbar 1989)

Unfortunately the science of microzoning ground conditions and predicting their likely performance in future earthquakes is relatively young and there are large uncertainties. Estimates of likely response characteristics of different ground types are only approximate, and detailed knowledge of the sub-strata underneath sites is difficult to obtain. There are only a few places where earthquakes have recurred and where detailed observations have been made of how the ground conditions affect the intensity experienced. In most other places, the detailed effect of ground condition on ground motion severity can only be crudely estimated.

Frequency Characteristics of Soils

The information provided by microzoning studies cannot predict very accurately the severity of shaking and the amplitudes of acceleration likely to be experienced in a future earthquake, but it can be much more reliable in determining the frequency content of vibration due to different local ground conditions. This is important because certain building types are more vulnerable to different frequencies of ground motion vibration than others. (See Section 7.5.)

Seismic microzoning can be used to ensure that a match does not occur between buildings vulnerable to certain frequencies of vibration and ground conditions that are likely to vibrate in that frequency range. This is chiefly a problem for taller high-rise buildings and soft soils that may amplify earthquake motions in the long-period range. To avoid buildings being damaged by *resonance effects* in zones where the ground is likely to vibrate in certain frequency ranges, buildings should be designed either to have frequencies of natural vibration well outside the critical range or, more problematically, for the much higher seismic forces they are likely to experience. An example would be a zone where restrictions might be imposed on building structures of 10 storeys high, likely to have a natural period of about 1 second, because the zone consists of deep deposits of soft soil that are also likely to have natural periods of vibration of about 1 second so resonance would occur.

Uncertainties about ground conditions and their likely performance in an earthquake may be too great for major decisions on location to be solely based on seismic safety considerations, but they can add useful information to help decision-making for protection.

Limitations of Land-use Planning

There are a number of other important restrictions to land-use planning as a tool of the earthquake protection planner. The first is that land-use planning is essentially opportunistic: there has to be a need for the location of new buildings (e.g. an expanding city), a choice between alternative areas in which location is possible, and a difference between the expected earthquake performance of the different areas. The second and possibly greater restriction is that land use has to be controllable. In many very rapidly expanding cities, principally in developing countries, urban planning authorities have almost given up attempting to control detailed land use, because the administrative framework for planning controls is impossible to maintain. The more stable cities, e.g. in the developed world, have well-established planning control mechanisms but the opportunity for changing their risk through land use is very limited because the city already exists and will largely retain its historical layout.

Land Price and Earthquakes

A major factor in shaping cities is land price. Earthquake risk may itself change the shape of the city to reduce future risk without planning measures. Earthquakes have been known to have marked effects on land price, changing the character of urban areas in the longer term: poor ground conditions in a district of a city, highlighted by concentrations of earthquake damage, are likely to make that district less desirable and suppress land prices there.[4] Land prices and commercial forces also change the nature of urban areas in other ways. Higher land prices tend to make high-rise buildings more economic and this has implications for urban form, occupant densities and safety levels in the event of future earthquakes. Control of land prices directly is not normally part of urban planning in democratic countries, but is strongly influenced by planning decisions, by zoning and by planning permissions. Provision of services affects how desirable an area is and residential densities may be influenced by levels of provision of utilities and other services. Understanding the dynamics of urban land price economics is often important in planning a safer city.

Deconcentration of Cities

The worst earthquake disasters have occurred in 'direct hit' earthquakes – an earthquake epicentre directly underneath or very close to a large town. The concentrations of people and buildings represent targets of high potential loss. Deconcentration of cities spreads the elements at risk by reducing densities and decentralising facilities. Deconcentration and density limitations are desirable in cities for other reasons too, including environmental improvements and limitations on service provision. Most urban plans already limit densities of development. Limitations of density, height restrictions, plot development regulations and other controls can all be used to limit concentrations of building stock. It is, of course, very difficult to change the densities of existing urban districts, and much easier to limit densities on areas of future development.

Reducing Densities in Existing Cities

The densities of existing urban areas can be reduced by city authorities buying up plots and demolishing to create open space among the blocks or redevelopments at lower densities. After some earthquakes in the past this has been achieved by the city authorities buying up the sites of collapsed buildings and

[4] After the 1985 earthquake in Mexico City, a number of banks relocated their office buildings from the badly damaged Reforma area to the more desirable and firmer ground condition of the nearby Polanco district to avoid problems of disruption to bank activities from future earthquakes. This had a significant effect on land price in the Reforma area and affected the development process.

making them into urban memorial parks.[5] Such urban parks, even if they are small, add greenery to the city, help with urban hydrology, humidity and microclimate, and provide areas for emergency facilities or population evacuation or temporary shelter housing in the event of any future disasters. Some cities now have large budgets for the re-greening of their built-up areas, buying up plots as they become available on the open market. In Japan, earthquake protection objectives (chiefly deconcentration for fire risk and the provision of refuge areas for the population) have been set at the provision of 3 square metres per person of parkland in all major cities. With the price of land in Tokyo currently the highest in the world, this is an expensive and long-term policy: Tokyo Metropolitan Government has achieved nearly 1 square metre per person so far, but other cities in Japan are closer to their target of 3 square metres per person.[6]

Limiting Densities in New Settlements

In the planning of a new town in a seismic area it is important to limit the size and potential for high-density over-concentration of development. Density controls include restrictions on building height, limitations on the plot ratio of allowable development for any site, and limitations on access to basic services.

Where direct density controls are not easily enforceable, other methods of achieving lower densities include the design of street patterns, wider streets and limiting plot sizes by physical planning means, using the design of the layout of the town and positioning of street furniture to maintain street frontages and to limit plot developments.

There are, however, no absolute levels or recommendations about density targets for earthquake safety. Urban population densities vary considerably from country to country and town to town, and the vulnerability of the building stock is the overriding factor in determining how much the population is at risk from earthquakes. In a neighbourhood of fairly vulnerable buildings (masonry, for example) the height and proximity of buildings, particularly buildings on a slope, should at a minimum be constrained to prevent one building collapsing onto or into a neighbour. The 'domino' collapse of buildings, particularly down a slope, has been one of the causes of high fatalities in earthquakes. Similarly street layout road widths, particularly major routes needed for emergency access, should be wide enough not to be made impassable by the rubble of a collapsed structure. Vitally important routes should be wide enough to survive the collapse of structures on both sides of the road simultaneously.

---

[5] After the 1985 earthquake, the sites of several collapsed buildings in Mexico City were turned into urban parks.

[6] Itoh (1985), Ashimi (1985).

It is also important to reduce densities by designing open spaces in the city, particularly spaces within the built-up areas. Such spaces also form safe congregation areas for the population, away from the possibility of injury from pieces falling from the façades of buildings and, in areas at risk from fires, provide some safety refuge in the event of multiple fires.

Deconcentration and Fire

Deconcentration is particularly important to reduce the risk of fire spreading from building to building in cities of flammable buildings. The danger of conflagrations following earthquakes is particularly acute with timber frame structures or those with combustible roofs: in such cases deconcentration becomes a major earthquake protection measure. The division of urban areas into small cells by wide roads, rivers, parks and other fire-breaks limits the potential for conflagration. The chief risk for fire or earthquake disaster in many cities is in squatter areas or informal housing sector developments. These are likely to be beyond conventional planning measures, but general programmes to upgrade squatter areas should include reductions of density, access routes for fire and other emergency service vehicles, and discouragement of siting on hazardous slopes.

Decentralisation of Major Cities

In many countries, there are efforts to decentralise capital cities and other major regional centres. There may also be programmes to reduce the rate of urbanisation generally and to discourage large-scale migration of rural populations to the cities. Both of these measures reduce earthquake risk in a seismic region. Decentralisation of major conurbations reduces earthquake risk by reducing the concentration of people and building stock, and earthquake protection is an additional argument for decentralisation. Decentralisation is commonly tackled using a number of methods including the development of 'satellite centres' (local services in the suburbs), 'necklace' development (suburban development beyond green belts), the promotion of secondary towns in the region, or moving ministries and other key facilities to other cities, or promoting relocation grants for industry and preferential provision of services in order to reduce development pressures on an over-centralised city.

After the city of Tangshan was devastated in 1976 by the most lethal earthquake of the twentieth century, the Chinese planners rebuilt the city as three separate smaller towns, several kilometres apart, partly in order to reduce the potential for an earthquake to cause another similar disaster.[7]

[7] Wu Liang Yong (1981).

Protecting Urban Facilities

Planning new facilities and managing existing facilities in cities are a vital part of the earthquake protection of the community. Facilities provided and managed by local authorities may include hospitals, schools, public housing, government buildings, museums and many other publicly owned elements of the building design stock. Other policies likely to be developed at city level include the conservation of historical buildings, and policies to maintain the cultural heritage of valuable building stock, or to preserve the overall townscape qualities of historic districts. In addition, urban planners are likely to be involved in the siting decisions for many privately owned, large-scale facilities, like major industrial plants, shopping malls, office complexes and other major private developments. The location and design of public services and utilities, transportation system networks, terminals and many other facilities are all a part of urban planning in its broadest sense.

A checklist of urban facilities is included in Table 6.1. These community facilities are important – some are critical – elements in the continued functioning of the urban society. Protecting them against failure in an earthquake insures against the breakdown of urban society and the economic damage caused by loss of urban services.

Decentralised Facilities

At a strategic level, services provided by one central facility are always more at risk than those provided by several smaller facilities. This principle applies equally to hospitals, government administration buildings and fire stations as it does to power stations, water treatment plants and airports.

The collapse of the central telephone exchange in the 1985 Mexico City earthquake cut nearly all telephone communications in the city for a vital 48 hours. In the reconstruction, the telephone system was redesigned using new technology and dispersed, mini-exchanges to make the system less vulnerable to earthquake disruption.[8]

Networks such as water supply, piped gas supply or electricity may also benefit from being compartmentalised into relatively independent zonal blocks, so that the failure of any part of the network is localised in its consequences.[9] The decentralisation of key services should be a primary objective for earthquake protection, or at least the protection against the failure of the service by the loss of one or two elements within it.

The creation of a robust system for each important urban facility listed in Table 6.1 should involve a vulnerability analysis of the facility itself. For example:

---

[8] Aysan *et al.* (1989).

[9] Tokyo Gas Company has subdivided the pipeline system of the entire Tokyo metropolitan area into zonal blocks as an earthquake protection measure (NLA 1987).

Table 6.1 Usage classification of elements at risk.

	Occupancy	Emergency Function	Loss Component and Role in Recovery
Residential			
Single dwelling houses	Daily cycle, low occupancy	Shelter	Large percentage of total building stock
Multi-dwelling apartment buildings	Daily cycle, high occupancy	Shelter	Significant percentage of total building stock
Public buildings			
Hospitals, clinics, nursing homes	Permanent high occupancy	Critical – medical facilities	Expensive to replace
Schools, colleges, universities	Weekly cycle, high occupancy, children at risk	Public congregation points/aid distribution centres/shelter	Expensive to replace
Churches, mosques or shrines	Occasional high occupancy	Public congregation points/aid distribution centres/shelter	Expensive to replace
Museums, galleries	Moderate occupancy	Non-essential	Cultural value and heritage. Exhibits and contents may be irreplaceable
Public administration offices			
Police station	Continuous level of occupancy	Critical – emergency services	Moderate financial loss. Possible coordination role in recovery
Fire station	Continuous level of occupancy	Critical – emergency services	Moderate to high financial loss, especially if equipment lost. No role in recovery
Ambulance station	Continuous level of occupancy	Critical – emergency services	Moderate to high financial loss, especially if equipment lost. No role in recovery

Table 6.1 (*continued*)

	Occupancy	Emergency Function	Loss Component and Role in Recovery
Public administration offices	Daily cycle, high occupancy	Important coordinating role	Important coordinating role in recovery
Commercial			
Offices	Daily cycle, high occupancy	No emergency role	Critical to the employment and continued income of a large sector of the community
Shops	Variable occupancy	No emergency role	Provides employment and sells products important for daily life
Shopping malls, markets	High occupancy, daily and weekends	No emergency role	Provides employment and sells products important for daily life
Hotels, guest houses, pensions	Permanent high occupancy	Temporary shelter for homeless	Economic generators (especially in tourist areas)
Cinemas, theatres, sports stadiums, etc.	Occasional very high occupancy	Emergency equipment storage/morgue	Public morale
Restaurants, night clubs, bars	Occasional moderate occupancy	No emergency role	Public morale
Warehousing, storage	None	Potential storage	–
Industrial			
Hazardous plant	–	Could cause secondary disaster	–
Factory (essential production)	Daily or permanent occupancy cycle	None	Critical to recovery phase
Factory (non-essential production)	Daily or permanent occupancy cycle	None	Not critical, but may provide employment and continued income for many people
Warehousing	Low occupancy	–	–

(*continued overleaf*)

Table 6.1 (*continued*)

	Occupancy	Emergency Function	Loss Component and Role in Recovery
Utilities and services			
Electrical network	–	Important to emergency operations	Power supply very important for industry and public safety
Water network	–	Critical for firefighting	Drinking water needed for public
Gas network	–	Short suspension of service acceptable	Important for industry and public comfort
Sewage and surface drainage network	–	Not important	Important for public health
Telephone network	–	Critical to emergency communications	Important for economic business
Road network	Variable traffic flows. Bridge collapses etc. could cause life loss	Critical. Paths needed for emergency vehicles	Critical
Rail network	Rail accidents are a serious threat	Possibly needed to import heavy equipment	Important
Public broadcasting, TV and radio	–	Important for public information	Important

Are the fire station buildings that house the vital fire tender trucks sufficiently earthquake resistant to remain serviceable when they are most needed? What failure rates can be expected on electricity cabling throughout the city network? An identification should be made of any weak links in the system. Where possible, decentralisation of all key services should be a primary objective for earthquake protection. Where it is not possible, the critical elements in the system must be protected to much higher standards. If the expense and loss of efficiency involved in setting up more than one specialist hospital or in having a dispersed government administration cannot be justified, then the single specialist hospital and the central government administration building should be strengthened if their continued function after an earthquake is essential.

Routing of networks – the piped services, electrical and communication systems cabling and the road and railway links that make up the transportation network of the city – is also important for earthquake protection. A grid network is more robust than a radial network because if one element fails, the same points can still be reached by another route. Compartmentalisation of facilities gives additional safety.

Prioritising Protection

Facilities can be prioritised for their level of protection. One level of prioritisation is life protection of building occupants. Buildings with large numbers of occupants in residence for a large proportion of the time should receive high levels of protection. The length of time that buildings have occupancy and the peak numbers of occupants are important considerations. Nursing homes have almost permanent occupation. Prisons are often forgotten as permanently highly occupied buildings.

It may be possible for a vulnerable building with high, permanent occupancy to have its usage changed – transferring the occupants to a less vulnerable building. Some categories of buildings may also be given a high priority – schools, for example, often receive high levels of protection because society instinctively protects children.

Inventories of the facilities of the city and an evaluation of their seismic vulnerability are an essential part of developing a plan for earthquake protection.

Street Safety

Urban planners are also responsible for the safety of the general public on the streets. A protection measure which can be undertaken relatively rapidly and effectively is a survey of street safety. In public places and routes most commonly travelled by foot and road traffic, any element of building façade, billboard, or street furniture shaken loose in an earthquake can have lethal results. It is a relatively straightforward exercise to identify such threats: parapets, unstable masonry, broken windows, poorly fixed street signs and any other potentially dislodged item can be fixed, bolted, strapped or demolished to make the street safe for the general public below.

6.4.2 Building Code Enforcement

The formulation of building codes and the training of the engineering profession to understand them are the responsibility of national governments and are discussed in Section 6.5. But their implementation and enforcement are normally part of the responsibility of the urban authority, and carried out in the department of the municipal engineer or the building control department.

If there is no effective system of checking that the code is applied, the level of code compliance is likely to be very low. In a number of countries separate regulatory agencies are judged too expensive or too restrictive, so a scheme of voluntary code implementation is adopted where a signed drawing by a registered engineer is accepted by municipal authorities as code compliance. Investigation after earthquakes has shown that in such circumstances there is a very poor rate of buildings achieving code standards.[10] This may be because engineers make mistakes while intending to comply with codes, or because designers intentionally ignore the code requirements judging them not to be important, or because buildings are not built as designed.

Proper code enforcement is likely to require a regulatory agency maintained by a local authority that is capable of checking drawings and calculations, capable of visiting buildings under construction and with powers to prevent unsatisfactory structures being completed. The regulatory agency needs to have sufficient competent staff to make a general check on the design of most buildings and to make a detailed check on a significant percentage of the building designs submitted for approval. The professional staff required for code enforcement have to be budgeted for adequately as part of the costs of a community's protection. In a city of half a million people, there may be several thousand engineered buildings under construction at any one time, and a staff of 20 municipal engineers would be stretched just carrying out simple checks of design drawings and calculations. The municipal engineer has been referred to as the front-line soldier in the community's battle for earthquake protection. As an investment in public safety, the employment of an extra municipal engineer may be one of the most cost-effective actions that a local government authority can take. The role of the municipal engineer is also important in giving advice and promoting earthquake protection concepts in addition to the role as a construction policeman. Building code enforcement is discussed further in Chapter 10, Section 10.2.

6.4.3 Building Stock Management

Building design codes on their own are limited in the extent to which they can reduce the vulnerability of the built environment, and in the speed with which they can increase earthquake protection.

When a new code is introduced it applies to all new engineered buildings built from then onwards (assuming that the code is well implemented), as shown in Figure 6.2. If the building stock is increasing owing to expanding population growth, population migration into the city, or increasing economic capability of the city's population, then only the additional buildings built each year can possibly comply with the new codes and have improved earthquake resistance. Over time, some old buildings in the city will also be demolished and replaced by new ones. The replacement rate of buildings depends on the useful lifespan of structures, the durability of construction, land prices and location, and other

[10] Estimates of percentages of urban buildings complying with seismic design codes vary considerably from country to country, but in some cases could be as low as 2% of new urban construction complying with code standards (Bayülke 1985).

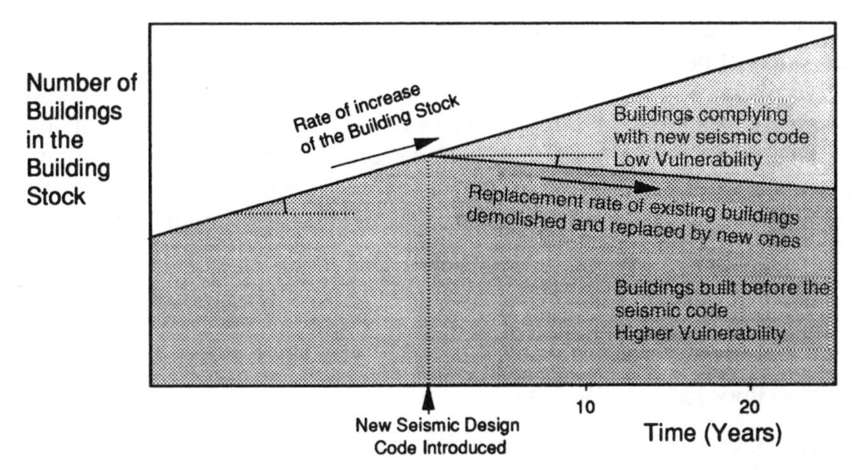

Figure 6.2 Effect of a new seismic design code in reducing the earthquake vulnerability of the building stock over time

factors like architectural fashions and economic affluence of the population. The reduction in vulnerability of the whole building stock as a result of the new code is highly dependent both on the rate of increase of the building stock and on the replacement rate of the building stock. It can easily be seen from Figure 6.2 that in the case of a static building stock – one with no increase and no replacement of buildings – or in the case of a declining building stock, then the introduction of a new seismic design building code will have little or no effect on reducing the vulnerability of the building stock. From this it can be seen that seismic design codes are most effective in cases of rapidly expanding and changing building stocks.

Where existing buildings will continue to be the main elements at risk for some time, a more comprehensive approach to building stock management may be required, where seismic design codes are just one element of a range of measures to reduce the vulnerability of the building stock as a whole. There are a range of possible measures to encourage the upgrading of existing buildings.

A building stock management plan for a city or a region, or for a country as a whole, should begin by identifying the classes of building stock most at risk and the characteristics of buildings with the highest vulnerability. A description of building stock in these terms would include construction types, age distributions, occupancy levels, ownership types and rates of increase and replacement. Most of the risk from earthquakes is to the houses, commercial buildings and other privately owned building stock that makes up most of the built environment. The proportion of the building stock that is in public ownership varies from country to country and with different political systems. The protection of publicly owned building stock by national and local authorities is much more straightforward than influencing protection levels in the privately owned building stock. Protection of publicly owned building stock is discussed later in this section.

Reducing the earthquake risk in privately owned building stock involves getting the owners of property to improve their buildings. In extreme circumstances, say a privately owned building in danger of imminent collapse onto a public highway, most local authorities have the power to serve a closure order on the building, to take remedial action themselves or to demolish it. Less extreme actions, other powers, and possibly more positive actions are also available to local authorities to influence changes in private sector building stock. Buildings that have a high vulnerability to earthquakes, or that would have serious consequences if they failed, can be targeted in a special programme to persuade their owners to upgrade them.

Building Improvement Grants

Offering incentives in terms of building improvement grants to the owners or subsidies for structural strengthening measures are established and relatively successful methods of upgrading building stock and require a significant budget and considerable administration and monitoring. In Japan, private buildings situated in zones along earthquake evacuation routes are eligible for improvement grants to improve fire resistance and to secure glass and cladding from falling into the street. Areas designated as Housing Improvement Areas, consisting of old, high-density housing vulnerable to earthquakes, are also eligible for a range of grants and redevelopment incentives.[11]

Development Incentives

Encouraging premature demolition and replacement of building stock, accelerating the replacement rate of the most vulnerable building types, is possible by allowing tax benefits or possible planning dispensations to land developers – the selective relaxation of planning requirements like urban plot ratio, urban densities, height restrictions or parking may make redevelopment of certain building types more attractive to their owners. Development taxes or land improvement waivers have been used to get private developers to fund the seismic upgrading of poorer quality buildings when building new structures elsewhere.[12]

Influencing Consumer Demand

In a situation where there is choice, the public choosing which type of house to live in and making demands on employers to provide a safe working environment will rapidly affect the building stock: market forces will make earthquake-resistant buildings more valuable than vulnerable ones and encourage upgrading

[11] Ashimi (1985).

[12] This has been used in the urban planning of Mexico City (Aysan *et al.* 1989).

and changes in the building stock. Methods of encouraging private owners to choose their own protection voluntarily include public awareness campaigns and education of the general public in what is an earthquake-resistant structure and what is not. Most people are unaware of how earthquake resistant their own house or workplace is. Where detailed campaigns have been mounted to explain which types of houses are most vulnerable to earthquakes, the general public themselves have proved well equipped to bring about building stock upgrading.

Identification of the most vulnerable structures by the local authority has been advocated as a method of using public opinion to pressurise building owners into doing something about their buildings, but the publication of vulnerability maps or seismic risk indices building by building has been resisted by local authorities for legal and logistical reasons.

Financial Penalties

Other methods can also be used by local authorities to reinforce the economic motivation for upgrading, by fining owners whose structures are excessively vulnerable or imposing other financial penalties. It has been argued that local property taxation, or some type of insurance premiums, should reflect earthquake vulnerability, with more vulnerable buildings paying higher contributions.

Building Certification

Particularly important structures may be required to obtain building code certification by local government. Buildings used commercially as workplaces for more than a certain number of employees, or for concentrations of members of the public, may be licensed for seismic safety and in many countries are likely to be licensed already for fire, safety at work and other public safety regulations. Licensing should involve some verification of structural vulnerability of the building and certain minimum structural criteria required, and possibly insurance, before a licence is granted. Public display of certification is an added aid to enforcement and reinforces public awareness of earthquake protection.

Targeting Weakest Buildings

Unless the probability of an earthquake is high, or the consequences of failure of a particular structure or class of structures are severe, it may be difficult to justify making structural interventions or forcing owners to carry them out to increase earthquake resistance. Costs of structural reinforcement of strengthening existing buildings are high – anything from 10% to 50% of the value of the building; generally it costs far more to increase the earthquake resistance of an existing building than it does to design a new building to a higher standard of earthquake resistance. For a building that may already be half-way through its useful life, this may be a poor investment, and on a larger scale, for the building stock of a city or region, it is rarely going to be an option to advocate reinvestment of sums equivalent to a significant percentage of the value of the current building stock on strengthening existing structures. The building stock that constitutes the main risk in most cities is the older, poorer quality building stock, not much of which may be worth strengthening. Instead a programme of identifying the worst structures and encouraging their gradual replacement by better structures over a realistic timescale (perhaps one or more decades) may be a practical approach. The means whereby replacement is encouraged depends on the powers, budgets and other means available to each local authority.

6.4.4 Low-income Communities and High-vulnerability Structures

The low-income communities most at risk from future earthquakes and whose buildings commonly make up the most vulnerable sector of the building stock are usually those who are least able to contribute to their own safety. Their abilities to make choices about where they live or what they live in are minimal and their priorities for food, income, housing quality and basic living standards may eclipse any concern for earthquake safety. The most vulnerable groups are inevitably the poor: those living in the least agriculturally productive areas of the region, or marginalised in the urban areas. The lowest income groups can afford the least to spend on their housing so end up with the poorest quality sector of the building stock, they have access to the least vulnerable land so live in the most hazardous locations and have minimal savings or economic resources so are least able to recover after a disaster. Locations for the poorest members of any community, rural or urban, are likely to be the marginal lands and include areas of high hazard: the steep hillsides likely to collapse in heavy rains or ground tremors, areas prone to flooding or rockfall, polluted or infested areas, and areas within the poorest levels of service provision.

Vulnerable Old Buildings

In cities many of the poorest and most vulnerable members of the community may not own their own houses but rent poor-quality accommodation from private sector landlords. Many of the oldest and weakest buildings in a settlement are increasingly occupied by the older generations of the community, as the younger generations and more economically productive members move out or build themselves new houses. Targeting such buildings for special assistance from the local authority is a way of helping to offer protection to those least able to protect themselves. Some earthquake protection programmes have involved enabling tenants' cooperatives to buy buildings from their landlords and to renovate and upgrade them using government grants,[13] others have obliged landlords to upgrade rental

---

[13] Aysan *et al.* (1989).

accommodation under licensing regulations. The fact that such buildings often form the historical buildings centre of their community can help to elicit political support for such assistance.

Informal Settlements

In many rapidly expanding cities, particularly in the developing world, the informal housing sector, the squatter settlements or shanty towns, represents some of the highest risks of life-loss, injury, homelessness and emergency needs in the event of a future earthquake. These areas are beyond the reach of the conventional planning process, the implementation of building controls or even of administrative jurisdiction, so efforts to impose earthquake protection measures or to extend planning measures into these areas are likely to be ineffective. Development experience has established that earthquake protection or hazard mitigation programmes in isolation are unsuccessful in these areas. Earthquake protection for these areas has to be part of a much more general upgrading strategy – the improvement of housing standards and services, legitimisation, land registration and income improvement.

6.5 National Risk Management

A major earthquake affects the national economy, is paid for through national taxation or national debt and so earthquake hazard tends to be a country-wide problem. Many aspects of earthquake protection can only be addressed at national level and while there are many things that local communities can do to protect themselves and that private and other organisations can do to bring about protection (discussed in later sections), ultimately if there are no national efforts, earthquake protection will be very limited. Governments establish the baseline level of risk that is acceptable by society generally by legislating building codes and setting safety standards. If the government takes the lead in demonstrating that earthquake protection is important, other people will take it seriously.

Conversely, and increasingly commonly in many countries, if the general public and other concerned lobbies demonstrate that earthquake protection is needed and possible, the national government will follow public opinion and implement safety measures. Political lobbying is a legitimate and often necessary part of instigating protection or improving safety standards. In the aftermath of a major earthquake, the need for protection is strongly demonstrated and political pressure at this time achieves government action which would not be possible at other times. Examples of political campaigns for earthquake legislation, like the prolonged campaign for legislation on unreinforced masonry in Los Angeles,[14]

[14] See Section 10.7.

demonstrate very clearly that the occurrence of a lethal earthquake, either in a neighbouring country or more significantly within the area of jurisdiction, is the main spur for government action. An earthquake is unlike many other political issues, in that no one will normally argue on the side of the earthquake, but in between earthquakes, many other issues tend to take priority on the political agenda. Experience has shown that pressure groups, such as community groups, scientists and engineers who maintain a continuous lobby to persuade government to adopt tougher legislation and have a prepared agenda for action, are better placed to implement action and achieve more results in a post-earthquake situation than an ad hoc lobby arising in the immediate aftermath.

National Disaster Preparedness Plan

It would be best if government actions to reduce future earthquake losses were taken within a broad-ranging overall strategy for earthquake protection – perhaps coordinated with protection strategies for other natural hazards in a national disaster preparedness plan. An integrated earthquake protection plan for a nation would decide on what levels of risk are acceptable (see Chapter 10), identify what are the priority areas for action and the role of private and public sectors in bringing it about, and coordinate legislation and budgeting within an overall timescale and set of objectives for achieving protection levels. Government committees, consisting of leading earthquake engineers, seismologists and other specialists, can help define such an integrated plan, but it is important that economists, community group representatives and legislators are also well represented on such committees to set the technical possibilities for protection within what is practically achievable, economically acceptable and socially desirable. Government committees that have been overreliant on technical specialists have, in the past, tended to be unbalanced and to propose over-ambitious recommendations that are ultimately unsuccessful in achieving an effective plan.

Long-term Planning

Planning for protection does not have to be instant. A 10-year plan or a 25-year plan may be more realistic, to envisage gradual changes in building stock as buildings come to the end of their lives and are replaced, to accommodate expansion of the population, to build up institutions, and to raise technical and educational standards. There are often problems in planning protection strategies for timescales beyond the lifetime of political administrations. A new administration often has new priorities and budget ideas. Programmes instigated by previous administrations may be downgraded, even if they are politically acceptable. One method to ensure long-term objectives is to institutionalise the reforms – to create bodies or institutions as independent as possible with responsibility for promoting seismic safety. Trusts, safety councils and academic or professional bodies may prove suitable vehicles for institutionalised national earthquake protection strategies.

The aspects of earthquake protection that can only be carried out at national level, e.g. the measures that are needed in legislation, financing, building code, professional standards, curriculum, etc., have to be implemented by the government. A national earthquake protection strategy would include a range of measures from construction controls, reconstruction and mitigation budget, hazard research, educational standards, public awareness and emergency preparedness. These are briefly discussed below.

6.5.1 Construction Controls

The best protection against earthquakes is to ensure that the built environment is a strong one. The quality of buildings, measured by their seismic resistance, is of fundamental importance. Minimum design standards and quality standards for earthquake resistance structures, legislated nationally, are an important first step in establishing future minimum levels of protection for important structures.

Many earthquake-prone countries now have national codes of practice and building regulations for seismic design. These codes are in constant review and the international engineering community continues to advance its knowledge of effective earthquake engineering design. Any major destructive earthquake normally provokes a review of the current seismic design codes in that country and in other countries that have similar codes. Field investigations are mounted to analyse whether the earthquake was stronger than expected for that part of the country, whether buildings designed to the code provisions performed adequately or whether damage revealed any gaps in the coverage of the code. The development of design codes for engineering structures is discussed in Section 8.6.

Code Levels

A new design code needs to be carefully considered and adapted for its particular application, and in particular gauged to the economic capability of the community to which it is to be applied. A building code that is insufficiently strict will result in buildings being damaged or causing injury in future earthquakes. But an earthquake design code that is too stringent may also cause problems. In developing countries where capital for development is precious, the level of earthquake protection aimed for is more critical than in countries more easily able to invest in higher cost infrastructure. Every 1% added to the cost of a structure by higher earthquake codes means that for every 100 hospitals, schools or houses that are built, one extra hospital, school or house has to be sacrificed to pay for the safety. A code requiring an increase of cost of several per cent to structures can seriously retard development and construction of the public and private facilities badly needed in many developing towns. The consequences of too severe a code, one considered unrealistic by the developers, contractors and engineers at large, are that the code is ignored, buildings are built without the considerations required and end up more vulnerable than they would have been with a lower, less ambitious code. This has happened in a number of rapidly developing countries where the expanding population and demand for facilities have outweighed the capability of both municipal authorities and people commissioning, designing and constructing new buildings to comply with seismic building codes.

Stricter codes – that is, ones requiring design for higher levels of seismic coefficient – are not always the best way to improve earthquake protection. Increased enforcement of even rudimentary seismic principles may be more effective than a new code of increased severity.

The implementation of a building code has to be seen in two parts:

(1) the definition of minimum design and construction standards, and
(2) the powers and implementation mechanisms for ensuring that minimum standards are achieved

Code Review and Consultation

The level of protection afforded by the code is likely to be taken as a benchmark of safety by other members of the community. Private companies, organisations and individuals are likely to take the protection levels stipulated in the national design codes as officially sanctioned objectives for everyone. The costs and consequences of the requirements stipulated in the code mean that the right level of protection needs to be judged very carefully. This balance between code strength and cost is best decided by a broad consultation process involving the practitioners, building industry, designers, client groups and planners in addition to the engineers drafting the building code. This review process may take some time, but should be thorough, soliciting the comments and taking representation from across the broad range of the building industry before passing a final version of the code into law.

Code Education

The implementation of building codes and design standards is often neglected or underestimated. Highly detailed building codes or complex calculation requirements may be difficult for some building designers to carry out correctly. Mid-career engineers may be unfamiliar with the latest design theory that the new codes are based on. Educational standards of practising engineers in provincial parts of the country may not be as high as those in the capital, for example, or the authors of the code, who are often eminent engineers at the top of their profession, may assume levels of training in their target audience that are slightly beyond the average engineering practitioner. Sometimes the legal phraseology of statutory codes is difficult to understand. Initiatives to explain seismic design codes in simpler language and with step-by-step calculation examples have proved popular with practising engineers and effective in improving seismic design capability in some countries.[15] Figure 6.3 shows one such example from Mexico City.

The proposal of new codes may need to be integrated with training initiatives for building designers and with support for the dissemination and clear understanding of what the codes are requiring them to do. Code enforcement is primarily a concern of urban authorities rather than national governments and is discussed in Section 6.4.2 above.

National Earthquake Insurance

Compulsory earthquake insurance for buildings has been considered in a number of countries as a solution both to financing reconstruction costs and as an incentive for protection measures, and the number of such countries is growing. A compulsory national earthquake scheme was introduced in Turkey in 2000 following the 1999 earthquakes,[16] which is discussed further in Section 7.6. In other countries which have tried to set up such schemes difficulties have been encountered in persuading commercial insurance companies to participate, not least because of the enormous financial risks involved. Local property taxes or insurance premiums only work in encouraging earthquake protection if the premiums reflect vulnerability levels – those improving the earthquake resistance of their building should benefit by reduced premium levels, for example – but the administrative cost in assessing premium levels to sufficient levels of detail may not be economically justified.

Disaster Protection and Economic Development

For the highly vulnerable, the linkage between being disaster-prone and economic development is clear.[17] Developmental programmes for the most vulnerable sectors of the community aimed at improving income levels, improving employment capability, supporting enterprise, giving access to credit and increasing economic security are likely to provide capability for that community to reduce its risk.[18] Such programmes may incorporate specific disaster mitigation measures to ensure that when the community becomes capable of choice, it exercises it in an effective way to provide protection against future hazards. Squatter upgrading, site and service schemes and housing programmes can all include elements for disaster and mitigation. Disaster mitigation measures incorporated as part of development programmes may include builder training programmes, site selection and

---

[15] A booklet explaining the seismic design codes for engineers in Mexico proved to be a popular and successful method of improving building code uptake in a United Nations project.

[16] Bommer *et al.* (2002).

[17] See for example Cuny (1983).

[18] Funding catastrophes and mitigation activities as development investment is explored in Freeman (2000).

Department of the Federal District of Mexico in Association with UNDP and UNCHS (Habitat)
Manual for Seismic Analysis of Structures

2.4 Seismic Coefficient

Federal District Regulations

ARTICLE 205
The zones of the Federal District as defined in Article 219 apply to this regulation.

ARTICLE 206
The Seismic Coefficient, c, is the component of the horizontal shear force acting at the base of the structure during an earthquake, proportional to the weight above that level.
The base of the structure is defined as the level from which the ground no longer provides lateral restraint to the structure. The total weight is calculated taking into account the total dead loads and live loads as defined in Chapters IV and V of this Regulation.
The seismic coefficient for structures classified as Group B in Article 174 is taken as equal to 0.16 in Zone I, 0.32 in Zone II and 0.40 in Zone III, unless:
a) The simplified method of analysis is used, in which case the coefficients of the Complementary Technical Code apply; or
b) In exceptional zones, where other values of c may be defined
For structures in Group A, the seismic coefficient is increased by 50%.

Commentary

It is not possible to predict the exact severity or direction of forces exerted by an earthquake on a structure. The effects produced by seismic motion are complex and the dynamic response of a building is difficult to analyze. Instead the Federal District Regulations use an empirical parameter that reflects a safe lateral load for the structure in each zone of the Federal District. This parameter is the Seismic Coefficient c, and is defined in Article 206. This defines the horizontal force acting at the base of a structure to be:

$$V_0 = cW$$

Where
V_0 = Horizontal force, i.e. the base shear
c = seismic coefficient as a proportion of gravity
W = The total weight of the building above ground level
(See examples 2.4, 2.5, 2.6 and 2.7)

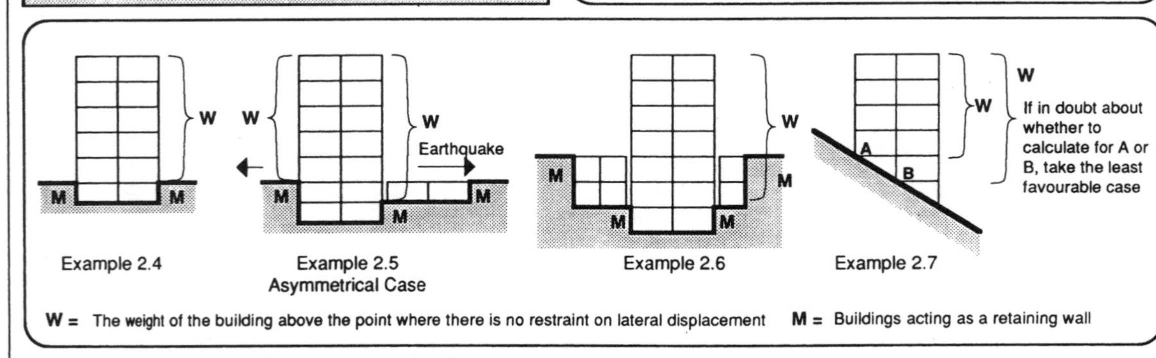

Figure 6.3 Illustrative example of an explanatory manual produced to accompany a seismic design code (Federal District of Mexico 1987 revision). Such manuals, aimed at practising engineers, help understanding and better implementation of building codes (after DDF 1987: authors' translation)

improvement projects, community action plans and others described in more detail in Section 6.6.

Traditional Buildings

The greatest earthquake risk throughout the world is faced by traditional rural communities that build their own houses from locally gathered materials. These houses, variously described as 'traditional', 'earthen', 'vernacular', 'owner-built', 'non-engineered', 'low-income' or 'low-quality', form a high proportion of the building stock of many developed countries. Their earthquake performance is notoriously poor, as outlined in Chapter 1 and touched on in each of the subsequent chapters of this book.

It is estimated that over 75% of the world's population lives in housing of this type.[19] Houses are built by the family itself, perhaps employing a skilled builder from within the community to direct operations. Traditional construction materials for this type of house are naturally occurring and used in building construction with only limited processing or quality grading, such as earth, stone, wood and fibre. Increasingly commonly used are modern building materials: cement, steel, concrete blocks, fired clay bricks, roof tiles, sheeting, processed timber and other materials bought for cash from nearby markets.

These types of communities similarly tend to be beyond the control of urban building regulations or planning requirements imposed from central or urban authorities. Instead development programmes based on capacity building and incentives for self-protection against earthquakes have been more successful. There have been a number of development projects focusing on improving the earthquake protection for traditional rural communities by increasing the capability of local craftworkers and the builders in each community to build earthquake-resistant structures with the skills and materials they have available.

These types of projects, discussed in Section 6.6, have been mainly pioneered by development agencies but increasingly adopted by governments as regional development initiatives. Government support can help extend the take-up rate of such programmes of education and training builders with building improvement grants, materials, subsidies and other incentives to establish earthquake-resistant construction techniques within the building traditions of the hazard-prone areas.

6.5.2 Education and Training

The overall level of competence of the design professionals and personnel in the construction industry has a major effect on the vulnerability of the built environment. In the longer term, the quality of the buildings that will be constructed in

[19] Razani (1981).

5–10 years' time depends largely on the standards of training being received by the students and apprentices of today.

Engineering Education

Engineers' curriculum and vocational training achievement levels are usually established by national authorities and in a seismic country all engineering students should have a thorough grounding in earthquake engineering as part of their core curriculum. The standard of earthquake engineering being taught is important and should be reviewed as an integral part of the longer term earthquake protection strategy. Mid-career training for engineers in practice is also important in order to increase awareness of earthquake issues, update them on recent developments in earthquake engineering and explain code revisions or regulatory procedures. Professional engineering institutions or colleges are useful vehicles for increasing education standards, and mounting mid-career training courses for practising professionals.

Education of Other Professions

It is also important to spread seismic design knowledge out wider than the engineering professions. Architectural education should also include earthquake design principles in the undergraduate course of student education and post-qualification training. Building surveyors, property managers, real estate agents and construction contractors could also benefit from a knowledge of earthquake-resistant design principles developed through college courses, further education and professional groups. Vocational training or on-the-job skills development for building supervisors, tradespeople and construction labourers also increases the quality of building construction and improves earthquake resistance. Requirements for certain skills to be represented in trade certification and basic training also help develop an earthquake awareness at every level of the building industry workforce.

6.5.3 Public Awareness

Public awareness of earthquake risks and support for the measures needed to be taken for protection are a necessary prerequisite for action to be taken. The support of the community and its participation in protecting itself and mandating its representatives to take actions to protect it are the essential elements of earthquake protection. It is clear from studies of perception of risk, presented in Chapter 10, that the actions communities take to reduce risk and the support they give to reduce unacceptable risk are related to the nature of the risk and perception of its degree and severity. Availability of information about the actual level of risk faced, demystification of the threat and familiarisation with protection capability are important aspects of motivating the community to protect itself. Studies of some communities at risk, particularly rural groups, developing societies and communities with limited access to information, suggest that their perceptions may distort and underestimate earthquake risk – individuals may be more at risk from earthquakes than they realise. Earthquakes are rare events and few people are likely to have personal experience of them. They may have an incorrect image of earthquakes being all-powerful and totally devastating.

Psychologists suggest that hazards like earthquakes which embody high levels of 'dread', which are perceived to be uncontrollable and that few of the community have personal experience of, are difficult to protect against. These hazards may be mentally rejected or perceived fatalistically or in other ways that reduce the motivation to take action to reduce them (see Section 10.6). It is clear that increased access to factual information can increase perception of risk, and affect what is considered safe.

Public information campaigns, increased exposure of earthquake issues in the media, including disaster safety in school teaching and encouraging civil protection to become more a part of public life, raise awareness in general. Some elements of public information campaigns for earthquake preparedness were discussed in Chapter 3, Section 3.6, including drills, practice emergencies and anniversary remembrances. Information on earthquakes can increase familiarity with the hazard and reduce its dreadfulness, and it can demonstrate that mitigation is effective and necessary. Public information campaigns can also address the more pragmatic issues of what to do in the event of an earthquake and other response activities that may save lives and reduce damage. But without a preparatory background programme to increase familiarity with earthquakes, the what-to-do type of information is unlikely to be well received or the need for it comprehended. The primary focus of public awareness campaigns is to motivate the community to protect itself as far as possible (see Section 6.2).

6.5.4 Earthquake Hazard and Engineering Research

At a national level, it is important to understand the nature of the earthquake threat. There are many countries that suffer repeated destructive earthquakes in which basic seismological data is poorly gathered, and national observatories under-resourced and understaffed. Hazard research (outlined in Chapter 7) can define the areas where earthquakes are most likely to strike, the rates at which earthquake activity can be expected, the characteristics and severity of future earthquakes and the probable consequences of seismic activity. Such research is best carried out and coordinated at a national level, or even at an international level where several neighbouring countries cooperate in hazard assessment. A national seismological observatory maintaining its own network of seismometers, or coordinating networks of different universities and research agencies, can observe patterns of seismic activity over time and across the whole country or region and contribute to everybody's understanding of the hazard they face.

Earthquake engineering may be an important area of research, as the construction types, preferred materials and design practices are different in each country and often differ from region to region within a country. The development of earthquake design building codes based on those of another country is common practice, but adapting them to the local building types needs research. Structures, e.g. concrete frame structures, are likely to be similar from one country to another and design methods may be transferable, but construction practices, e.g. infill construction, construction of engineering movement joints, etc., and the local building forms and typologies vary considerably from one region to the next. It is clear from earthquake engineering research that local construction techniques affect seismic performance significantly and studies are needed to optimise earthquake safety measures for local building types. Traditional building types and the non-engineered building stock that makes up most of the risk of earthquake damage tend to be very poorly researched and largely dismissed by the engineering community in most countries. Research is needed to develop earthquake-resistant techniques, design principles and construction details that are appropriate to the types of buildings normally built in that country.

Coordination and support for a broad programme of independent research, involving universities, public utility companies, government research institutes, private companies and other research establishments, may also be an area of government initiative or budgeting. Research activities are a vital part of national earthquake protection efforts.

6.5.5 Budgeting for Losses and Mitigation

The degree of influence that national or local authorities can bring to bear on improving earthquake safety is likely to be related to the budget available.

Many measures available for earthquake protection require the commitment of significant resources. The establishment and sustaining of institutionalised safety councils, the implementation and policing of adequate building codes, the construction of earthquake-resistant public buildings, and the use of grants for building improvement, establishing research institutes and many other measures advocated here, require adequate funding. These measures, as an integrated package, represent the cost of public safety against earthquakes.

The spending of public finances to improve public safety is justifiable on its own but financial costs of earthquakes are themselves high and there is additional justification for spending on earthquake protection in reducing these costs. In addition to the unquantifiable but considerable costs to society, intangible losses of injury and the loss of human life, there are costs in the destruction of property, both public and private, costs of emergency mobilisation, relief and recovery, the disruption to the economy, loss of earnings and lost production, and costs of lost opportunity and delayed developmental progress. The few studies that have been made of the costs and benefits of spending on earthquake protection have shown that well-targeted investments in areas of high hazard can be cost-effective in reducing losses – that is, the financial savings can exceed the costs of investment. There are therefore economic arguments for earthquake protection measures quite apart from civil protection and saving human life. Chapter 10 presents the use of cost–benefit analysis and other methods of calculating the value of earthquake protection measures.

Budgeting for Earthquake Losses

Few governments cost future earthquake losses systematically, but a number of countries maintain some disaster budget or contingency account used mainly for relief and emergency needs. Most disaster losses are funded through borrowing and there is a convention that disaster losses are unforeseeable, and so are not planned for on the budget sheet. By not budgeting explicitly for earthquake losses, it is harder for the economics of earthquake protection to be shown or to be argued for. Systematic costing of earthquake losses is both possible and necessary in a country that has repeated earthquakes. Earthquakes, of course, happen irregularly and there can be many years between major events. When a large earthquake occurs it can cost billions of dollars. Smaller earthquakes occurring more frequently add smaller costs to the damage total. Averaged out as a cost per year, the losses due to earthquakes can be significant, and seismological hazard studies, historical experience and future risk analysis (see Chapter 9) can predict with a fair degree of accuracy what the annual average loss is likely to be over a 10-, 20- or 100-year period. Such studies cannot, however, predict *when* this loss is likely to occur – the need to budget on an annual basis, or even for a multiple year administration, means that the probability is low that the expenditure will come within that particular, short time frame. This tends to lead to it being ignored in short-term accountancy. Longer term accountancy and policy development, such as that being increasingly practised by government administrations in areas like environmental policy, energy and mineral resource exploitation and transportation policy, is needed for national protection policies against earthquakes and other natural hazards.

Deciding on levels of budget that are appropriate to commit to protection against earthquakes is a matter of the severity of the risk, the prioritisation of that risk against other calls on the resources available, and the social and political judgements that each community uses in making decisions. In Chapter 10 there is further discussion of decision-making on risk, perception of risk and comparable levels of risk that communities find acceptable.

Protection Fundraising

Apart from direct treasury allocation, government budgets can be raised in a number of ways to fund protection measures. In some countries, specific lotteries or an added tax on tobacco or consumables have been used to raise a budget specifically for disaster measures. Disaster budgets, however, are not always spent on disasters, and if not spent are liable to be used for other things. Civil protection programmes that are allowed to draw from a range of budget sources are more sustainable and should be one of the main beneficiaries of the disaster budget.

Inevitably the funding for protection programmes will be disbursed piecemeal, with hazard observatories funded from higher education and research budgets, code administration funded from local administration, building improvement grants from capital funds, and so on. Some individual initiatives to raise local funds for building improvement from development tax and to encourage partnerships of public and private sector funding have been discussed above. The main key to funding is motivation and belief that earthquake protection is possible and desirable. If the community at large and the individuals at funding level believe that earthquake protection is a valid activity, then funding will become a higher priority. The role of national and local government authorities is central to establishing earthquake protection as a credible, achievable and essential part of life in a hazard-prone country.

6.5.6 Supporting the Design Professions

The design of buildings and facilities to protect them against earthquakes (described in Chapter 8) is a skilled job and requires a thorough understanding of the destructive power of an earthquake and the mechanisms that operate. Specialist earthquake engineers in the fields of civil, structural and geotechnical engineering have a vital role, both in the structures they design and in promoting earthquake protection to be adopted more widely. Some of the strongest advocates for seismic legislation, community preparedness and earthquake protection measures are the earthquake engineering specialists. But earthquake-resistant design has to become a broader, general skill within the general engineering profession for a truly safer community to result. The standards of earthquake-resistant design of the average engineering practitioner are what determine the safety of our towns and cities.

Understanding the principles of earthquake-resistant design also has to be widened to include all the other design professions: architects, surveyors, services engineers, interior designers. It is now well established that a good engineer cannot make a bad architectural design earthquake resistant. All the members of the design team need an understanding of earthquake issues to make a safe building – the architectural form of the building, the location of the service runs, the materials specifications and the non-structural fittings all need to be designed with an understanding of how they affect earthquake safety.

Most countries have professional institutions that represent each of the various design disciplines, regulate professional qualification standards and lobby for their interests. A full endorsement by professional institutions of the role played in earthquake protection by their members will enable that profession to move more fully into earthquake protection activities. The professional institutions may require competence in design for earthquake safety as a qualification or membership requirement. Continuing professional education or mid-career training should also include earthquake-related topics, particularly following the introduction of a new building code or in the aftermath of a destructive earthquake which has had lessons for design professionals. All these activities depend on the support of national governments.

Persuading Clients to Protect

The client, when commissioning the design of a building or other facility, commonly relies on the advice of the design professional for a range of matters. Structural safety and functional reliability are factors in which clients are likely to be influenced by professional recommendations. The professional engineer, architect or other designer is an advocate for earthquake protection and has an important role to play in educating the client about the risks involved and persuading the client to take earthquake protection seriously in the design process.

A client may be unwilling to pay for the additional costs involved in incorporating appropriate levels of earthquake resistance unless convinced of the necessity and benefit of doing so. The design professional may be able to convince the client of the need for design safety by demonstrating the hazard faced and the risks of earthquake damage. The client should be encouraged to protect the facility to the fullest extent practicable. The client also should be made aware of the protection levels afforded by the statutory minimum design requirements and encouraged to protect the facility to higher standards where this is appropriate or justifiable.

6.6 International Aid and Development Organisations

Earthquake disasters frequently reach international proportions. The scale of a major disaster often exceeds the capabilities and resources of a national government and the international aid community is usually quick and generous in its response. Aid from the richer countries to the poorer is commonly given for emergency relief to assist recovery after a major earthquake or other disaster.

At other times and in non-disaster circumstances assistance is given by the countries of the developed world to help other countries generate economic development and to improve the lifestyle and safety of communities. It is being increasingly realised that assistance to developing countries to help them reduce the effects of future disasters before they occur is more effective than providing aid afterwards: prevention is better than cure.

Disasters are closely related to economic development. The great majority of casualties and disaster effects are suffered in developing countries. Development achievements can be wiped out by a major disaster and economic growth reversed. The promotion of earthquake protection in the projects and planning activities of development safeguards development achievement and assists populations in protecting themselves against needless injury.

There are many organisations and operations devoted to development assistance and these have an important role to play in helping countries and communities protect themselves against future earthquake disasters. Organisations representing multi-lateral economic assistance (i.e. funded by contributions from several countries) include the many United Nations agencies and regional organisations of the globe including the European Community, Pan-American, Pan-Arab, Pacific Cooperation, and other multi-nation technical, cooperation, economic and development organisations. International organisations like the Red Cross and Red Crescent have significant involvement in most disasters. Bilateral aid programmes – provided by a single country directly to another – make up a very large proportion of the economic assistance that passes between countries. Most of the industrialised nations, the members of the Organisation for Economic Cooperation and Development, maintain a government ministry or department responsible for development assistance to other countries, and maintain departments, attachés or representatives in the embassies and consulates of the countries to which they disburse aid. Other organisations that are instrumental in international development include non-governmental organisations (NGOs), agencies like CARE, Oxfam, GTZ, and very many other private organisations which are humanitarian, religious or developmental in nature. Often these NGOs are the channel for implementing projects in the recipient areas funded by the bilateral or multi-lateral aid organisations. In addition to the international NGOs there are also large numbers of NGOs within each developing country that implement development projects and assist in humanitarian activities.

The number and types of development organisations are considerable. Each can, if it directs its efforts in the right direction and is motivated to do so, bring about effective, sustainable achievements to make communities safer against future earthquakes. Incorporation of disaster protection into the activities of development organisations can be achieved without major shifts of emphasis in their work, providing the issues of protection are understood.

Earthquake Relief

Many development agencies have an extensive involvement in disaster relief and any major earthquake is likely to involve development organisations working in that country or region in helping the worst affected communities to recover from its effects. Increasingly such organisations are implementing programmes to reduce the effects of future earthquakes as part of their operations. Relief and reconstruction programmes can contribute to the future safety of the affected community if they are orientated towards long-term revitalisation of the economy and sustainable development as discussed in Chapters 4 and 5. Short-term needs, such as shelter and food, are undeniable, but provision of emergency assistance will not result in any sustainable community recovery unless this is followed up with help to the community to reknit the social structure, re-establish the economic activities and regenerate the damaged buildings and structures through the normal construction processes and building workforce.

The objective for any reconstruction programme in a development context is to assist the community to rebuild its own economy, houses and workplaces. Reconstruction programmes have rarely been successful where outside agencies have made the major decisions for the affected community or have built houses for it or relocated damaged settlements, or introduced new, rapidly built building types in order to accelerate the reconstruction process. Only by allowing the affected community to maintain control over its own reconstruction can an outside agency assist a recovery which will be sustainable and seen to be beneficial 10, 20 and more years later and that will result in a community less vulnerable to a future earthquake than it was. In practice this means consultation and community-led decisions on issues like priorities for the assistance that is available, location of new facilities, labour available and timing. Community consultation and decision-making can be a lengthy process and may appear to place undesirable delays on the reconstruction operation – delays that it is tempting to short-cut with centralised planning – but the benefits of having a committed and participating community will be seen in the sustainability and developmental achievements of the project.

In housing, for example, the argument has been made in Chapter 5 that rebuilding damaged houses should be carried out by the normal building construction industry, expanded if necessary to meet the large-scale demand. In many rural areas and developing regions, houses are built by householders themselves or by village craftworkers or small-scale contractors. Assistance projects by development agencies to help these builders meet the reconstruction need is far preferable to the donation of several-thousand housing units.

Protection Beyond the Reconstruction Area

It is important in a reconstruction that the processes of building more strongly are established as well as the houses being built strongly. The next major earthquake in the region is likely to be nearby – in the next valley or in a neighbouring district. The opportunity should be taken to use the earthquake to introduce protection measures to as broad an area as possible in the neighbouring seismic region. Builders trained in strong construction may be encouraged to use their skills and qualifications to build for other clients in the neighbouring region, or to train other builders in nearby villages. Development agencies can promote earthquake protection over a broader region, using the earthquake reconstruction as the initiating opportunity.

Establish Long-term Protection

Apart from broadening the scope of protection activities geographically beyond the reconstruction area, it is important that the processes of building more strongly are well established so that they are sustained over time. Earthquakes generally have long periods between occurrence, and protecting against them through improving the quality of the built environment is a long-term process. The buildings likely to be affected by a future earthquake may not be this generation of structures, but those that exist in 20 or 50 years' time. Many more buildings are likely to be built during the period between earthquakes and if the process of building them has been improved, then future earthquakes will result in lower damage levels. But the improved construction skills will need to be maintained throughout the next 20 or 50 years.

Development organisations involved in post-earthquake emergency and reconstruction operations can help to instigate protection against future earthquakes by establishing regional and sustainable building improvement programmes as part of the community-based reconstruction.

Building Improvement Programmes

The most vulnerable parts of the built environment are the non-engineered buildings constructed by householders, craftworkers or small-scale contractors, from a wide range of locally available and purchased building materials. The poor earthquake performance and lethal consequences of these building types are well known. Research, analysis and testing of these building types has identified their behaviour in earthquakes and the vulnerability of their construction practices.

Technical methods of improving the construction of these types of buildings to make them less vulnerable to earthquakes are described in Chapter 8. The greatest priority for development organisations concerned with reducing earthquake disasters is to implement improved construction techniques in the many thousands of houses being built across the seismic regions of the developing world. There are a variety of ways, described below, to encourage a community to improve its construction techniques. The most appropriate measures will depend on the type of community, its normal construction processes and the resources and technical capability of the assisting agency.

Incentive Programmes

Building improvement programmes may be instigated by encouraging the community to build to improved building designs and higher technical standards by offering grants or preferential loans or other incentives to anyone building a new structure. In a grant-based building improvement programme, householders are given building plans or construction specifications and are eligible for financial benefits if their new structure conforms to these standards. Financial benefits may include provision of building materials, labour or mechanical assistance, or other incentives, including services provision or credit. Staged assistance is common with incentive programmes providing one instalment of grant or building materials when the foundations are dug, another when the walls are complete, and again after the roof is fitted, and so on. The advantage of such programmes is that if they are administered correctly with each structure being monitored by a competent official, there can be a high degree of confidence that buildings are achieving the technical levels of construction and design aimed for. The disadvantages of incentive programmes are that they require large resources – both to provide the materials or grants and to administer – with each new structure requiring at least one visit by a technical specialist to check that it conforms. A further disadvantage is that it is seen as prescriptive – the external agency is specifying what is required and the community is not in charge of its own building process.

Communal Building Programmes

Learning-through-doing can be implemented by instigating community projects which require the builders and labour force of the community participation groups to work together under skilled supervision. An example of community-building programmes may be the construction of a public building – a school, a meeting hall, clinic or religious building. The community provides labour and perhaps land. The development organisation pays for the capital project costs, including building materials, provision of equipment and may also pay wages or provide food for the community workforce. Some community projects have involved housing projects where housing is built by the community and allocated to individual families when the project is complete. The community-built project is built to a high standard of earthquake protection using techniques easily reproducible by the community when it builds its own houses or additional buildings at some time in the future. The project is designed with this reproducibility as a deliberate aim, using only the materials, construction methods and skills normally available within the community. Experience has shown that working under skilled supervision is an effective way of teaching less skilled builders improved techniques that are reproduced in their later projects, unsupervised. Community projects also have additional benefits of reinforcing social cohesion, improving decision-making and community participation. Disadvantages with community projects are that they are only successful in certain circumstances, and require a relationship to be established between the development organisation and the community which may not be achievable on a large scale or with very large numbers of communities.

Technical Assistance On-site

Building improvement can be achieved by providing technical assistance and advice at the point it is most needed and effective – during the construction of a new building and provided on-site while the builder is working (Figure 6.4). A mobile technical team or skilled supervisor travels through areas during the building season, stopping at sites where construction is happening. Advice can be offered and discussions held about the need for earthquake protection and the best way to go about it. This type of travelling advice can be effective if the advisor is familiar in the village, passing through regularly, and a respected figure, perhaps an experienced master builder from the region itself. Considerable coverage of a large area can be achieved, including many villages and acting as a catalyst for improved construction at relatively low cost. The impact of any advice on-site is likely to be limited to construction detail rather than broader design decisions and any completed construction will be unaffected. As an additional element in other types of building improvement programme, e.g. as technical support for builder training off-site, or as follow-up on training and involving recruitment for community projects, the provision of technical assistance on-site can be a highly effective use of additional resources.

Builder Training Off-site

In many rural areas, most houses are built by the families that occupy them, and most families would have some building skills in addition to being farmers, animal keepers and the many other occupations that make up the rural family economy. But there are often craftworkers within the villages who spend more of their time building, who have a reputation for building well and may well be paid by other families to assist with the building of their house. In more affluent villages these craftworker builders may be professional, earning their entire living by building to commission. In larger villages and towns, building may be fully commercial, with small teams of contracting builders owning their own construction machinery and operating on a much more formal basis. In places where specialist or craftworker builders operate, they can be very influential in the construction styles and housing quality built by the community.

In communities where specialist builders build to commission or help householders build their own homes, selectively training builders is an effective way of improving the construction process in a community. Builders trained in earthquake-resistant construction can be highly influential within a community, persuading house owners to invest more in earthquake safety and setting examples

Figure 6.4 Building improvement can be achieved by providing advice where it is most needed – on the site of a new building under construction. Here a mobile technical support team provide advice on earthquake-resistant construction to builders engaged in the reconstruction after the 1982 Dhamar earthquake, Yemen Arab Republic

for other builders to emulate. Builders can be recruited and taken for training at a local training centre if the project is well publicised and presented in such a way as to be prestigious for the trainee. Training is carried out over a number of days by engaging the builder in practical exercises, building sample buildings under skilled supervision. The core of practical exercises can be supplemented by classroom teaching, group discussion and educational literature or training films. The objective is to raise the awareness of the builder of earthquake risks, to convey a limited number of technical messages about earthquake-resistant building and to develop a pride in quality of construction generally within the class.

Any building improvement programme of this sort is likely to need a general context of a supportive community in order to flourish. Without a general awareness of the risk and demand for protection by the community, no building improvement will be able to succeed on its own. A complementary part of the building improvement project may involve public information campaigns.

Public Information Campaigns

Simple messages to the population of earthquake areas to generate support for protection measures can be disseminated in many ways. The most effective is usually through the people: the opinion-makers, the leaders or other influential members of the community itself. If the leaders of the community are convinced of the need for protection, or the schoolteachers or the respected craftworker builders, then other members of the community are likely to agree. The most effective public information campaigns are where peer pressure operates. Where the community believes that earthquake protection is sensible and desirable, 'safe is smart', there will be support for community measures to reduce risk and individual action to protect itself.

Information is transmitted through many media and the most effective medium is likely to be different from place to place and for each different society. The channel for authoritative information, i.e. information that is believed and taken notice of, perhaps from radio messages or proclamations from the village council, may be different from channels that are perceived as entertainment media, e.g. TV and comic-books. Public information campaigns on earthquake risk channelled through the wrong media may be misconstrued and ignored.

Visual material is a strong reminder of messages that people have already been acquainted with. A poster on earthquakes or a booklet may be a useful reinforcement of the message in a community that has already discussed the problem, but it is rare that printed materials on their own can convey a new message. A coordinated campaign of identifying and informing the key opinion-makers in a community, subtle messages in familiar information channels and supplementing with carefully designed printed materials can bring about public support for the earthquake protection measures being undertaken in development projects.

Development Programming by International Agencies

International development organisations, multi-lateral aid organisations and bilateral aid programmes can also assist in building up protection against future disasters through technical assistance. Protection from disasters is an international concern. Disasters are, with a few notable exceptions, infrequent and any individual country is unlikely to have regular experience or build up expertise in dealing with all of the wide range of hazards it is likely to experience. That expertise is available on an international level – countries that have recently experienced a major earthquake may be best placed to assist another country instigating seismic protection programmes, for example. International development organisations are important vehicles for facilitating international exchanges of expertise and developing an international approach to disaster mitigation.

Earthquake Protection in Development Projects

When development projects, like any other projects, are undertaken without regard for the risks of future hazards, the investment level considered adequate for the programme may be insufficient to protect it during its lifetime.[20] Cost–benefit analysis (CBA) and risk assessment techniques for use in development project planning are outlined in Chapter 10.

It is not just the engineering content of development programmes that needs to build in safety factors and protection, the entire project needs to be designed with a level of risk awareness. Investments in development projects have been lost repeatedly in hazard-prone areas wiped out by a major earthquake – often a hazard that should have been foreseen. Perhaps more common is the occurrence of a disaster interrupting an ongoing project and diverting resources from their original intended use.

One important procedure that has been proposed[21] is to include disaster potential in the economic analysis of the design of all development projects. Some major institutions currently require a disaster potential analysis in the project formulation of any major development funding. Capital projects are required to build in earthquake protection at the design stage of the project. The extra costs of protection, it is sometimes counter-argued, would make some development projects not economically viable. However, the basic argument for integrating disaster awareness into development planning is that it is wasteful not to do so.

Building up Skills and Institutions

One of the most important long-term, sustainable aspects of disaster mitigation is for the development of skills and technical capacity in-country. Professional development and a pool of expertise in disaster mitigation techniques will allow longer term development of the issue. Helping to build national institutions and formal structures that perpetuate the mitigation programme is an important element of development assistance. In a number of countries, the response to any individual disaster is to set up a special disaster committee to handle the emergency. At the end of the emergency or reconstruction, the committee or government department has the advantage of retaining skills and experience

[20] Anderson (1990).

[21] Anderson (1990).

and also allows some emphasis to be switched from post-disaster assistance to pre-disaster preparedness.

Further Reading

Aysan, Y., Clayton, A., Cory, A., Davis, I., Sanderson, D., 1995. *Developing Building for Safety Programmes*, Intermediate Technology Publications, London.

Cuny, F., 1973. *Disasters and Development*, Oxford University Press, Oxford.

FEMA publications (from www.fema.org).

Jaffe, M., Butler, J. and Thurow, C., 1981. *Reducing Earthquake Risks: A Planner's Guide*, American Planning Association, Planning Advisory Service, Report Number 364, 1313 E. 60th Street, Chicago, IL 60637, USA.

Lagorio, H., 1991. *Earthquakes: An Architects Guide to Non-Structural Seismic Hazards*, John Wiley & Sons, New York.

Maskrey, A., 1989. *Disaster Mitigation: A Community-Based Approach*, Oxfam Development Guidelines, No. 3, Intermediate Technology Publications, London.

UNCHS, 1989. *Human Settlements and Natural Disasters*, (UNCHS), PO Box 30030, Nairobi, Kenya.

UNCHS, 2001. *Cities in a Globalizing World: Global Report on Human Settlements*, Earthscan, London (Chapter 13).

Wolfe, M.R., Bolton, P.A., Heikkala, S.G., Greene, M.M. and May, P.J., 1986. *Land Use Planning for Earthquake Hazard Mitigation: A Handbook for Planners*, Special Publication 14, Natural Hazards and Application Information Center, Institute of Behavioral Science #3, Campus Box 482, University of Colorado, Boulder, CO 80309-0482, USA.

7 Site Selection and Seismic Hazard Assessment

7.1 Choice of Siting

The selection of a suitable site is a crucial step in the design of a building or planning a settlement in an earthquake area. There are a number of earthquake-related hazards which should always be considered when choosing a site, together with the influence of the ground conditions at the site on the ground motion which the building may experience in a future earthquake. An assessment of the extent of the earthquake hazard should always form a part of the overall site assessment and of the specification for the design of any structures to be built there. No site can be expected to be ideal in all respects, so the choice of site will often involve a judgement about relative risks and the costs of designing to protect from them. But there are some sites which are so hazardous that they should be avoided if at all possible, since the cost of building safely is likely to be prohibitive.

The factors which need to be considered as a part of a site assessment include:

- What active faults or seismic source zones are close enough to the site to give rise to potentially damaging earthquake ground motions.
- What the pattern of earthquake occurrence is on these faults, or in the region generally (in terms of size and nature of event, recurrence period), and what ground motions these are likely to cause at the site.
- Whether any active fault can be identified which passes through the site.
- How the ground motion effects at the site are likely to be affected by the subsoil conditions at the site.
- What other earthquake-related hazards need to be considered (such as landslides, soil liquefaction, settlement or subsidence, tsunamis).

The site assessment should provide information which will be used to determine design loading and other inputs for the structural design of buildings. In most cases this process is greatly facilitated by using the zoning map and other procedures specified in an existing code of practice. Where no code exists, or a more rigorous analysis is required than required in the code, the task is more complex and some sort of seismic hazard assessment for the site has to be carried out.

This chapter first reviews the range of site-related earthquake hazards (Section 7.2), and then discusses the methods available to assess the ground shaking hazard (Section 7.3). The effect of site conditions on seismic hazard is discussed in Section 7.4. The use of microzoning techniques to map the variation in the principal earthquake hazards within an administrative region is discussed in Section 7.5, and the chapter concludes with a consideration of risk mapping and its uses in insurance (Section 7.6).

7.2 Site-related Earthquake Hazards

7.2.1 Large Ground Deformation

Large, permanent ground deformations often occur at the surface breaks associated with fault ruptures in earthquakes. Vertical and horizontal displacements of one side of the fault break relative to the other of a number of metres have occurred; where this relative movement occurs under a building catastrophic damage can result. Local deformations sufficient to cause severe damage can occur up to a few hundred metres from the fault.

Fault breaks are known to occur repeatedly at the same location and it is therefore advisable not to locate buildings in the immediate vicinity of known previous fault breaks, although avoiding these does not guarantee protection from new surface faulting. It is particularly important to avoid such locations for sensitive installations such as power stations, chemical plants or major hospitals, the loss of which could be catastrophic for the whole community. For sub-surface pipes, roads and railways it may be impossible to avoid the network crossing a fault, and building to resist rupture may not be feasible. In such cases, the best protection strategy is to ensure that alternative routes are available, and that the flow of liquid or gas in the pipelines can be rapidly shut off in the event of a rupture.

7.2.2 Liquefaction

Earthquake-induced soil liquefaction has been the cause of catastrophic damage in a number of earthquakes. Certain types of soils, when they are saturated with water and then suddenly shaken by an earthquake, completely lose all shear strength, and flow like a liquid. The support to the foundations of buildings built

Figure 7.1 Building failure caused by liquefaction. This recently completed building in Adapazari, Turkey, overturned when the soil below its shallow foundations liquefied in the 1999 Kocaeli earthquake. The building, otherwise little damaged, is nevertheless a complete loss

on such soils then disappears, and they can plunge into the ground or overturn as shown in Figure 7.1, or be carried sideways bodily on unliquefied masses of soil. Liquefaction is most likely to occur in loose cohesionless soils, such as fine sand or silts; these are most commonly found in sea or river-deposited sediments laid down within the last few thousand years. Simple *in situ* soil testing using a cone penetrometer has been shown to be a good indicator of potential liquefaction susceptibility in a soil layer, and it is possible to establish magnitude and intensity thresholds below which liquefaction is not likely to occur.[1] Clearly sites which may be subject to liquefaction should be avoided if possible for any massive structure; alternatively foundations should be designed to bear on stable soil layers below the layers that may liquefy.

7.2.3 Landslides

Sloping ground or rock masses which are stable under normal loading can lose their stability during an earthquake causing effects ranging from a slow

[1] EERI (1986), p. 29.

progressive creeping of the ground to a dramatic landslide, rockfall or flow failure. Slope failures are particularly likely to occur when the ground is saturated following rainfall. Whether sudden or slow, such slope failures are liable to cause complete destruction of any building founded on them or in the path of the slide. Slope failures can contribute a high proportion of the losses from earthquakes in mountainous terrain.[2] Earthquakes in mountainous terrain can also trigger rockfalls and mudflows large enough to engulf whole settlements. Landslides and lateral spreads can also cause extensive property damage.[3]

The only effective means of protection from the landslide hazard is to avoid building on sites which may be affected. Sites on or at the top of steep slopes, or where there is evidence of recent instability, are those most obviously at risk. Known landslides can sometimes be stabilised through drainage, excavation, retaining structures or other geotechnical work, but while this may protect structures below the slide, it is unlikely to make the site safe for building. In some areas maps of previous and potential landslide areas may be available.[4]

7.2.4 Tsunamis and Floods

Tsunamis are sequences of long-period sea waves generated by earthquakes, often those which occur in the sea bed (see description in Section 4.5). They travel long distances at high speed, and when they reach the shore, they may under certain conditions result in huge waves a number of metres in height, which can surge well inshore. Low-lying coastal areas on the margins of the large oceans, especially the Pacific Ocean, are most vulnerable. Considerable damage can be caused by tsunamis[5] and many coastlines such as those of North America, Japan, Hawaii, Peru and Chile are vulnerable. Some warning of the arrival of a large tsunami is usually available, enabling the vulnerable population to evacuate. Low lightweight buildings may be severely damaged by the high-velocity water impact, but more substantial structures can survive.

Flooding following earthquakes may also result from *seiches* (oscillation of the water in enclosed bodies of water such as reservoirs) or from the failure of reservoirs or embankments. The probability of such flooding hazards is not easy to determine. They need to be acknowledged in selecting a site which is vulnerable, but the risk of damage or life-loss is probably not great enough for the site to be avoided altogether, except for very sensitive facilities.

---

[2] Slope failure was a major cause of building damage in the 1980 southern Italy earthquake and the 1986 El Salvador earthquake. In the small hill town of Calitri, the 1980 southern Italy earthquake reactivated an old landslide, causing a relative ground displacement of about a metre over a period of several months, and resulting in severe damage to dozens of houses located on the slide.

[3] The Alaska earthquake in 1964 caused more than $300 million of property damage (EERI 1986, p. 28).

[4] For example, California – see Lagorio (1991), p. 81.

[5] Damage from the tsunami following the 1964 Alaska earthquake occurred as far south as California.

7.2.5 Ground Shaking Amplification

Choice of siting should also take into consideration the probable effect of the siting on the extent of ground shaking which will be experienced in an earthquake. It has frequently been observed that earthquake damage is greater in settlements sited on soft soils than in those sited on hard soil or on rock sites. This is mainly due to amplification of the ground motions in transmission from bedrock to surface through the soil layer, but additional factors which may be involved include the destructive effect on foundations of subsidence which may have occurred on soft ground prior to the earthquake and the effect of ground deformations during the earthquake. Generally rock sites are to be preferred, and where siting on soft soil is unavoidable, provision should be made in the design of the building and the foundations for the more severe movements which will be experienced. Most building codes include provision for the effects of subsoil conditions. A full geotechnical investigation of the site is needed to consider the likely consequences of the subsoil conditions for the design of buildings.

Settlements located on deep deposits of soft soil types or compressible deposits are a special case. Such deposits can have a strongly defined natural frequency of vibration, amplifying that part of the bedrock motion which is of similar frequency, and filtering out the rest. Buildings will be affected selectively according to their own natural frequency of vibration (see Section 7.2). Such amplification will be particularly strong for distant earthquakes for which filtering of the high-frequency component of the motion has already occurred. Low-frequency components of ground motion have caused damage to medium- to high-rise buildings on a number of city sites located on deep soft soil deposits.[6] In settlements founded on such deep alluvial deposits it may be necessary to restrict the height or mode of construction of buildings so that their natural frequency of vibration is not of the same order as that of the underlying soil deposits. Avoiding such sites altogether is rarely an option, since the pattern of urban development may have already been established.

Ground motion amplification can also occur as a result of topographical effects; in particular, buildings sited on ridges may be vulnerable.[7] However, the extent of this effect and the factors influencing it are not yet sufficiently well understood for any clear rules to be formulated. Again, it may well not be possible, for economic reasons, to avoid building on ridges.

---

[6] The selective damage to buildings between 6 and 15 storeys high in Mexico City in the earthquakes of 1957 and 1985 is explained by this phenomenon. Other cities which have experienced the same phenomenon are Caracas, Venezuela, in 1967, Bucharest, Romania, in 1977, Bursa, Turkey, in 1970, and Istanbul, Turkey, in 1999.

[7] These topographical effects are described in the EEFIT report on the Chile earthquake (EEFIT 1988), and were confirmed by later microtremor studies described by Celibi (1990).

7.3 Estimating Ground Motion Hazard

For most earthquake protection measures, the critical factor is the probability or likelihood of a damaging earthquake occurring. The higher the probability of an earthquake, the more important it is to protect against it, the more cost-effective protection measures become and the more public support can be generated for the measures. From what we know about earthquakes and data from the past century or more of earthquake records, it is now possible to make a reasonably accurate estimate of the rate of earthquake activity in any area of the world. For government authorities or agencies responsible for a region, the concern is how often earthquakes are likely to occur within that region and how severe they are likely to be. *Regional seismicity* is defined by the occurrence rate of earthquakes of given *magnitude* within a geographical area. *Site-specific hazard* is the probability of the site experiencing a certain level of ground motion specified in a way that enables design loadings for buildings to be calculated. Both are elements of a seismic hazard assessment.

7.3.1 The Framework for Seismic Hazard Mapping

Today's seismic hazard maps, developed through many years of collaborative work among the world's leading seismology groups, are powerful tools to support earthquake protection. The general procedure on which they are based involves the following steps:

- Compilation of a regional catalogue of earthquakes and their effects
- Definition of a set of earthquake source zones
- Determination of the expected earthquake recurrence rate for each source zone
- Definition of attenuation relationships for the required parameters of ground motion
- Synthesis at each location of the effects of all earthquakes.

Within this general framework, different methods have been adopted for each of these steps depending on the extent of historical and scientific information available and the analytical resources available. Some of the issues affecting the approach to each stage are discussed in the following sections.

7.3.2 Compilation of Earthquake Catalogues

For any seismic hazard assessment, a knowledge of past earthquake occurrence in the region concerned is an essential first step, and a good catalogue is needed of all significant events of which there are records. For many areas good catalogues already exist, but the data on which they are based needs to be understood. Catalogues of earthquakes are often available from national seismological observatories, or can be obtained from international seismological institutions for the region of interest.[8] Earthquake catalogues are compiled from two distinct sets of data: recordings from seismometers and historical data.

Instrumental Catalogues

Instrumental catalogues of major earthquakes during the twentieth century are relatively complete. With progressively better instrumental detection of events from seismic stations around the world, few significant earthquakes since 1900 are likely to have gone undetected. The accuracy of their determination, the assessment of magnitude and location of epicentre, from instrumental records, may be extremely variable, however, and it is important to understand the uncertainties in instrumental catalogues when compiling a hazard assessment of a site or a region.

The accuracy of instrumental determination has improved over time, notably since 1963 following the introduction of the World Wide Standard Seismograph Network. Prior to 1920, some instrumental locations of epicentres, calculated from intersecting the distances estimated by a number of different stations, are known to be in error by as much as hundreds of kilometres.[9] The detection of small-magnitude events and the accuracy of instrumental determination depend largely on the proximity and positions of seismic stations around the earthquake sources. On plots of earthquakes recorded in successive decades, it can generally be seen that greater numbers of small-magnitude events are detected as instrumental capabilities improve over time. Locations of important events in the instrumental catalogue may be assisted by macroseismic information from post-earthquake investigations.

Historical Earthquake Data

The period for which instrumental data is available is extremely short compared with the geological timescale over which seismic activity occurs. The recurrence interval of larger earthquakes predicted from such studies is generally much longer than the return period for which data exists. A longer term database is needed to assess the accuracy of the estimates of long return-period events, to estimate maximum magnitude events with greater confidence, to study changes in seismicity over time and to compile more complete zoning maps.

Catalogues of earthquakes occurring before the twentieth century are difficult to compile with certainty. There are many inaccuracies in modern catalogues of historical events, which often present second- or third-hand information uncritically.

[8] A number of seismological institutions provide regional catalogues of earthquakes on request, e.g. the International Seismological Centre, Newbury, UK, United States Geological Survey, Reston, VA USA, British Geological Survey, Edinburgh, UK.

[9] Ambraseys (1978).

The review of primary, contemporary sources for information is complex and requires the skills of an historian rather than a seismologist, but the information gleaned can provide much more accurate assessments of present-day seismic hazard and can give extensive insights into the historical perspective of seismicity.[10] Determining the magnitude of a historical earthquake depends on drawing an intensity map. Intensity maps of earthquake damage are of particular interest because they indicate the geographical extent and spatial distribution of building damage from individual events. These maps are complex to compile because each map represents a compilation and synthesis of much data about a single event from a variety of sources. Because of the approximate nature of intensity definitions, intensity assignment is uncertain, but it does give an indication of extent and general degree of damage. The highest intensity level observed, the epicentral intensity (I_0), represents the severity of the earthquake and maps of I_0 can be used as an indication of the location of the most severe earthquakes which have occurred historically. Both the epicentral intensity I_0 and the felt area can be used to assess the magnitude of earthquakes in the historical record.

7.3.3 Seismic Source Zones

Any earthquake catalogue with a sufficiently long period of observation is likely to show a distribution of seismic events by time, space and size. A range of different magnitudes is likely to have occurred unevenly across a region and at various times during the duration of the catalogue. An example of a plot of the seismic activity of a region is given in Figure 7.2. There are a number of ways of analysing the catalogue to derive patterns of seismicity across the region: in terms of magnitude recurrence relationship, as a time sequence and in terms of a rate of strain energy released. A first step is usually to divide the region into source zones which have an approximately uniform seismicity.

Knowledge about the tectonic processes causing the seismic activity is used, together with geological information, geomorphological understanding, topography and other data, to define areas within which an approximately uniform level of earthquake activity can be expected. In rare cases, singular, linear faults can be identified along which large earthquakes can be expected, but in most cases earthquakes occur within broader 'fault systems': areas of multiple faults, some visible, others not, where earthquake occurrence is clustered. These areas can be defined as seismic source zones, within which seismicity can be assumed to be roughly uniform. One possible subdivision of the area in Figure 7.2 into separate seismic source zones is given in Figure 7.4 below. The seismicity of each source zone in terms of its magnitude recurrence relationship can then be obtained, counting all the earthquakes within its boundary.

---

[10] Ambraseys (1978) discusses the unreliability of modern sources in cataloguing pre-1900 events. An example of a catalogue derived from historical study is the study of Persian earthquakes by Ambraseys and Melville (1982).

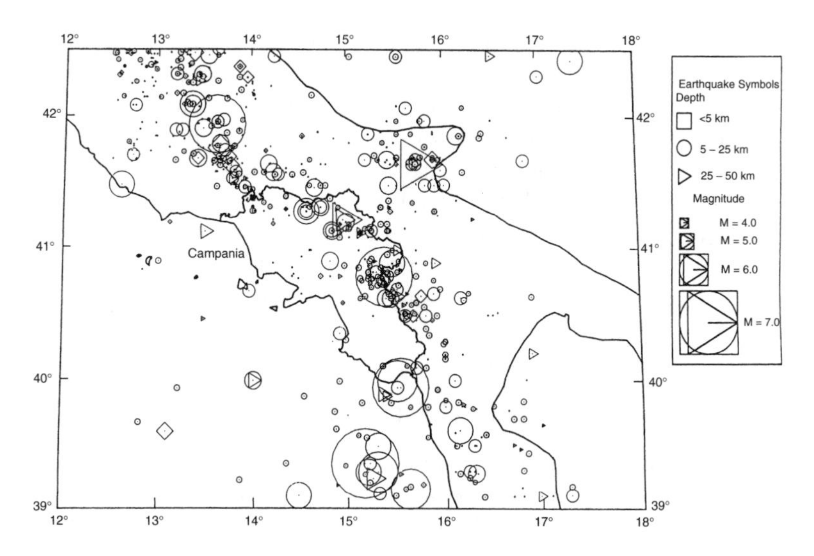

Figure 7.2 Map of earthquake epicentres in Southern Italy, 1900–1980

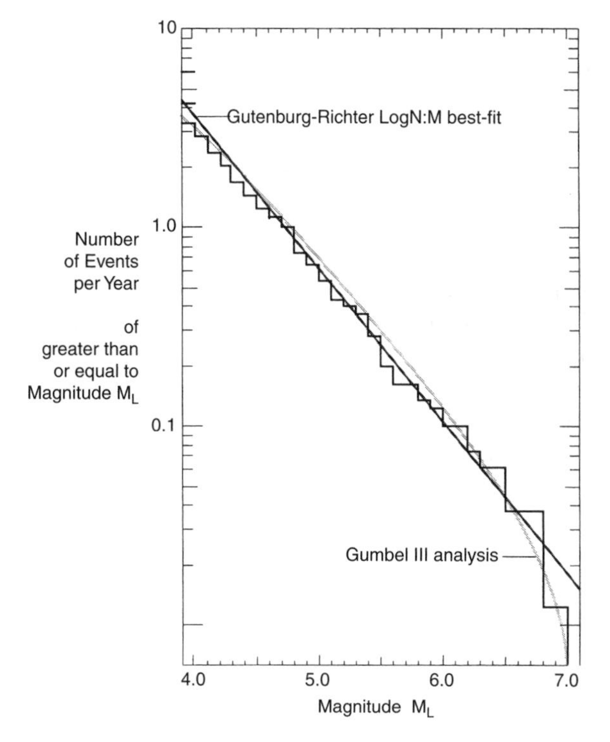

Figure 7.3 Seismicity of southern Italy, 1900–1980, expressed as cumulative frequency (log N vs M) of earthquake occurrence. The area covered is shown in Figure 7.2

The definition of source zones involves a review of all available information on the earthquake history of the hazard influence area, from both instrumental and historical and macroseismic records. The location of known active faults should be taken into account. The definition of the location of the boundaries of source zones can have a significant effect on the estimated hazard at sites near those boundaries. It is also somewhat artificial to delineate exact boundaries between zones, so it is advisable to consider a range of possible boundaries for source zones in a full analysis. More recent studies have avoided the problems of source-zone boundary definition by using spatially smoothed measures of source-zone seismicity.[11]

7.3.4 Magnitude–Recurrence Relationships

Within any seismic source zone, there will, over a period of time, be a large number of detectable earthquakes with a range of different sizes or magnitudes. Smaller earthquakes will always be more frequent than larger earthquakes, and the recurrence frequency of smaller earthquakes tends to be related to the recurrence frequency of larger earthquakes. Thus, a graph showing the number of earthquakes of each size against the size of the earthquake will tend to have a fixed shape. Gutenberg and Richter[12] postulated, on the basis of statistical recurrence laws, that the relationship between $\log N$ and M should be linear, where M is the magnitude and N is the number of earthquakes in the zone over a given time period with magnitude greater than M.[13] Figure 7.3 shows the earthquake data for Figure 7.2 plotted this way, indicating that, over the range of earthquakes for which there are an adequate number of recordings (M_s between 4.0 and 7.0), the linear relationship is reasonably good. The comparative seismicity of each separate seismic source zone (as defined in Figure 7.4) is presented in Figure 7.5, divided by the area of each zone to derive seismicity per unit area. Information from geological studies, such as the maximum magnitude event expected in each zone, can be added to define probabilistic extrapolations to higher magnitudes.

A relationship based on such a short period of observation is of course very approximate. A particular problem is that the larger earthquakes, which are the most important from the point of view of earthquake protection studies, are the most infrequent, and assumed recurrence periods based on extrapolation of the linear relationship derived from observations of lower magnitude earthquakes are likely to be unreliable. Extrapolation of the linear relationship would

---

[11] Frankel *et al.* (2000).

[12] Gutenberg and Richter (1954).

[13] The relationship can therefore be described by an equation of the form $\log N = A - bM$, where A and b are constants characteristic of a particular seismic region.

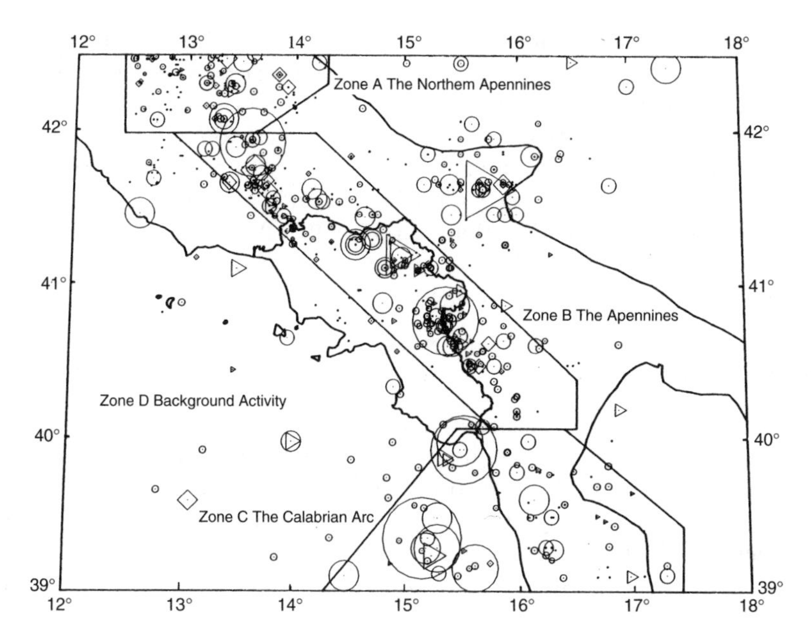

Figure 7.4 Division of earthquake activity in southern Italy into seismic source zones

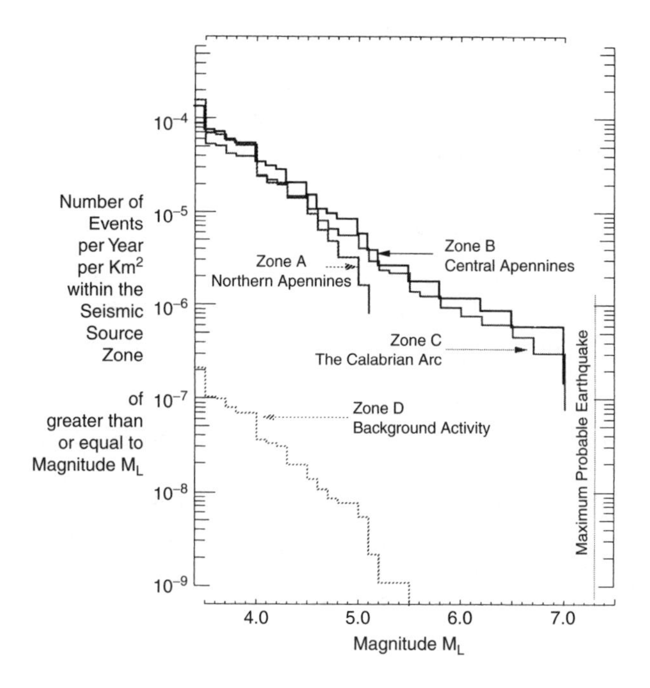

Figure 7.5 Seismicity of the separate source zones shown in Figure 7.4

give infinitely large magnitude values, which is unrealistic. For each earthquake region, there is in effect a limit on the maximum size of earthquakes which could occur, deriving from the geological nature of the faulting. To deal with this, various modifications of the Gutenberg–Richter formula have been proposed, such as the use of a curved or truncated linear relationship. Figure 7.3 compares the Gutenberg–Richter formula with an alternative formula[14] for the data for earthquakes in southern Italy, showing that the curved relationship with a definite upper bound is much more useful for predicting the recurrence of earthquakes of magnitude greater than 7.0.

Time Sequence Analysis

These analyses generate 'expected' return periods for events, i.e. the average rates of occurrence of earthquake activity. It is clear from most earthquake catalogues that earthquake activity does not occur uniformly in time – it is sporadic and unevenly spread over the years. An administration responsible for a region or an organisation working across an area may well be concerned to estimate the numbers of earthquakes of different sizes likely to occur within that region in any given period of time. An example of a typical time sequence of earthquakes of different magnitudes occurring across a region is given in Figure 7.6.

The number of earthquakes occurring within a given time period, e.g. 10 years, can be derived from the data for successive time intervals (1900–1910, 1901–1911, etc.) as presented in Table 7.1. Analyses like this give the range of observed behaviour in the past, indicate confidence limits for any prediction of average activity rates and identify any obvious patterns in the seismicity, such as cycles of quiescence and activity.

Rate of Strain Energy Released

A further alternative way of presenting the recurrence of earthquakes in a region, which derives more directly from an understanding of plate tectonics, is as a plot of cumulative strain energy released with time. If the plate boundaries slip at a constant rate, it would be expected that energy would be stored in the rocks at a constant rate, and that over a long period of time, energy would be released at a rate which, over a period of time, would be roughly constant. This type of analysis can be a means of indicating at any time whether there is a significant amount of stored energy, and also the size of earthquake which would occur if it was all released. For areas where the history of energy released on a fault is known, this type of information can be used in the estimation of present-day hazard, and in the compilation of seismic hazard maps.[15]

---

[14] Based on Gumbel's extreme value analysis, as proposed by Burton *et al.* (1984).

[15] Frankel *et al.* (2000).

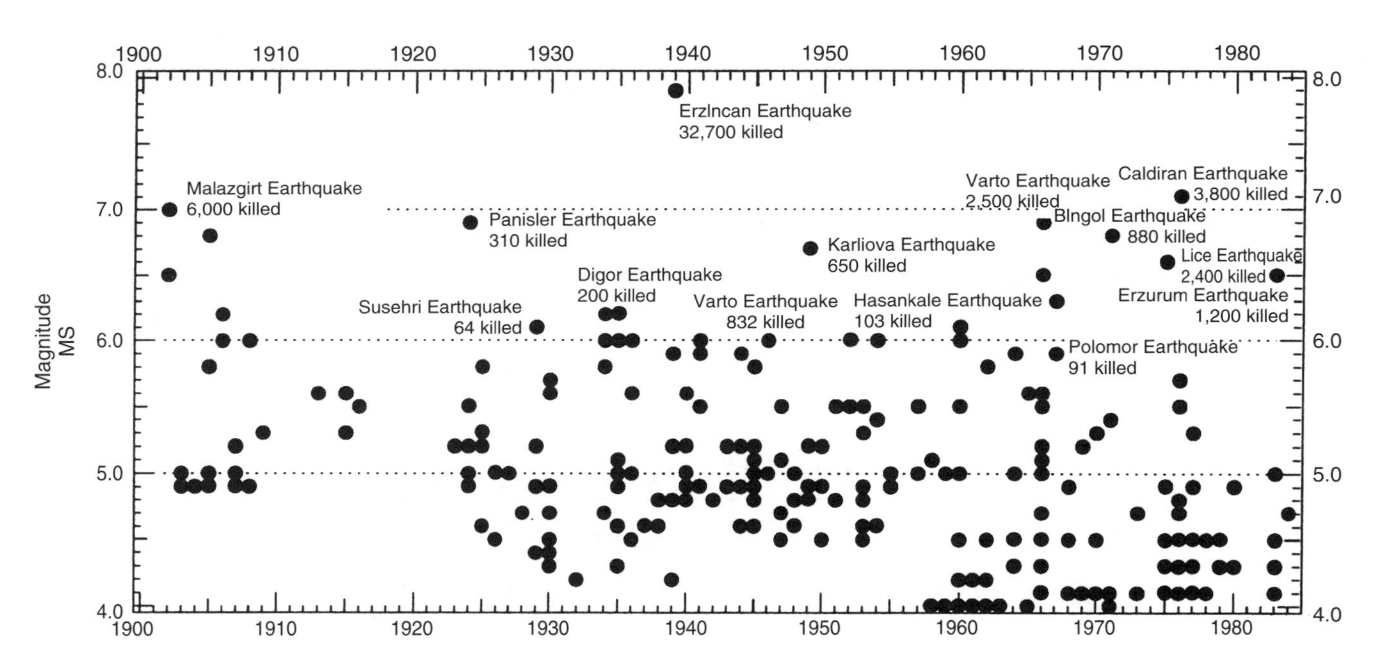

Figure 7.6 Earthquake recurrence over time. The recurrence of damaging earthquakes in eastern Turkey, 1900–1984

Table 7.1 Time sequence analysis of earthquake occurrence in a 10-year period (earthquakes magnitude $M_s \geqslant 6.0$, recorded in eastern Turkey, 37°–41.5°N, 38°–45°E, 1900–1984).

N Number of earthquakes $M \geqslant 6.0$	No. of 10-year periods in which *N* earthquakes occurred	Probability of *N* earthquakes in a 10-year period (%)	Probability of *N* or more earthquakes in a 10-year period (%)
0	6	8.33	
1	12	16.67	91.67
2	6	8.33	75.00
3	16	22.22	66.67
4	13	18.06	44.45
5	9	12.50	26.39
6	5	6.94	13.89
7	2	2.78	6.95
8	3	4.17	4.17
9	0	–	–

Average number of $M \geqslant 6.0$ earthquakes in a 10-year period = 3.29.
Return period of an $M \geqslant 6.0$ earthquake = 3.04 years.
The number of earthquakes likely in any particular period for a known average activity rate can be estimated mathematically, assuming a Poisson distribution, see Section 9.9.

7.3.5 Attenuation Relationships

Ground motion attenuation relationships give estimates of various parameters of ground motion, as a function of the magnitude and depth of the event and distance from the site to the epicentre (or fault rupture), with a known uncertainty. They may also take account of variations in ground conditions as explained below. They are based on probabilistic mathematical models of the earthquake source mechanism, and the earthquake wave transmission process, calibrated by the actual available data from strong motion recordings.

Where substantial data is available, e.g. in western North America, there is now a rather good agreement between the various published relationships, examples of which are shown in Figure 7.7.[16] For other areas of lower seismicity (or data availability) there is greater uncertainty, but it is clear that the attenuation relationships for parameters of spectral response are very different in different areas, and that a distinction needs to be made between regions of higher and lower seismicity. Ground motion tends to attenuate faster in areas of high seismicity than in areas of low seismicity. Attenuation relationships suitable for areas subject to intra-plate earthquakes have also been developed.[17]

[16] Atkinson and Boore (1990), Boore *et al.* (1997).

[17] Dahle *et al.* (1990), Toro *et al.* (1997).

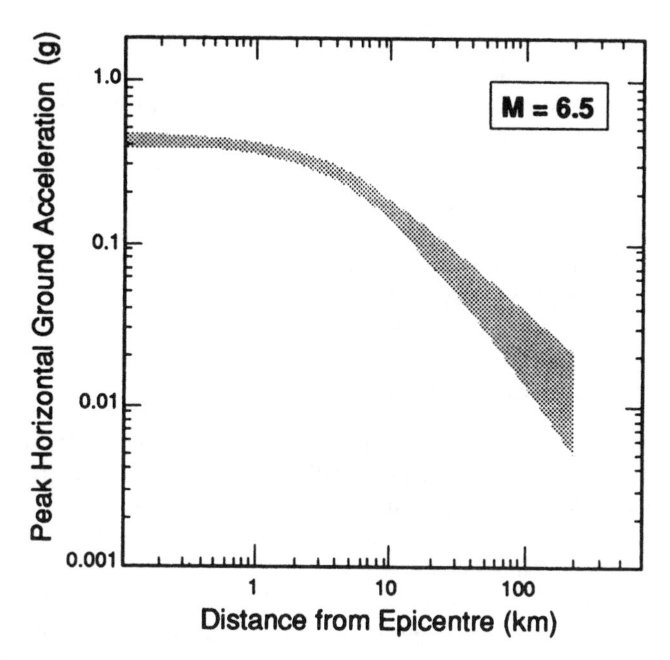

Figure 7.7 The range of published average attenuation relationships for acceleration with distance from an earthquake of magnitude 6.5 in western North America (after Atkinson and Boore 1990)

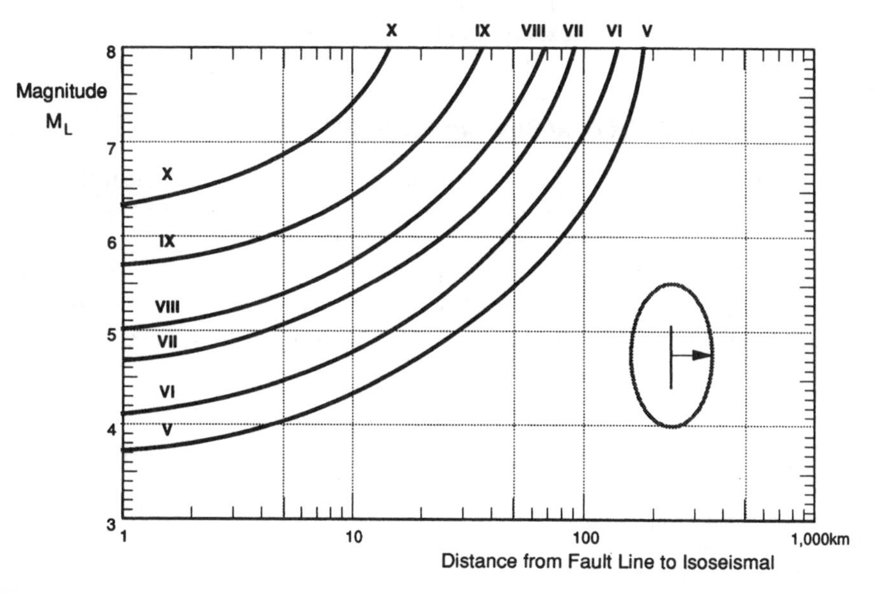

Figure 7.8 Average EMS intensity attenuation relationships from analysis of isoseismals of 53 earthquakes, southern Italy, 1900 to present (after Coburn *et al.* 1988)

In areas of lower seismicity where ground motion data is limited, it may be possible to derive attenuation relationships for macroseismic intensity based on historical records. Ambraseys[18] has derived intensity–attenuation relationships for the low-seismicity north west European area, and also derived appropriate magnitude–intensity relationships which can predict magnitude from the use of one or more isoseismal radii. Grandori[19] has given intensity–attenuation relationships from Italian earthquakes in terms of the epicentral intensity (I_0). Figure 7.8 shows intensity–attenuation relationships for southern Italy derived from the analysis of isoseismals of past earthquakes in the region.[20]

7.3.6 Computational Procedure

Using the recurrence relationships and other data relevant to earthquake occurrence, and the appropriate attenuation relationship for the relevant ground motion parameter, the hazard at any site can be determined. This now involves aggregating the effects at that site of earthquakes originating in each relevant source zone at each of a series of increments of distance from the site, up to the maximum distance at which the largest possible earthquake can have any significant effect. Appropriate and widely used algorithms for this are available[21] and computer programmes incorporating these algorithms have been published.[22] Since there is often uncertainty about which of several alternative earthquake occurrence models and attenuation relationships is appropriate, hazard maps are often synthesised by blending the data from different sources, using weightings for each source which are based on expert scientific judgement.

7.3.7 The USGS National Seismic Hazard Maps

The US national seismic hazard maps produced by the US Geological Survey[23] are amongst the most advanced maps produced to date. Separate maps show peak horizontal ground acceleration and spectral response at 0.2 and 1.0 second periods with 10%, 5% and 2% probabilities of exceedance in 50 years, corresponding approximately to recurrence times of 500, 1000 and 2500 years. The reference site conditions for the maps is firm rock with an average shear wave velocity of 760 m/s in the top 30 m.

---

[18] Ambraseys (1985).
[19] Grandori *et al.* (1988).
[20] Derived as a part of the analysis of site hazard in Campania, Italy (Coburn *et al.* 1988).
[21] The algorithm described by Cornell (1968) is commonly used.
[22] For example, that of McGuire (1978).
[23] Frankel *et al.* (2000).

The maps are based on the combination of three components of the seismic hazard:

(1) spatially smoothed historical seismicity, assuming that future damaging earthquakes will occur near areas that have experienced such earthquakes in the past;
(2) large background source zones based on geological criteria with maximum magnitudes of 6.5 to 7.5 for areas with little historical seismicity; and
(3) the hazard from 450 specific fault sources on which geological slip rates (observed or estimated from palaeoseismic data) were used to determine earthquake recurrence rates.

Hazard curves were calculated at a site spacing of 0.1° for the western United States and 0.2° for the central and eastern United States, a total of 150 000 sites. Several separate attenuation relationships were used and the results combined with equal weightings. Disaggregation plots for major cities (New York, Chicago, Los Angeles and Seattle) have also been produced to show what proportion of the total hazard at that location derives from different bands of magnitude and distance.[24] The maps of spectral acceleration at periods of 0.2 seconds and 1.0 second with 10% exceedance probability in 50 years are the basis of the maps of maximum credible earthquake (MCE)[25] used in the new 2000 International Building Code (Figure 7.9).[26]

7.3.8 The Global Seismic Hazard Assessment Project (GSHAP)

GSHAP was one of the major international achievements of the International Decade for Natural Disaster Reduction (1990–2000). It aimed to produce regionally coordinated and homogeneous seismic hazard evaluations and regionally harmonised seismic hazard maps. One key output was the world seismic hazard map of peak horizontal ground acceleration shown in Plate I.[27] This was produced by the integration of separate regional maps produced by 10 separate groups, each a collaboration between the major seismological groups active in the areas.

To some extent methods adopted and outputs produced varied from region to region. In Region 3 for example, which covers the 29 countries of central north and north west Europe, the work had as an additional goal the production of consistent maps to support the seismic zonation needed for application of

[24] Frankel *et al.* (2000).
[25] Leyendecker *et al.* (2000).
[26] ICBO (2000).
[27] GSHAP (1999).

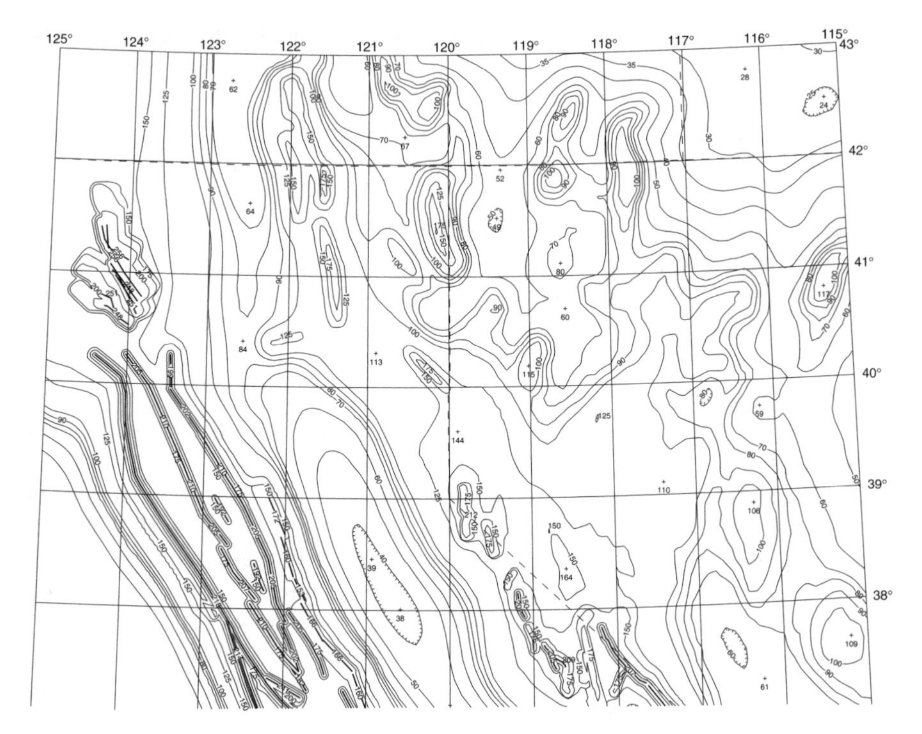

Figure 7.9 Maximum considered earthquake ground motion for region 1 of 0.2 sec spectral response acceleration (5 percent of critical damping), site class B

the European Building Code, EC8 (see Chapter 8). Key tasks involved in the production of the European seismic hazard map[28] were:

- Integration of separate earthquake catalogues covering over 20 different countries or regions and in some cases extending back more than 1000 years, and conversion of many different forms of magnitude measure into a single homogeneous, moment magnitude (M_w) measure.
- Definition of a single set of seismic source zones – in all 196 separate source zones were distinguished – and estimating characteristic focal depths, upper bound magnitudes and magnitude–recurrence relationships for each zone.
- Defining appropriate ground motion attenuation relationships to adopt and weighting coefficients to use where several separate attenuation relationships were relevant.
- Performing hazard calculations for a grid size of 0.1° latitude by 0.1° longitude (except in northern Europe), a total of 59 217 separate points.

[28] Grünthal *et al.* (1999).

The resulting regional map of horizontal peak ground acceleration with an exceedance probability of 10% in 50 years is shown in Plate II. The information shown on this map can be used directly in design to define a spectral response curve, and will also inform the national maps produced in the National Application Documents which accompany EC8.[29]

7.3.9 Defining Earthquake Design Loads

For the designers or owners of individual buildings, or for urban planners or city authorities, the issue is how likely a specific site is to experience earthquake forces of a certain severity. Building design codes adopt one of two alternative procedures for specifying the geographical distribution of design loads:

(1) seismic zonation or
(2) contour mapping of expected ground motion.

Most national codes of practice use the seismic zonation concept. The country (or region) covered by the code is divided into a small number (usually no more than four or five) of separate source zones, within each of which the lateral loading requirement for earthquake-resistant design is constant, and is specified by a zone coefficient. The zone coefficient relates to the expected peak ground acceleration within a predefined return period, but this information does not need to be known by the designer. The Turkish seismic zonation map (Figure 7.10) is a typical example. In this code the zone coefficients are 0.1, 0.06, 0.04 and 0.2 for Zones 1, 2, 3 and 4 respectively, corresponding roughly to the peak ground acceleration (as a proportion of the gravitational acceleration g) with a 10% probability of exceedance in 50 years. These coefficients are converted into a response spectrum for design using further coefficients for local soil type and building importance.

The advantage of this method for specifying design loads is its simplicity for designers. The zones, although defined from knowledge of regional seismicity, are not given a formal definition in terms of expected ground motion. Their significance derives from the use of the zone coefficient in the formulae in the accompanying code, so they have a semi-legal character, like district boundaries.

However, the approach also has disadvantages. One disadvantage is that the seismic zonation is coarse, and is unable to take into account the effects of local features such as fault zones. Another is that only a single parameter is defined, whereas it is now accepted that at least two independently varying parameters are needed to take adequate account of the variations in regional seismicity.[30] These two disadvantages are overcome through the use of contour maps such as those

[29] CEN (1994), Lubkowski and Duian (2001).

[30] Leyendecker *et al.* (2000).

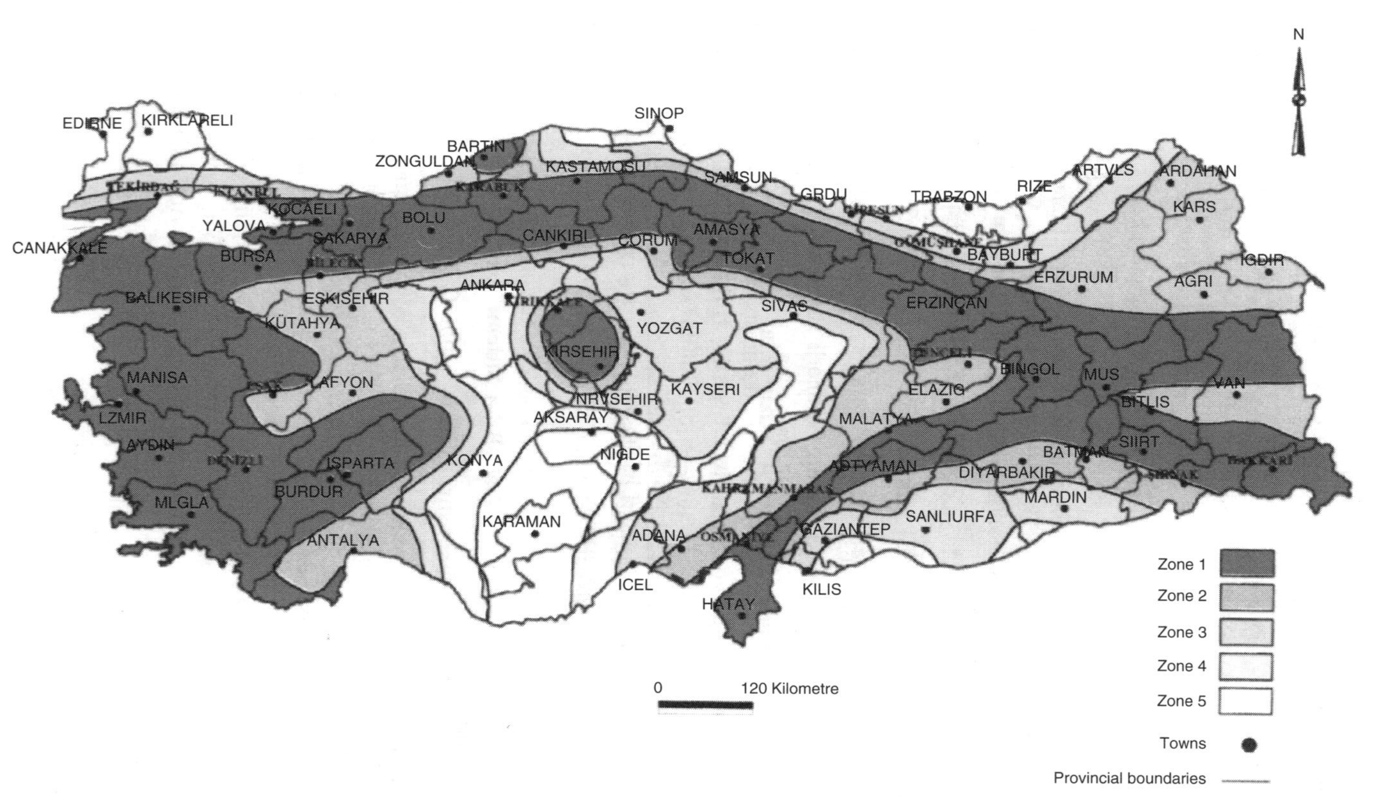

Figure 7.10 Seismic zoning map of Turkey from the 1996 earthquake code. Each zone is associated with a zone factor to be used in the design of structures. The darkest shaded area, in which 40% of the country's entire population lives, is the zone with the highest risk, with the highest zone factor (Reproduced by permission of Willis Consulting Ltd.)

accompanying the 2000 International Building Code.[31] The code specifies that the design loading should be that associated with the *maximum credible earthquake* (MCE) at the site. Contour maps of the entire United States indicate the values of two key design parameters to be used to construct the design ground motion response spectrum at that site: the spectral acceleration values at 0.2 s and 1.0 s periods. The value of these parameters is derived from the US Geological Survey's hazard maps which are contour maps of the 0.2 and 1.0 s spectral accelerations with a 10% probability of exceedance in 50 years, but with modifications for some parts of the United States to take account of the effects of known local faulting on design loads, and with variations for different classes of site defined by soil conditions.[32] Figure 7.9 shows, for example, the MCE ground motion map of a small part of Western United States for the 0.2 s horizontal spectral acceleration (% of g), for Site Class B. Site effects are discussed in Section 7.4. Maps such as these represent a considerable step forward in defining appropriate design load coefficients and are likely to become the standard approach for future codes in other countries.

7.4 Effect of Site Conditions on Seismic Hazard

Whichever approach to hazard estimation is used, the influence of site conditions needs to be taken into account. It has been shown that amplification of peak ground acceleration (PGA) by a factor of 5 or more is possible in a particularly unfavourable site,[33] while studies of macroseismic intensity[34] indicate that an intensity increment up to three steps on the EMS scale is possible owing to ground conditions.

The site conditions giving rise to ground motion amplification have already been discussed in Section 7.2. Variation in subsoil surface geology is the principal cause of variation, though site topography can be a significant factor as well. Ideally, the effect of these factors on the site hazard will have been determined by a detailed microzoning study as discussed in Section 7.5. Methods based on records of micro-tremors can be used to determine relative amplification factors over an area for comparison with a particular reference site.

The type of subsoil condition also affects the shape of the site response spectrum – on soft sites low ground motion frequencies are often amplified and high frequencies filtered out, for instance. Thus different amplification factors may need to be defined for different frequency ranges. The influence of soil

[31] ICBO (2000).

[32] Leyendecker *et al.* (2000).

[33] Such as Mexico City, see Singh *et al.* (1988).

[34] For example, in the USSR, see Medvedev (1965).

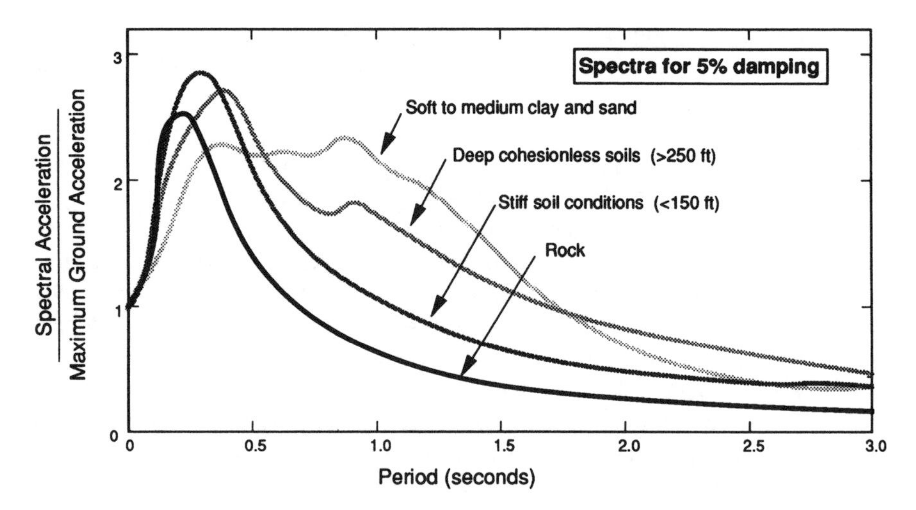

Figure 7.11 Influence of soil conditions on average acceleration spectra experienced at a site (after Seed *et al.* 1974)

conditions on average acceleration spectra is illustrated by Figure 7.11, which is based on the shape of the spectra for 104 records.[35]

The effects of ground conditions have been incorporated into some published attenuation relationships.[36] These attenuation relationships divide ground conditions into mainly three types, namely rock, shallow soil and deep soil. The influence of ground conditions on peak ground velocity and displacement appears to be stronger than on peak ground acceleration, and the influence is much greater for spectral values at low frequencies than at high frequencies. However, very little of the data on which such relationships depend is derived from ground motions strong enough to cause building damage.

7.5 Microzoning

Microzoning is a developing technique which promises in the future to bring very important benefits for earthquake protection. Its aim is to identify and map the variation in earthquake hazards within a limited area – typically a city or municipality – as a result of variation in ground conditions or other characteristics. Microzoning maps can be used in conjunction with larger-scale hazard mapping to inform urban land-use planning and decide on allocation of resources for strengthening existing buildings.

[35] Analysed by Seed *et al.* (1974).

[36] Such as those of Joyner and Boore (1981), Campbell (1981) and Toro and McGuire (1987).

The local site conditions which may have an influence on the suitability of the site for a building or settlement are discussed in Section 7.1. They include:

- soil conditions which can amplify ground motions, either generally or selectively in certain frequency ranges;
- susceptibility to liquefaction and other types of ground instability;
- topographical variations which can cause ground motion amplifications;
- sites which could experience large permanent ground deformations such as those associated with surface or shallow faulting;
- low-lying coastal sites vulnerable to tsunamis.

Other characteristics of an urban area which affect the expected consequences of an earthquake are related to the buildings and their occupants; such characteristics as highly vulnerable building types, fire potential and social deprivation can be usefully mapped.

7.5.1 Amplification of Ground Motion

An ideal microzoning would consist of the determination of relevant ground motion parameters with specified probability of exceedance for given return periods for all points in the study area, taking account of local effects, and presented as contour maps. In practice, simplified methods are needed to make microzoning feasible. One method is based on micro-tremor survey, another is based on calculating the response of the site based on subsoil survey data.[37]

Micro-tremor Methods

The first type of approach uses instrumental records of micro-tremors and other low-amplitude ground motions to determine and plot ground motion amplification factors across the zone. The method[38] makes a valuable contribution to an understanding of the parts of the zone that can expect the greatest ground motion in future earthquakes, and its validity has been demonstrated by the close correlation found between these areas of high amplification and areas of extreme damage in several earthquakes.[39] Figure 7.12 shows a microzoning map of Mexico City derived in this way. The method can also be applied making use of aftershock measurements following a major earthquake. However, the method has been criticised on the grounds that the characteristics of low-amplitude vibration may be in many cases quite different from those of damaging ground motion, in ways which it would be impossible to assess in advance of an earthquake.

---

[37] Vaciago (1989).

[38] One version is referred to as the Nakamura method (Mucciarelli *et al.* 1996).

[39] Singh *et al.* (1988).

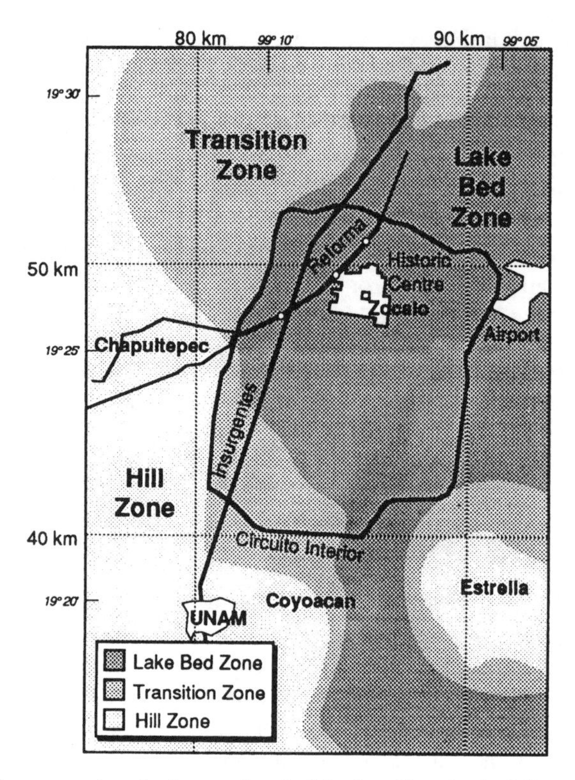

Figure 7.12 An example of microzoning in Mexico City for building design to resist earthquakes

Subsoil Modelling

A second approach uses a detailed geophysical model of the subsoil in the area for microzonation, and calculates expected ground motions at the surface for a series of reference earthquakes. Three-dimensional modelling of the subsoil defined as a series of layers with defined boundary conditions is combined with an input motion for the reference earthquake (or a combination of earthquake scenarios) to produce maps of surface effects. These can in turn be used to define soil amplification coefficients appropriate to different zones, suitable for use in codes of practice for design of buildings. The approach depends on substantial information on the subsoil in the top 30 metres or so being available. It has been extensively used in Italy.[40]

A related method, which has been used for some years in Russia,[41] is based on estimation of the effect of the subsoil characteristics on the site intensity. For

[40] Ansal and Marcellini (1999), Marcellini *et al.* (1999).

[41] Developed for use in the USSR, see Medvedev (1965).

each site a site *impedance* parameter is determined, which is the product of shear wave velocity and density for the subsoil. The intensity increment at the site is assumed to be proportional to the difference between the seismic impedance for the site under consideration and that for a reference site located on granite.

An alternative way to obtain an understanding of the relative ground amplification over an urban area is by a survey of the felt effects of an earthquake. This approach has been used in Japan to develop a microzonation map of the city of Sapporo in Hokkaido Province following a damaging earthquake.[42]

7.5.2 Scenario Zoning

A simplified non-quantitative approach to microzonation for geological hazard factors is based on seismic hazard 'scenarios'.[43] Four groups of seismic hazard scenario are defined, each of which is easily identified from field observations and existing geological maps:

(1) active or potential situations of slope instability;
(2) amplification of ground motion due to site morphology;
(3) amplification of ground motion due to ground behaviour;
(4) potential for significant permanent deformations.

Particular scenarios falling into each of these groups are illustrated in Figure 7.13. This method gives only qualitative information, so is only suitable for preliminary microzoning to identify areas of special risk. But maps derived from it can readily be incorporated into urban planning procedures.

7.5.3 Risk Zonation Mapping

Another type of microzonation map which can be of great value in planning emergency response is a mapping of the vulnerability of structures, lifelines and vulnerable inhabitants. In Japanese cities maps have been produced showing the proportions of old timber-framed buildings – which have a high vulnerability – in different localities.[44] For the Tokyo metropolitan area, where more than 400 000 houses were destroyed by fire following the Great Kanto earthquake of 1923, a microzonation of the city has been prepared to identify, for each 0.5 km grid square, the relative risk of building collapse, fire outbreak, casualty generation, and evacuation potential.[45] In San Francisco, where earthquake-related fire is also a serious risk, possible fire damage potential following a major earthquake has

[42] Kagami and Okada (1986).

[43] This was developed in wide-scale application following Italian earthquakes in the 1980s (Bressan *et al.* 1986).

[44] Kobayashi and Kagami (1972).

[45] Watabe *et al.* (1991).

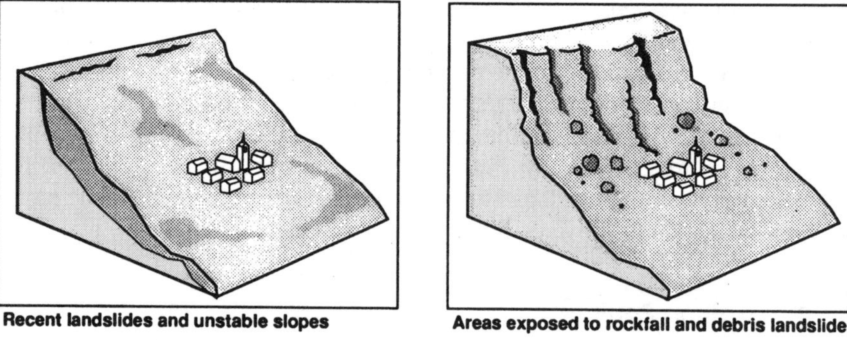

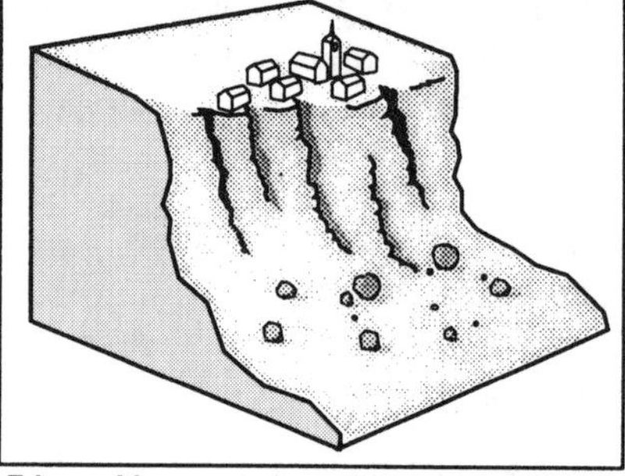

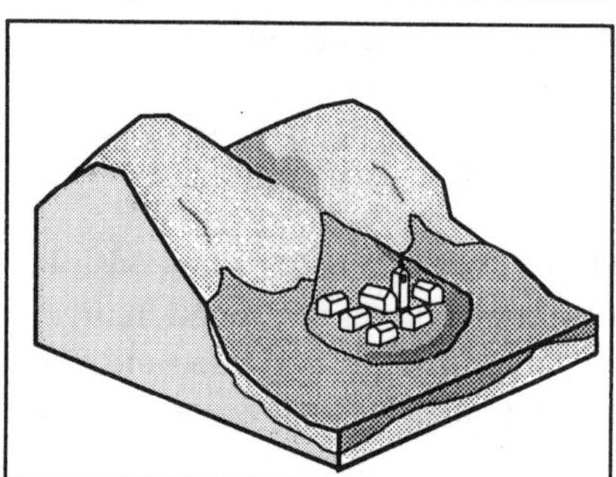

Figure 7.13 Scenarios for microzoning

been mapped.[46] Mapping of the potential indirect effects of future earthquakes can also be carried out by considering the distribution of population by income and the distribution of economic assets.[47]

7.5.4 Use of Microzonation Maps

Once microzoning maps are available, they may be used in conjunction with the already-existing urban and land-use planning instruments of a local authority to assist in earthquake protection in a variety of ways. For example:[48]

- by defining *where* to implement building controls;
- by specifying *how* to control building stock management, e.g. by defining how both the level and detailed construction rules for seismic upgrading of existing buildings will vary geographically depending on microzoning;

[46] Scawthorn *et al.* (1988).
[47] See Abolafia and Kafka (1978) for Los Angeles and Kagami and Okada (1986) for Hokkaido.
[48] Vaciago (1989).

- by influencing urban exposure, by planning *what* and *how much* may be built and where;
- by specifying *when* plans may be realised, by specifying timetables for compliance with upgrading, with priorities defined by microzoning.

The use of microzoning by urban planners is discussed in more detail in Chapter 6. Microzonation maps can also be valuable tools to support the planning of the emergency services.

Microzoning is beginning to be accepted practice in a number of countries, notably Italy, Japan and the United States. Italy is a country with a large building stock of high vulnerability and great historical importance. Here efforts towards microzonation are stimulated by the need for planning tools to assist in the allocation of resources for reconstruction following earthquakes as well as to assist in the allocation of public funds for seismic upgrading.

7.6 Mapping of Insurance Risks

Insurance companies were amongst the first to produce risk maps, and their fire risk maps dating from the early part of the century have continued to be a valuable source of information on the building stock insured. In recent years, the developing science of risk evaluation and increasing competition have driven insurance companies to attempt to define their risk geographically in more detail in a manner which enables them to set appropriate insurance premium rates, and also to estimate the total loss to which their portfolio of risk could be subjected. In California, the California Earthquake Authority offers insurance to all householders based on zipcode mapping of the earthquake risk. An insurance quotation can be calculated instantly over the internet, based on information supplied on zipcode location and age and form of construction of the house[49] with tariffs varying from $1 to $5 per $1000 sum insured.

Where insured risks can affect large areas, it is important for both local and international insurers and reinsurers to limit their overall exposures, to ensure that these do not exceed their reserves. This is achieved by subdividing each insured territory into separate accumulation zones, and ensuring that all insurance risks which each insurer undertakes in each zone are systematically reported. CRESTA (Catastrophe Reinsurance Evaluating and Standardising Target Accumulations) is an international organisation to standardise the reporting and recording of earthquakes and other natural hazards by accumulation zone. It also collates and makes available to insurers a range of information about the major perils, particularly earthquakes, country by country, including, where available, standard insurance tariffs charged for different types of buildings in different zones.

[49] California Earthquake Authority (www.earthquakeauthority.com).

CRESTA zonation maps and earthquake-related information at different levels are available for a total of 62 different earthquake-prone countries in the CRESTA manual.[50] The zonation for Japan, identifying 12 separate primary zones, with sub-zones, is shown in Figure 7.14. For some countries the CRESTA manual not only defines separate accumulation zones, but distinguishes also standard tariff zones. In Mexico, for instance, the 48 CRESTA zones are grouped into 12

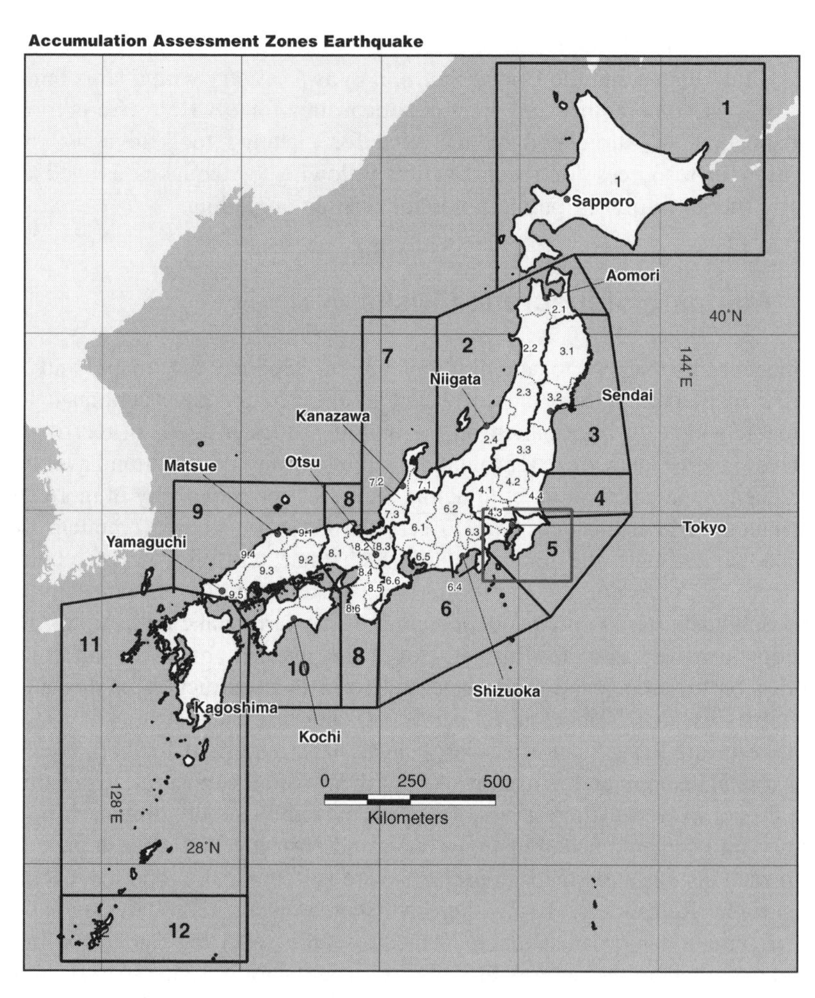

Figure 7.14 CRESTA zones: the insurance industry's standard accumulation assessment zones for earthquakes in Japan (Reproduced by permission of Infotech Enterprises Europe Ltd.)

[50] The CRESTA manual is available at www.cresta.org.

separate tariff zones, for each of which a basic tariff is reported. Tariffs differ for low-rise and high-rise buildings, and range from $0.28 to $7.27 per $1000 insured in the lowest and highest tariff zones respectively, and varying requirements for deductibles and co-insurance are also reported.

In Turkey, to support the introduction, in September 2000, of a compulsory national earthquake insurance scheme, a national mapping of earthquake risk was carried out, district by district, with the building stock (derived from a national building stock survey) divided into 14 separate vulnerability classes.[51] Plate III shows the variation of earthquake risk, aggregated at province level, defined by the average annualised loss to residential buildings as a ratio of complete rebuilding cost. The very wide variation across the country partly derives from the seismic hazard (see Figure 7.10), but also relates to the higher concentration of vulnerable building types (weak masonry and poorly constructed apartment blocks) in some parts of the country. High values of damage ratio are concentrated in the south western parts of Turkey and the provinces which were affected by the 1999 Kocaeli and Düzce earthquakes.

Further Reading

Cornell, C.A., 1968. 'Engineering seismic risk analysis', *Bulletin of the Seismological Society of America*, **58**, (5), 1583–1610.

CRESTA, 1998. CRESTA Manual, CRESTA, Munich Reinsurance, D-80791, Munich (www.cresta.org).

Dowrick, D.J., 1987. *Earthquake Resistant Design for Engineers and Architects* (2 edition), John Wiley & Sons, Chichester.

EERI, 1987. *Reducing Earthquake Hazards: Lessons Learned from Earthquakes*, Publ. No. 86-02, Earthquake Engineering Research Institute, California.

Gutenberg, B. and Richter, C.F., 1954. *Seismicity of the Earth and Associated Phenomena*, Princeton University Press, Princeton, NJ.

Tiedemann, H., 1992. *Earthquakes and Volcanic Eruptions: A Handbook on Risk Assessment*, Appendices 22 and 23: *Catalogue of Earthquakes and Volcanic Eruptions*, Swiss Reinsurance Company, Mythenquai 50/60, PO Box, CH-8022, Zurich, Switzerland.

[51] Bommer *et al.* (2002).

8 Improving Earthquake Resistance of Buildings

8.1 Strong and Weak Building Types

The earthquake resistance of buildings plays a central role in earthquake protection. The overwhelming majority of deaths and injuries in earthquakes occur because of the disintegration and collapse of buildings, and much of the economic loss and social disruption caused by earthquakes is also attributable to the failure of buildings and other human-made structures. The principal cause of failure of buildings in earthquakes is ground shaking, and improving the resistance of buildings to ground shaking is the subject of this chapter.

A small percentage of building failures is caused by secondary earthquake hazards, such as landslides, tsunamis and gross deformation of the ground. The protection of buildings from these hazards by appropriate siting is therefore an essential first step; measures to improve earthquake protection through siting have been discussed in Chapter 7.

Studies of earthquake damage show that some types of construction tend to be more vulnerable than others. The form of construction of the main vertical load-bearing elements is one of the main determinants of vulnerability: a building with unreinforced masonry walls can be expected to be much more vulnerable than a timber frame building, for instance. Table 8.1 shows a general classification of the construction types found in many seismic areas of the world. The vulnerability of these construction types, on average, can be expected to decrease from the top to the bottom of the list, i.e. earth and rubble stone are the most vulnerable and would be expected to suffer the most damage in

Table 8.1 Classifications of structural types by seismic vulnerability.

	Construction type classification	Main structural classification	Building type	Other vulnerability parameters
Non-engineered buildings	Masonry Type A Weak masonry	AR Rubble stone	AR1 Rubble stone masonry in mud or lime mortar	Roof type (heavy/lightweight), craftwork quality, age, condition, structural deterioration
		AE Earthen	AE1 Rammed earth construction, earth cob, pise or solid soil	
			AE2 Composite earth with timber or fibre, wattle and daub, earth and bamboo	
		AA Adobe (earth brick)	AA1 Adobe sun-dried earth brick in mud mortar	
	Masonry Type B Load-bearing unit block masonry	BB Unreinforced brick	BB1 Unreinforced fired brick masonry in cement mortar	Roof type (heavy/lightweight), number of storeys, plan shape and room sizes, mortar strength, masonry bond, construction quality
			BB2 Brick masonry with horizontal reinforcement	
		BC Concrete block	BC1 Concrete block	
		BD Dressed stone masonry	BD1 Stone masonry, squared and cut, dimensioned stone, monumental	
	Building Type C	CC reinforced concrete (RC) frame cast *in situ*	CC1 Reinforced concrete frame, *in situ*	Column and beam sizes, spans, structural form, regularity
	Frame structures	CT Timber frame	CT1 Timber frame with heavy infill masonry (e.g. Bagdadi)	Roof type (heavy/lightweight), number or storeys, age, jointing quality, connection to foundations
			CT2 Timber frame with timber cladding, lightweight structure	

Engineered buildings	Building Type D Engineered structures	DB Reinforced unit masonry	DB1 Reinforced brick masonry	Design for earthquake resistance
		DC *In situ* RC frame	DC1 *In situ* RC frame with non-structural cladding	Conformity to earthquake design code (data of construction, and code revision in force at that time)
			DC2 *In situ* RC frame with infill masonry	
			DC3 *In situ* RC frame with shear wall	
		DP Precast RC structure	DP1 Precast RC frame with infill masonry	Design quality, structural detailing
			DP2 Precast RC frame with concrete shear walls	Quality of construction
			DP3 Precast large-panel structure	
		DH Hybrid or composite steel/RC structures	DH1 Composite steel frame with *in situ* RC casing	
		DS Steel frame structures	DS1 Light steel frame (portal frame, steel truss, low rise)	
			DS2 Steel frame, moment resistant	
			DS3 Steel frame with infill masonry	
			DS4 Steel frame, braced	
			DS5 Steel frame with RC shear wall or core	

an earthquake, with steel frame structures suffering least. However, different areas of the world have their own building styles and construction methods and the form of construction is not the only significant factor. Indeed a potentially strong building type can be very weak if the configuration or design details are badly considered, and conversely, basically weak systems can be greatly strengthened by careful design and good construction. For any particular area, a classification of the local building types needs to compiled.[1] This chapter discusses the factors that influence a building's earthquake resistance. It starts by looking at the way buildings are shaken in earthquakes and how they respond. It then describes the principles of the design of buildings to resist earthquakes. It is well established that the configuration of a building – its size, plan layout, shape, height and mass distribution – has an important influence on its performance in an earthquake, whether or not it has been designed in accordance with an earthquake-resistant building code. The choice of materials is equally crucial. A sensible overall building form and appropriate choice of materials should therefore be the first consideration of designers. It has rightly been said that 'the structural engineer cannot make a building of poor structural form behave well in an earthquake'.[2] The following sections of the chapter therefore discuss the influence of materials choice and structural configuration on resistance to earthquakes. This leads to a consideration of design codes which have been drawn up nationally and internationally to assist in the design of earthquake-resistant buildings.

But the vast majority of the ordinary dwellings in the poorer earthquake-prone countries are built using local materials and building traditions which are not regulated by such codes. A major problem for earthquake protection is how to reduce the often extreme earthquake vulnerability of such dwellings. The occupants of houses of rubble stone masonry for example are many thousand times more likely to be killed in an earthquake, given the same severe ground shaking, than the occupants of a reinforced concrete structure designed and built to modern code standards. Most people in developing countries build without the help of professionals, without submitting plans for approval, and often without the assistance of trained builders; thus they are outside the reach of formal building codes. Different ways need to be found to reduce the vulnerability of such dwellings. The chapter therefore discusses the means to improve the earthquake resistance of these buildings.

It is easier to improve the earthquake resistance of new building than to upgrade existing ones, yet most of the world's existing buildings will continue to be inhabited for many years to come. Moreover, as buildings age their vulnerability tends to increase. Thus earthquake protection, especially in old cities, requires consideration of how to upgrade and strengthen existing buildings at modest

---

[1] See, for example, Applied Technology Council, ATC-13 (1985).

[2] Dowrick (1977), p. 80.

cost. Buildings of historical importance are a special problem. The last part of the chapter therefore discusses the strengthening (and post-earthquake repair) of existing buildings.

8.2 Building Response to Earthquakes

Large earthquakes cause violent ground motion shaking, with simultaneous components in horizontal and vertical directions, and accompanied by rocking, twisting and distortion of the ground. These ground movements set up forces in the structural elements of any building attached to the ground, giving rise to complex stresses and deformations. Buildings designed only for anticipated gravity and wind loading may well be unable to withstand these forces resulting in significant damage or collapse of the building.

8.2.1 Description of Ground Motion

To understand how buildings respond to earthquakes, we must look first at the nature of earthquake ground motion. The energy released in earthquakes travels through the ground in seismic waves, somewhat similar to sea waves, which can be clearly seen by an observer in a large event, and the way in which these waves are triggered and travel from the earthquake source has been discussed in Chapter 1. During an earthquake the ground will move rapidly in a complex way in both horizontal and vertical directions simultaneously. A modern *strong motion instrument* records three components of acceleration at a point: one in the vertical direction, and two in perpendicular horizontal directions.[3] Figure 8.1 shows a typical strong motion record. From such records the key features of ground motion can be identified.

The *peak ground acceleration* (PGA) is the maximum value of acceleration of the ground itself reached at any instant during the ground motion. This is a commonly used parameter of ground motion severity,[4] and its use has been discussed in Chapter 7. Peak horizontal and vertical accelerations are usually identified separately; the vertical acceleration is often (but by no means always) smaller than the horizontal acceleration and the two horizontal components are often of a similar amplitude. In some earthquakes there has been a marked directionality to the horizontal acceleration, which may be associated with the direction of 'throw' on the earthquake fault or direction of travel of the seismic waves.

[3] Acceleration is a more direct measure of the effect of an earthquake on buildings than displacement. From the acceleration history the velocity and displacement time histories of the point can be deduced if needed.

[4] Ground acceleration is measured either in cm/s^2 or more commonly as a ratio of the acceleration due to gravity ($9.8\,m/s^2$). Thus an acceleration of 10% of g is close to $1\,m/s^2$.

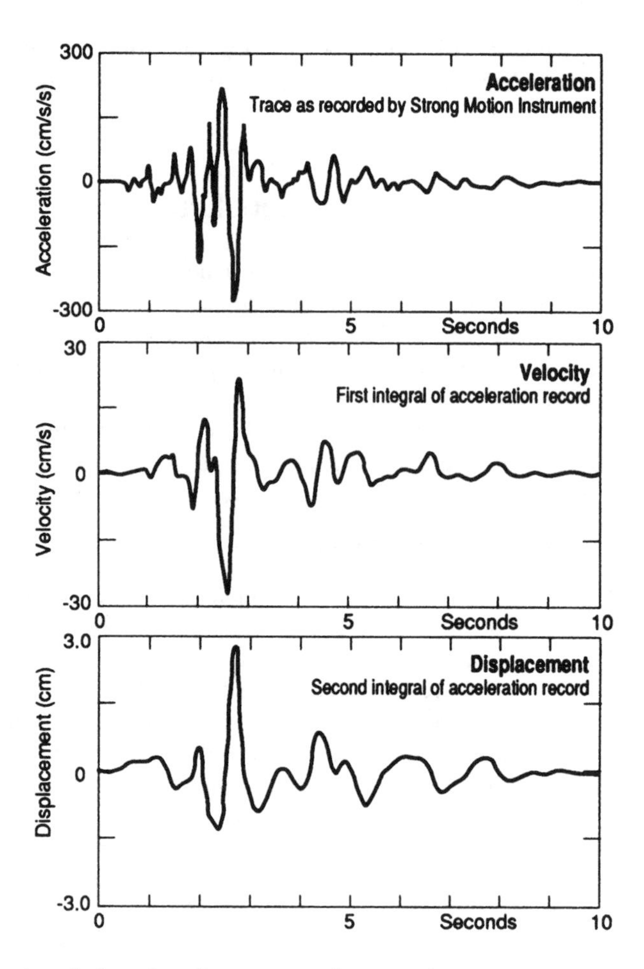

Figure 8.1 A typical earthquake strong motion record

The *duration* of the earthquake shaking is a measure of the length of time during which the acceleration peaks exceeded a certain amplitude. The duration of strong ground shaking can vary widely from a few seconds to a minute or more. The longer the strong shaking continues, the more destructive the earthquake will be.

The *frequency* of the motion is a measure of the number of times the ground moves backwards and forwards per second. Earthquake motion rarely exhibits regular cycles of motion, but an earthquake record can be interpreted as containing motions of many different frequencies and of different amplitudes superimposed on one another. Mathematical analysis is able to determine the contribution of motions of different frequencies, shown in the *power spectrum* of the earthquake, identifying which are the predominant frequencies present. However, from the point of view of building response, an even more important measure of frequency content is the *response spectrum*, which is discussed below.

8.2.2 Building Response to Ground Motion

When a rigid object is shaken, so-called *inertia forces* act on it which increase according to the acceleration of the object, and to its mass. If an absolutely *rigid* building is firmly tied to the ground, and shakes with the ground, then the inertia forces are transmitted from the ground into the building: the magnitude of the force is proportional to the mass of the building and varies with time in the same manner as the acceleration.

This simple model is unfortunately inadequate, however, because no real building is quite rigid. All buildings deform to some extent as they are shaken, and the deformation of the building substantially alters the force distribution. Small, massive buildings are relatively stiff, but as buildings become taller and lighter they tend to become more flexible. When a flexible building is shaken, the force acting on any part of it is still proportional to the mass and acceleration of that part, but the distribution of forces within the building depends on the way the building itself deforms. Depending on the mass and flexibility of the building, the accelerations within the building may be greater or less than the ground accelerations, and thus the forces may also be greater or less than if the building was a rigid body. The consequences of this for building design are of great significance.

The property of a building which principally determines its dynamic response to earthquake ground motion is its *natural frequency*. Because all buildings are flexible they will vibrate when jolted, and they will then sway backwards and forwards in a regular way. Taller buildings have lower natural frequencies (they sway more slowly) than lower buildings. A building 10 storeys high may take about a second to sway backwards and forwards in one cycle, i.e. its *natural period* is 1 second. A building of two storeys will take about one-fifth of a second: its natural period is 0.2 seconds. (A rough guide is that each storey adds about one-tenth of a second to its natural period.) Three- to five-storey structures are likely to have a natural period in the order of 0.3 to 0.5 seconds. High-rise frame buildings of 10 to 20 storeys have periods of between 1 and 2 seconds. And very high buildings can have period up to 4 seconds or more.

If the disturbance is a short one, the swaying will continue after the disturbance has finished, but it will gradually die away. The rate at which the swaying decays after the end of the disturbance is a measure of the *damping* in the building's structural system.

If the building is shaken by regular ground oscillations (like the effect of a rotating machine), its response will depend on the relationship between the frequency of these oscillations and the natural frequency of the building. For ground motion frequencies much less than that of the building, the building will simply move with the ground, and deform very little; as the frequency of the ground motion increases, so the deformation of the building will increase, and when the two frequencies are equal the building deformation will reach a peak which may be many times greater than that of the ground. The ground motion and the building are *in resonance*. For frequencies of ground motion still greater than the natural frequency of the building, the deformation of the building will be less the further away it is from the resonant frequency. The relationship is illustrated diagrammatically in Figure 8.2.

When a building is shaken by a real earthquake, which has a ground motion consisting of a mixture of frequencies all added together, its response will depend both on the natural frequency of the building and on the frequency content of the earthquake. A 10-storey building, with a natural frequency of 1.0 cycles per second, will be particularly affected by the component of the ground motion with this frequency, but much less by the components with higher and lower frequencies.

The effect of a particular earthquake ground motion on a range of buildings is shown by the *response spectrum*. The response spectrum for a particular ground motion shows what the maximum response would be to that ground motion for buildings[5] of different natural frequencies. Its shape depends on the

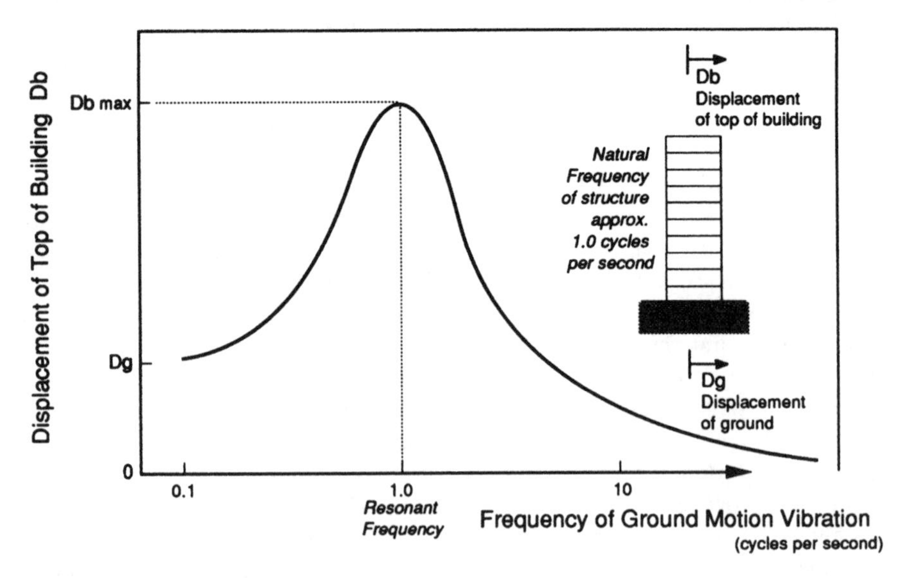

Figure 8.2 Diagrammatic representation of the response to a 10-storey building to the frequency of ground motion vibration

[5] Or, more strictly, to damped mass–spring systems, which are a useful mathematical idealisation of building structures.

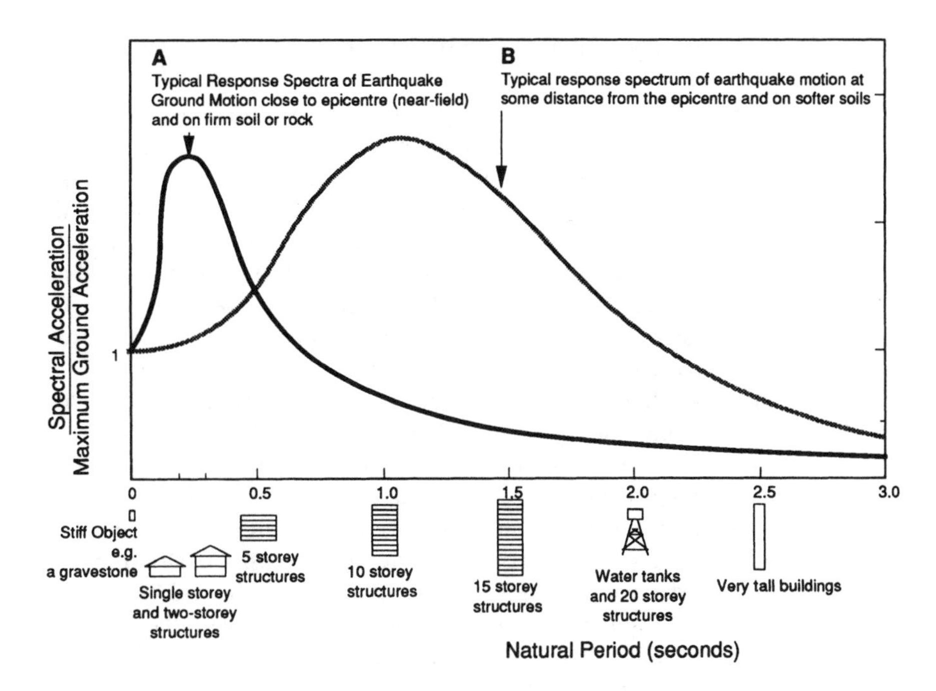

Figure 8.3 Typical response spectra and the building types they affect

frequency content of the earthquake and on the degree of damping of the building. Figure 8.3 shows some typical examples of response spectra. Example A is what the response spectrum might look like for a site close to the epicentre of an earthquake on firm soil or rock. It has a peak value of around 3 cycles per second. It would therefore be most damaging to low-rise buildings, but less so to taller structures, which would experience smaller forces. Example B shows a typical shape for a site at some distance from the epicentre and on a soft soil, with a peak value at about 1.0 cycles per second. This event would be especially damaging to the taller structures, but would be felt much less strongly by the low-rise structures.

An example of this second type of behaviour was the 1985 Mexico City earthquake which caused ground motion in the lake bed area of the city with a period strongly concentrated around 2 seconds; the earthquake caused particularly serious damage to recently constructed 10–20-storey apartment blocks, while leaving much of the older, weaker, low-rise masonry much less severely damaged.

The response spectrum is commonly used in building design codes to define the design earthquake which buildings should be able to resist without damage. Codes are discussed further in Section 8.6.

8.3 How Buildings Resist Earthquakes

Many small buildings are so stiff that they can be assumed to be rigid in a first estimate of earthquake forces. If a horizontal shaking occurs, the forces on each element of the building can be found by assuming that it is static, but has a horizontal force acting on it (through its centre of gravity) proportional to the ground acceleration and to the mass of the element, but in the opposite direction. This is what is referred to as the *inertia* force. The effect of vertical shaking is similar. The resistance of this stiff building is principally determined by the ability of the structure to transmit these large and rapidly varying inertia forces to the ground without failure.

Consider first a single-storey building, consisting of four walls (with window and door openings) and a flexible roof which sits on two of the walls, but does not tie them together, see diagram A in Figure 8.4. The effect of a primarily vertical ground shaking will be to increase or decrease the vertical forces, but as the structure is capable of carrying substantial vertical gravitational forces under normal conditions, it can usually accept extra vertical forces without difficulty.

The effect of a horizontal shaking parallel to two of the walls will be to set up horizontal inertia forces on each wall in proportion to their mass: the forces on the walls parallel to the direction of the shaking (the in-plane walls) will be along their length, while those on the perpendicular walls (the out-of-plane walls) will be at right angles to them. The force on the roof will also cause an additional horizontal force to be transmitted on to whichever wall supports it. The principal effect of *out-of-plane forces* is to cause the walls to bend (i.e. deform out of

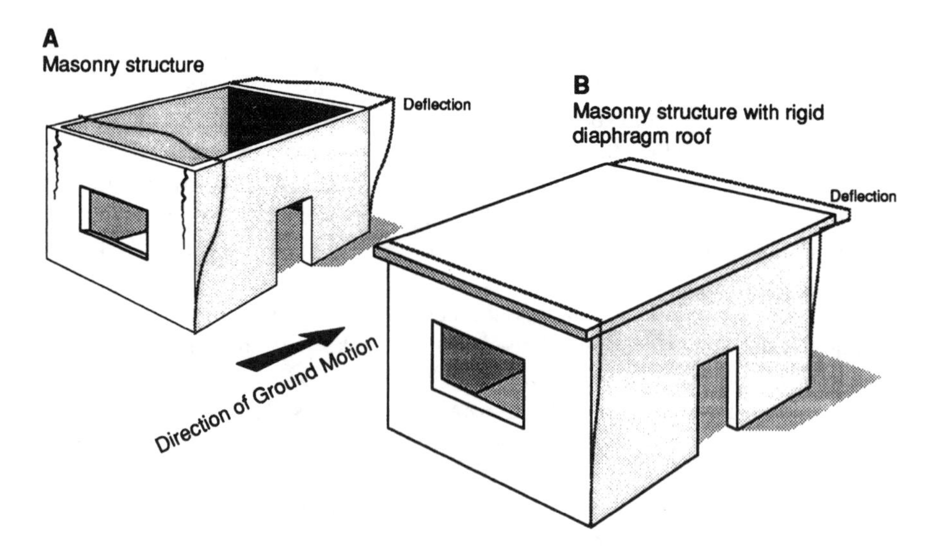

Figure 8.4 Response of single-storey masonry building to earthquake ground shaking

their plane), which can cause damage to brittle masonry structures even under low levels of loading. Wall elements tend to be stronger under *in-plane forces*: these cause in-plane shear forces which are easier to resist in a solid wall or can be provided for by bracing or other means.

In the same building the effect of a horizontal force in the direction perpendicular to that just described would be to exchange the responses of the walls, those previously out-of-plane becoming in-plane and vice versa. Thus under a real earthquake shaking, with horizontal shaking in all directions, all walls are subjected to both out-of-plane bending and in-plane shear simultaneously. This type of building tends to have little resistance to earthquake forces.

If instead the roof is constructed in such a way as to tie the tops of the walls together as a rigid *diaphragm*, the behaviour will be different, as in diagram B in Figure 8.4. The unresisted out-of-plane bending of diagram A will be prevented, as the out-of-plane wall will be connected to the roof diaphragm member, which is then able to transfer the forces involved to the tops of the stiffer in-plane walls, and then to the ground. In addition the continuity of the roof will also tie the corners together, inhibiting corner cracking. Under shaking in the other plane, the behaviour is the same in reverse.

Thus these elements – the stiff vertical *shear wall* in each direction, to carry the loads to the ground, and the stiff *horizontal diaphragm*, to transfer the earthquake forces at this level to the appropriate wall – form the basis of an effective earthquake-resistant structural system. The same system can be used as effectively in multi-storey construction, in which case the horizontal loads to be transmitted by the shear walls increase (as do the vertical gravitational loads) from top to bottom of the building, so that the ground floor walls are required to transmit to the ground the horizontal forces acting on the whole building.

However, the use of extensive shear walls can often create serious limitations on the planning of a building, and the equivalent shear strength can also, in some cases, be achieved by means of alternative vertical elements such as *braced frames* and *moment-resisting frames* (Figure 8.5).

In the *braced frame*, the bracing members transmit the horizontal forces in tension and compression; such frames can be very stiff but are often appropriate only on the external walls of a building. In the *moment-resisting frame*, the horizontal forces are transmitted by bending moments in the columns and in their framing beams. The moment-resisting frame can be designed (using steel or reinforced concrete) to be as strong as required, but frame structures will tend to be rather more flexible than braced or shear wall structures.

Similarly it is not always necessary (especially in a small building) for a fully rigid diaphragm to be provided at each level. Cross-bracing of a framed floor (steel or timber or trusses), along with the provision of a ringbeam in concrete or even timber, may in some cases be an adequate alternative.

Where a flexible, moment-resisting frame is to be used, care also needs to be taken with the additional bending moments in the columns which arise from

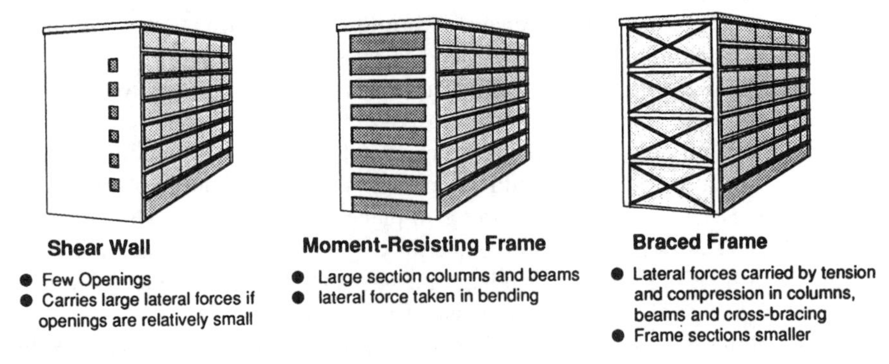

Figure 8.5 Alternative earthquake-resistant structural forms: shear wall structures, moment-resisting frames and braced frames

the relative displacement of their ends. This so-called *P-delta* effect can be the cause of rapid material breakdown and collapse if adequate provision has not been made for it.[6]

8.4 Structural Form and Earthquake Resistance

The simple elements of an earthquake-resisting structure described in the previous section can be provided in a great variety of ways. But simply providing these elements is unfortunately not sufficient to guarantee good performance in an earthquake. The static force analogy presented above fails to explain the complex behaviour of real structures subjected to the unpredictable, large and rapidly varying forces of real earthquakes. In addition, there are certain principles of overall structural design which need to be observed. Structures should be *symmetrical, continuous, small in plan, not elongated in plan or elevation.*

Experience has repeatedly shown that simple structures symmetrical in plan perform much better in earthquakes than complex and unsymmetrical ones. The force distribution in complex and unsymmetrical structures under earthquake loading is extremely difficult to predict; torsional forces are liable to be set up if the centre of mass is not coincident with the centre of resistance, and this can cause local failures.

Adequate design of members and details with complex arrangements and under complex force systems is much more difficult than for simple cases. The same applies to re-entrant plan shapes even if they are symmetrical. Uniformity and continuity of structure are of equal importance, because changes in cross-section, either in overall elevation or in one particular element, cause concentrations of stress which are very damaging.

[6] For further details see Dowrick (1987), Penelis and Kappos (1997) or Booth (1994).

Experience has shown[7] that a structure will have the maximum chance of surviving an earthquake if:

- the load-bearing members are uniformly distributed;
- the columns and walls are continuous and without offsets from roof to foundation;
- all beams are free from offsets;
- columns and beams are co-axial;
- reinforced concrete columns and beams are nearly the same width;
- no principal members change section suddenly;
- the structure is as continuous (redundant) and monolithic as possible.

The concept of redundancy implies that any applied load can find many alternative routes (load paths) to the ground. Given the unpredictable nature of earthquake motions and the real chance of local overload, a structure designed so that if one element fails others will be able to carry its load must evidently have a better chance of survival in an earthquake.

Avoid Soft Storeys

One particular type of discontinuity is worth elaborating on. Very commonly multi-storey frame buildings are provided with cross-walls or frame infilling in residential upper storeys, but these are omitted or partially omitted on the ground floor to provide open commercial or car-parking space; this is often the cause of a serious weakness on this floor. This has been the cause of the disastrous failure of the ground floor of many buildings such as that illustrated in Figure 8.6. The effect of setbacks in elevation is similar and these should also be avoided for the same reason.

Plan Size and Slenderness Limitation

Limiting the size of a building size in plan is important because earthquake forces vary rapidly in both time and space and a long building is likely to have different ground movements applied to it at each end, coupled with ground distortion along its length. Where a long building is needed for planning reasons, it is likely to perform better if subdivided into separate short lengths of structures with movement gaps between them.[8]

The slenderness of a building should also be restricted to limit horizontal deformations: a height/width limitation of 3 or 4 has been proposed,[9] although this can be exceeded with good design.

---

[7] Dowrick (1987).

[8] The legendary survival of Frank Lloyd Wright's very large Imperial Hotel in the 1923 Kanto (Tokyo) earthquake has been partly attributed to its separation in this way.

[9] Dowrick (1987).

Figure 8.6 Collapse of reinforced concrete buildings in Adapazari, Turkey, in the 1999 Kocaeli earthquake

Columns Stiffer than Beams

In framed buildings, additional important rules of design must be observed. One requirement is that columns should be stiffer than the beams which frame into them. If this is the case, the beams will fail before the columns, limiting failure to the area supported by the beam and enabling the beams to be used as energy absorbers; where the columns begin to fail first, failure tends to occur very rapidly, under their vertical load.

Infill Panels

The use of stiff infill panels in framed buildings as cladding or as internal or external partitions presents serious problems: often they are not treated as a part of the structure and are themselves weak. However, in an earthquake they tend to attract load initially, because of their stiffness. When they fail, this will be a brittle type of failure, which can cause serious damage to the main structure, as well as injury to occupants, and result in serious economic loss to the building, even if the main load-bearing structure is unharmed. Thus infill panels *either* should be treated as a fully integral part of the structure (making it a shear wall not a frame structure) *or* should be totally separated from it by movement joints which allow the frame to move independently. This latter approach presents detailing problems if the structure is expected to support the infill panel under normal conditions. Infill panels can have an equally disastrous effect if they are discontinuous in either elevation or plan. The effect of this is to create regions of high stress concentration in a structure for which it was not designed, causing local failures.

Separation Between Buildings

Individual buildings need to be provided with adequate separation, to prevent damage caused by *pounding* when they deform in earthquakes, which has been a serious cause of damage, even of collapse in recent earthquakes. The minimum separation gap depends on the height and flexibility of the building. The gap between buildings should exceed the expected cumulative maximum drift (lateral displacement) of all storeys added together with an extra allowance. Separation can be a particularly difficult problem to deal with where a tall building of complex or large plan is divided into smaller separate structural elements for reasons discussed above. The gaps created then generally need to be bridged to preserve functional continuity, but it is essential that any bridging should be designed not to transmit forces, so as to maintain structural separation.[10]

Alteration to Existing Buildings

Stress concentrations are very frequently caused by supposedly non-structural alterations carried out on existing buildings when their function changes. Not only the addition or removal of partitions, but also the positioning of windows, doors and staircases can significantly affect the earthquake performance of a building. Vertical or lateral building extensions, particularly where new materials are to be used, can be equally damaging.

Non-structural Elements

Finally, to achieve good earthquake performance, it is essential to pay attention to the non-structural elements of a building.[11] In recent earthquakes a high proportion of the damage was unrelated to the main structure of the building. Heating and cooling plant, fuel, electricity and water supply mains, elevator equipment, etc., need to be secured to resist earthquakes, otherwise serious damage including fire outbreaks can occur. Heavy furniture and equipment such as bookstacks need to be properly secured. Flying glass is a serious hazard in urban areas and for flexible high-rise buildings detailing for movement is needed. Cupboards and bottles containing hazardous chemicals have to be specially designed to avoid spillages.

---

[10] Solutions have been discussed by Arnold and Reitherman (1982) and Dowrick (1987).

[11] Lagorio (1991).

The jamming of doors in buildings as a result of deformation is a serious hazard as it may prevent escape or rescue; doors should be detailed so that some movement of the structure can occur without causing them to jam.

Foundations

Foundations, particularly for large buildings, should equally be kept as simple as possible. Only one type of foundation should be used for the whole building (or any structurally independent part of it). Separate column or wall foundations should be interconnected so as to achieve an integral action, and should all rest at the same level. Foundations should be loaded approximately uniformly under vertical load, and where possible sites with large variations in subsoil conditions should be avoided.

8.4.1 Engineering Techniques for Improving Earthquake Resistance

Some new engineering techniques for modifying the structure to achieve better earthquake resistance are available, and can be expected to become more widely used in the future. The most important of these techniques are *base isolation*, and the use of *energy absorbers*.

Base Isolation

The principle of base isolation is to introduce some form of flexible support at the base of a building so that earthquake forces transmitted to the building are much lower than if the building is firmly fixed to the ground. The simplest form of base isolation is a frictional sliding layer, which will slip if the force exceeds a certain proportion (perhaps 3–5%) of the weight of the building. As such slip is likely to result in permanent displacements, a spring system is normally preferable. Spring systems will transmit forces proportional to the relative movement of the ground and the building, and incorporating them will increase the natural period of vibration of the building, hence (for most earthquakes) considerably reducing the forces the building experiences.[12] Laminated rubber springs are the materials most widely used; they have a much lower stiffness in the horizontal direction than in the vertical direction, and thus are effective only to reduce the damaging horizontal forces. A lead core is incorporated to provide energy absorption through damping, thus further reducing the earthquake loads experienced by the building.[13]

---

[12] Key (1988), p. 70.

[13] Base isolation techniques have been discussed by Key (1988) and by Buckle and Mayes (1990).

Energy Absorbers

The function of energy absorbers is to absorb energy by deforming if the structure experiences a large earthquake, thus protecting the main supporting structure. The energy absorbers will experience permanent deformation (i.e. they will be damaged), but they can be replaced at a much lower cost than that of repairing damaged structures. A common location for energy absorbers is in cross-bracing for framed (particularly steel frame) structures.[14]

Both base isolation and energy absorption techniques are at present relatively expensive and their current use is mainly in high-rise buildings, But there is a growing body of evidence to demonstrate their effectiveness.

Active Control

Base isolation and energy absorbers are referred to as *passive control* or energy dissipation systems. A further range of techniques referred to as *active control* systems is currently under development. In these systems mechanical devices are incorporated into the building which actively participate in the dynamic behaviour of the building in response to measurements of its behaviour moment by moment during the earthquake ground motion. Only a few such systems have been installed to date and there is little experience of their effectiveness in real earthquake events, but they hold promise for the future.[15]

8.5 Choice of Structural Materials

The choice of materials for building materials in seismic areas is to a large extent dictated by questions of availability and cost. The essential material requirements for earthquake-resistant structures are strength and ductility, and these properties are closely interrelated. *Ductility* refers to the ability of a material to deform after its maximum strength has been reached, without losing its ability to carry load. Structures made from materials which have this property can survive short-term accidental overloads because, rather than breaking, they can deform during the overload and absorb a large amount of energy without losing strength, instead of simply breaking. Steel is an inherently ductile material, and is thus very suitable for building in earthquake areas.[16] California and Japan make extensive use of steel in large buildings of all types. Concrete and all types of masonry, without reinforcement, are brittle materials, but by means of embedment of steel

[14] There are a wide variety of techniques which have been discussed by Key (1988) and Hansen and Soong (2001).

[15] Soong and Spencer (2000).

[16] Although welded joints can be a source of weakness and have resulted in some failures in recent earthquakes.

reinforcement, suitably placed, they can be made to perform in a semi-ductile manner, making them suitable for earthquake-resistant construction.

Since the extra forces resulting from an earthquake are proportional to the mass of the structure, structural materials which are strong and ductile but light (i.e. have a high strength-to-weight ratio) are particularly suitable for earthquake-resisting structures. Timber has the highest strength-to-weight ratio of all structural materials, and steel is also good in this respect. Reinforced concrete is not so good, and masonry is poor in terms of strength-to-weight ratio.

Nevertheless, in many parts of the world economics dictates that reinforced concrete frame structures are used for mid-rise and high-rise buildings; to make such structures earthquake resistant requires careful attention to continuity and to ductility requirements. Such frames should be cast in *in situ* concrete; adequate ductility and continuity are much more difficult to achieve with precast concrete.[17]

Reinforced masonry, of brick, block or dressed stone, is a good material for low-rise structures, since it combines high shear strength with some ductility, provided that certain important rules are observed. Unreinforced brick masonry is less suitable in areas of high seismicity. As Figure 1.1 shows, the great majority of earthquake deaths over the last century have been caused by the collapse of unreinforced masonry buildings. But unreinforced masonry may be used with acceptable safety in moderate- or low-seismicity areas if the elements are properly interconnected as described in Section 8.4.

Low-strength unreinforced masonry materials (such as rubble, stone and adobe) have an extremely poor seismic performance and should be avoided whenever possible. However, where there is no economic alternative to their use, even these materials can, with suitable reinforcement of timber, steel or reinforced concrete, be made to behave in a semi-ductile fashion which will significantly improve their performance in moderate earthquakes. Techniques are discussed in Section 8.7.

Composite structures of steel or concrete frame with masonry infill panels are today much used in seismic areas because of their cheapness, and they need a lot of care in their design and construction. Just as an addition of reinforcement can help to enable brittle materials to achieve some ductility, the ill-considered incorporation of brittle infill materials can cause ductile materials to behave in a non-ductile way.[18]

Timber frame is mostly used only in low-rise structures: it has generally performed very well in earthquakes as a framing material owing to its high strength, low weight, and the ductility provided by flexible but energy-absorbing joints. The majority of domestic construction in New Zealand, Japan and southern California uses timber frame construction. However, in older structures where the timber

---

[17] Precast concrete structures have a poor record in earthquakes, as shown by recent experience in the 1994 Northridge earthquake in the United States (EEFIT 1994) and the 1999 Kocaeli earthquake in Turkey (EEFIT 2002b).

[18] Some rules for detailing such structures have been proposed by Smith (1988).

is not well preserved, its deterioration can be a problem, and it contributes to the risk of fire damage in earthquakes. Heavy roofs supported on old timber with poor connection to masonry walls or unbraced timber frames have been responsible for many deaths.[19]

8.6 Codes of Practice for Engineered Buildings

8.6.1 Philosophy

In seismic areas buildings will collapse or be seriously damaged unless they are specifically designed to withstand the expected loads from future earthquakes, and rules are needed to guide designers on how to achieve this safety. Both the level of protection to be aimed for and the means of achieving it will vary from place to place, according to the level of seismic risk, the resources available, the type of construction being considered and the capability of the building industry. Some of the world's most technologically and industrially advanced areas, such as Japan and southern California, are located in regions of high seismicity. Both experienced a series of devastating earthquakes during the twentieth century, and continue to experience regular shocks. In these areas, codes of practice for the design of new buildings have been in place for most of the last century; they are almost universally understood and adopted by designers of large buildings, and they have become models for the codes of practice used in other countries.

In California, the level of resistance aimed for in design has, since the late 1970s, been based on the concept of an 'acceptable risk'. The objectives are:[20]

1. To resist minor earthquakes without damage.
2. To resist moderate earthquakes without significant structural damage, but with some non-structural damage.
3. To resist major or severe earthquakes without major failure of the structural framework of the building or its component members and equipment, and to maintain life safety.

It is also recognised that certain critical facilities should be designed to remain fully operational during and after an earthquake.

Within this general framework the scope and reliability of the codes have been able to develop in recent years in step with the rapid developments in the scientific knowledge of earthquake hazards and the engineering understanding of the effects of earthquakes on buildings. The most recent US codes, incorporated in the

[19] The majority of the deaths in the 1995 Kobe earthquake were reportedly caused by the collapse of poorly maintained, timber-framed houses supporting heavy tiled roofs.

[20] As defined by Applied Technology Council, ATC-3-06 (1978).

2000 International Building Code,[21] are based on state-of-the-art understanding of ground motion characterisation and zonation, performance-based design concepts, and consideration of the non-linear response of structures. They also extend to considerations of the problems of assessing the condition of, and the strengthening of, existing buildings.

Similar codes have been developed in other seismically active areas such as Japan, New Zealand, the European Union and Mexico. The detailed provisions of earthquake codes are discussed below. Each country or group of countries has its own code related to construction types and to a form of organisation of the building process appropriate. At least 36 countries or regions now have distinct earthquake codes.[22]

Typical of the newest generation of codes is EC8,[23] designed eventually to be applicable throughout the countries of the European Union. The code is written in terms of the limit state philosophy now widely used in codes of practice for other aspects of structural design. The basic concept is that in the planning, design and construction of structures in seismic regions the following requirements should be met with an adequate degree of reliability under earthquake loading (in addition to non-seismic loads):

- a no-collapse requirement and
- a requirement limiting susceptibility to damage.

Different levels of reliability are envisaged according to the consequences of failure. Adequate reliability against collapse is ensured if certain specified detailing rules are observed, and if verifications of the structure's strength, ductility and overall stability are performed, while adequate reliability against damage is ensured if specified deformation conditions are satisfied. Extensive general design rules (including rules for structural regularity and for design loadings) and specific rules for different materials and elements are given. Separate documents relate to buildings, bridges, towers and masts, tanks and silos, and foundations. A feature of the Eurocodes is that many of the numerical quantities (safety factors, design loads and so on) required for design are not specified, but left to be decided by national authorities through a National Application Document. An important principle is complete compatibility between this code and the independent Eurocodes for steel, concrete, timber and masonry.

Similar codes have been formulated for application in many countries, but in some of the poorer earthquake-prone countries, codes have sometimes been found difficult to apply because of lack of resources and difficulties of enforcement, coupled with a lower level of public awareness of the risk. For such areas, which

[21] ICBO (2000).

[22] Paz (1994).

[23] CEN (1994), Lubkowski and Duian (2001).

include many areas of equally high seismicity, there is an inevitable trade-off between the desirability of improved earthquake resistance and the extent to which scarce resources can be committed to this particular risk, when so many other types of risk are faced.

8.6.2 Typical Requirements

Codes of practice for engineered buildings have been in a stage of rapid development in recent years, in scope, precision and complexity. Codes will normally do three things:

(1) define the earthquake loads to be used in the structural design of buildings;
(2) define criteria for overall structural performance;
(3) provide guidance on detailing the building structure appropriate to the common materials and structural systems in use.

For ordinary buildings the earthquake load is generally defined in terms of a base shear coefficient, which is multiplied by the applicable weight of the building to define the horizontal load which the building must be designed to resist at its base. Base shear coefficients are defined by means of seismic zoning maps (or from other seismic hazard maps as discussed in Section 7.3) for different regions of a country according to the historically experienced pattern of earthquakes. Most countries define no more than three or four zones with fairly large differences between the required coefficients.

The basic coefficient is then scaled by a series of modification factors which take account of:

- the fact that structures of particular importance need to be provided with a higher level of protection;
- the effect of the subsoil on the ground shaking which will be experienced;
- the degree of ductility and other aspects of the earthquake-resistant quality of the building (regularity of form, low eccentricity, good damping, redundancy);
- the reduction in base force with increased natural frequency of the building.

Figure 8.7 shows the general shape of the standard elastic spectral response curves for design of structures defined in the 1997 US Uniform Building Code. The actual curve for any location is generated by a pair of constants C_a, which defines the (peak acceleration-related) plateau level, and C_v, which defines the (peak velocity-related) shape of the descending branch of the curve,. The pair of values C_a and C_v to use in each case depends on a seismic zone factor Z (for the seismic zone applicable) and a soil profile type.[24] The class of the soil is derived

[24] Bachman and Bonneville (2000).

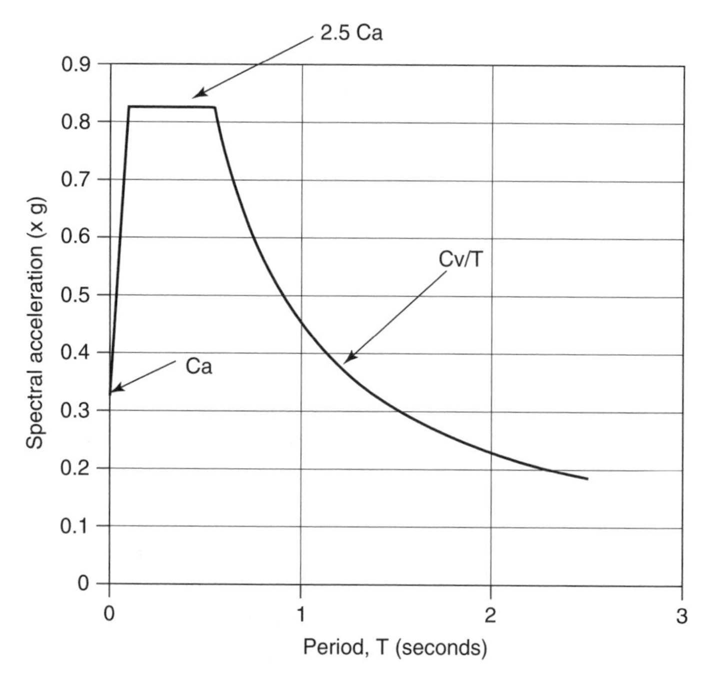

Figure 8.7 A typical response spectrum for use in design (using design response spectral formula given in 1997 Uniform Building Code for soil class C – dense soil with zone factor 0.3)

from the stratigraphy of the soil and its shear wave velocity or penetrometer test values in the top 30 metres. Design response spectra for the codes applicable to other countries are structured in a similar way.

Codes also specify procedures for distributing the total load thus computed between the different levels of the building, and for combining the earthquake forces with the other types of loading which the structure experiences (dead load, live load, wind load and so on).

In addition to these loading requirements, codes commonly specify limits in the overall form of the building plan layout (length, building height/width ratio, degree of eccentricity, extent of changes in structural resistance from one floor to the next). These may be absolute limits, or the code may require the building to be designed for higher forces if certain limits are exceeded.

Codes also specify, for each material, numerous detailing requirements needed to provide the structure with the necessary ductility and continuity, and may specify further that buildings provided with increased levels of ductility can be designed for a reduced level of loading. Codes will also commonly specify limits to the interstorey displacements, and provide for minimum spacing between buildings to avoid pounding.

Current building codes are by no means perfect and will continue to be extended, refined and adapted as more knowledge becomes available, but each additional requirement has to be carefully considered in relation to the expected benefit and the cost involved. Recent experience has shown that buildings designed to modern codes (i.e. those in place since the early 1980s) have in most cases a satisfactory degree of resistance to the level of earthquake which they are intended to protect against.

However, the usefulness of earthquake codes depends crucially on the extent to which they are actually applied. Several recent earthquake disasters in countries where modern earthquake design codes exist have shown that many buildings have in the past been built without reference to the code in force at the time, and many buildings collapsed as a result. Building control – the process by which local authorities and building owners ensure that the code of practice is adopted in both design and construction – is now recognised as a vital aspect of improving earthquake protection. Measures to improve building control are discussed in Chapters 6 and 9.

8.7 Improving the Resistance of Non-engineered Buildings

Some of the earthquakes of the recent past have offered ample evidence of the weakness and high vulnerability of the traditional dwellings of many earthquake regions, particularly those where rubble stone masonry or earthen construction is used. Figure 1.2 shows the destruction caused to a rubble stone masonry village by the Dhamar, Yemen, earthquake of 1982. Figure 8.8 shows the damage to masonry buildings in the 2001 Gujarat, India earthquake in India. Better construction methods are essential if disasters such as this are to be avoided, but improved methods can be expected to be implemented only if they can be afforded by people with very low cash incomes. In addition, the improved methods need to be within the capability of the existing builders and to be culturally acceptable to the villagers.

Programmes of rural upgrading to reduce future earthquake disasters were discussed in Section 6.6. Many proposed building improvement programmes of rural upgrading to mitigate disasters have been unsuccessful through failure to take each of these limitations properly into account. Within the constraints that exist, the replacement of existing dwellings with 'earthquake-resistant houses' is neither feasible nor, perhaps, desirable.[25] The fate of many such schemes has been to be rapidly abandoned, while their intended beneficiaries continue to build and live in houses of traditional form. It has been found more realistic to think, rather, in terms of low-cost upgrading of traditional structures, with the aim of

[25] Aysan and Oliver (1987).

Figure 8.8 Destruction of rubble stone masonry in Bhuj, in the 2001 earthquake. Buildings of this type have been found to be highly vulnerable to earthquake damage in many countries

limiting damage caused by normal earthquakes and giving their occupants a good chance of escape in the once-in-a-lifetime event of a large earthquake.

8.7.1 Removing Defects

Detailed study of the construction of weak masonry has indicated that many of the common defects in this type of construction, which lead to earthquake damage, can be eliminated at no cost or very low cost using a level of technology within the capability of owner builders or craftworker builders. Figure 8.9 shows, for example, some of the defects common at wall–roof junctions in the rubble stone masonry construction with flat earth roofs typical of traditional practice in many areas of the Mediterranean, Middle East and Asia. In Table 8.2, the symptoms, causes and effects on earthquake damage of common problems in walls and roofs are identified, and no-cost or low-cost preventative measures proposed.[26] Some of these are illustrated in Figure 8.10.

A fundamental weakness of much low-strength masonry construction is that it allows water penetration into the walls, causing the ends of timber roof joists to rot, and softening soil mortars which causes deformation and even instability in the walls; the incorporation of improved drainage and overhanging eaves at roof level and improved surface water run-off at ground level are very low-cost measures to deal with these problems. Another weakness with direct consequences for earthquake resistance is poor bonding of the stonework or other

[26] Coburn and Hughes (1984).

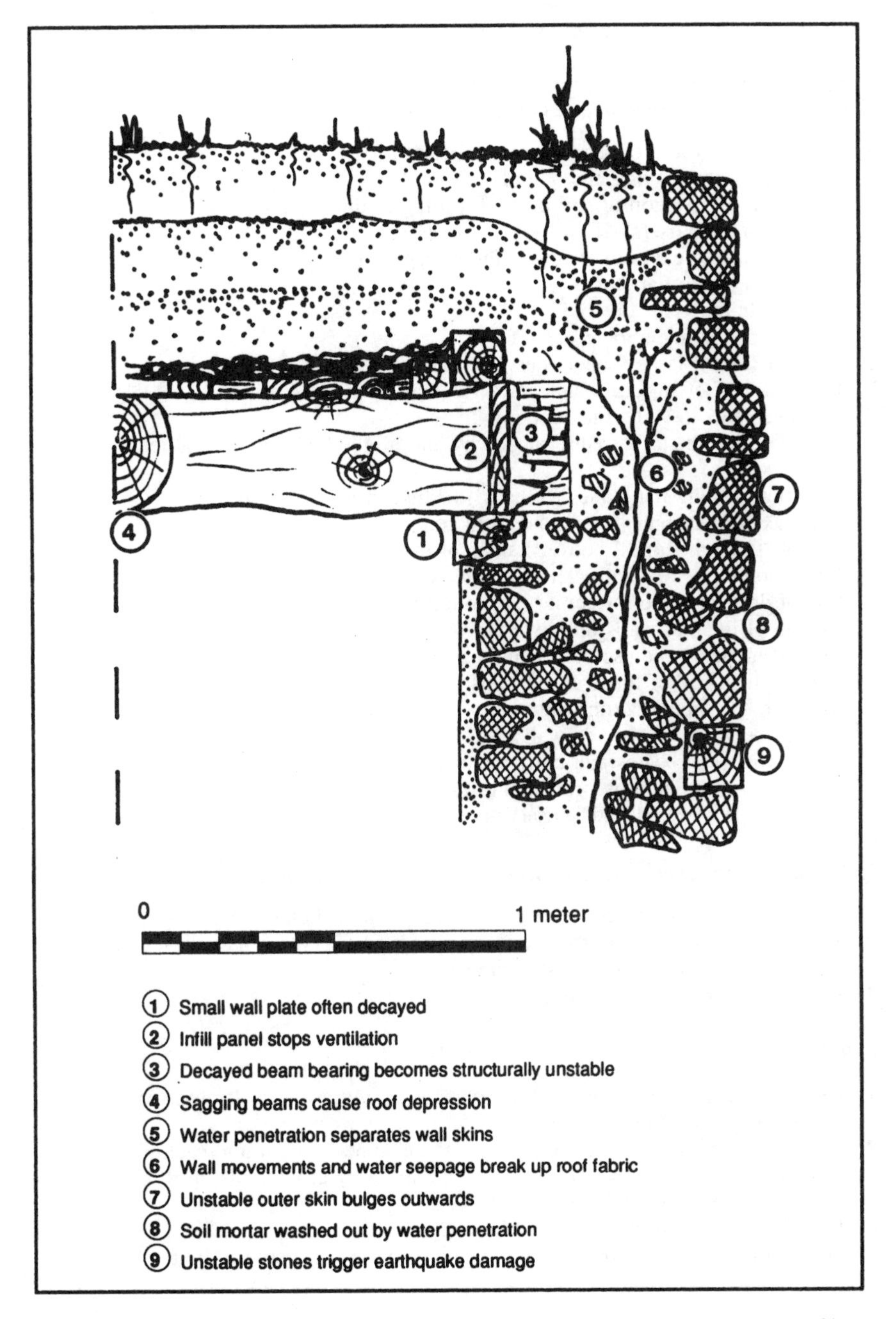

Figure 8.9 Common defects in traditional stone masonry construction at wall–roof junctions that increase earthquake vulnerability (after Coburn and Hughes 1984)

Table 8.2 Examples of low-cost and no-cost measures to reduce earthquake vulnerability in traditional construction.

a. Preventative Construction and Good Building Practice

1. Prevention of water penetration through roof:
   Grading soil by sieving and adding clay
   Extensive compaction of soil on roof, rolling in thin layers
   Use of soil stabilisers
   Construction of eaves overhang with drainage spouts
   Paving roof with flat stones
   Internal drainage in soil roof
   Non-permeable membrane in roof
2. Construction of large foundations, deep and wide.
3. Improvement of surface water drainage away from building.
4. Improvements to wall construction:
   Stones laid with stable centre of gravity and in good contact with at least two other stones
   Selection of long flat stones ('blocks') for use in wall construction
   Care taken to stagger jointing between courses
   Use of 'through stones' between skins of masonry
   'Dog-toothing' stones of each skin together
   Use of dressed stone for corner construction
   Keying of internal walls into external walls
5. Improvements to the use of structural timber:
   Even distribution of loads onto beams
   Beams to be supported on wall plates
   Fixings between beams and wall plates
   Fixings between beams and cross-members of roof
   Beam ends ventilated by provision of air gap
   Use of timber treated with preservative

b. Annual Maintenance to Reduce Deterioration and Weakening

1. Roof stripped and top soil level of roof recompacted before each winter.
2. Walls repainted or sections of masonry rebuilt where cracks have developed.
3. Replacement or rotten or infested timbers.
4. Reconstruction of site drainage.

c. Reinforcement Measures in Local Materials

1. External buttresses at corners of masonry walls and at intervals along long walls, preferably out of dressed stone.
2. Dressed stone corners extending two or three stones' length into each wall.
3. Horizontal double courses of dressed stone masonry at intervals up wall height, e.g. ground, cill, lintol and eaves level.
4. Long bearings on roof beams and lintols.
5. Duplication or triplication of fixings (nails, straps, etc.) between timber members.

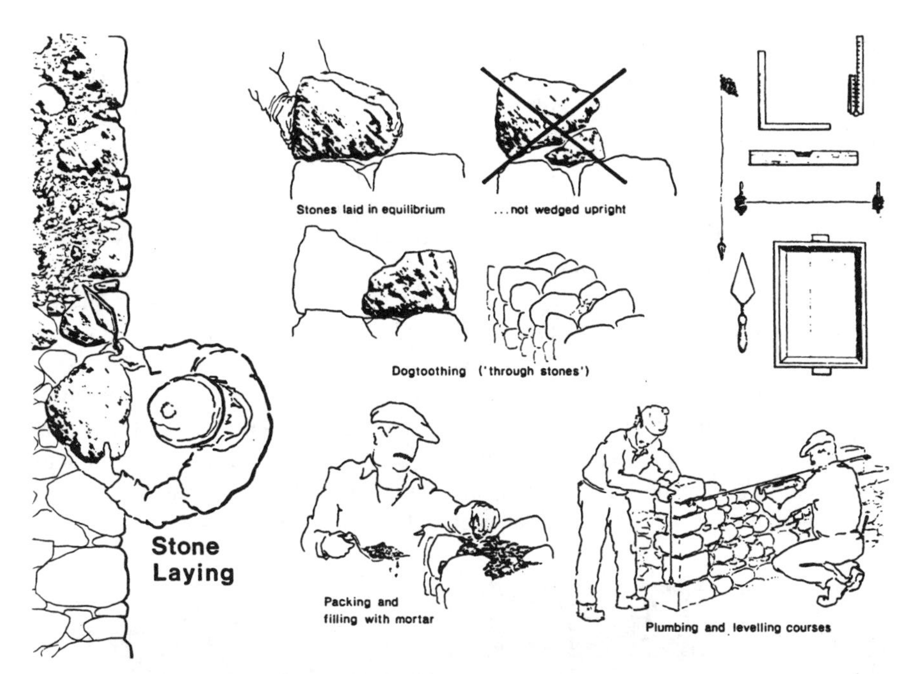

Figure 8.10 Buildings can be made more earthquake resistant without any additional cost, by careful construction. Measures such as careful placing, bonding and levelling of rubble masonry courses can result in a cohesive wall which is less vulnerable to earthquake vibration

masonry, leading to the separation of inner and outer layers of masonry, to the formation of vertical cracks in walls, particularly at corners, and to local wall failures.

In rubble stone masonry the improvement of bonding through the use of 'through stones', more frequent use of squared stones, especially at corners and openings, and greater attention to the equilibrium of individual stones during laying would also substantially improve earthquake performance at very little cost. Equally, improved building maintenance would often be a cheap way to reduce earthquake vulnerability significantly.

8.7.2 Low-cost Modifications

The improvements described above will eliminate the weaknesses which earthquakes tend to exploit. Adding further strengthening is also possible for small additional costs in the construction of a building. Section 8.4 explained the importance of roof diaphragms and ringbeams to transfer horizontal earthquake forces into in-plane forces in the walls, and to provide corner continuity. One method of strengthening masonry construction is to incorporate a reinforced concrete ringbeam of thickness up to 200 mm and width equal to that of the wall at the level of the eaves to tie the walls together and provide continuity.[27] This is effective but fairly expensive by comparison with the cost of the unstrengthened construction. Another method used traditionally in some areas is the use of timber wall ties at the level of the eaves; these can also be introduced at the level of the lintel and window cill, which has the additional benefit of improving the through-bonding of the wall and of reducing the effective height of each panel of masonry acting in shear. The same strengthening effect can be achieved by means of thin, lightly reinforced masonry ties at these levels. Figure 8.11 shows strengthening measures of this type being tested for their effectiveness; the results showed a dramatic strengthening effect from the addition of either timber or concrete wall ties.[28]

Monolithic action of the roof is not easy to achieve in traditional construction without the incorporation of a reinforced concrete slab; this is effective but involves a level of construction technology normally appropriate to contractor construction. A cheaper and simpler, though not so strong, alternative is to nail timber sheeting cross-members or a sheet material such as plywood or corrugated

Figure 8.11 The effectiveness of earthquake-resistant construction can be demonstrated using destructive experiments on full-sized buildings. Here a full-size rural masonry building is shaken apart on a simple 'impulse table' where a concrete slab vibrates on rubber bearings

[27] IAEE (1986), Daldy (1972).

[28] Spence and Coburn (1987a).

steel sheeting between the main roof beams to increase the degree of interconnection and redundancy of the roof structure; the use of wall plates to which all roof beams are firmly fixed is a minimum provision to spread and redistribute the earthquake loads which is not always present in traditional masonry construction.

The use of timber or concrete ringbeams as described above provides substantial wall strengthening at a low cost to both stone and adobe masonry. The strength of stone masonry walls can also be increased by improvement of mortar quality. Stone and adobe walls tend to be thick, 50 cm to 60 cm in width, and the volume of mortar used is high, so the use of a 1 : 6 cement–sand or 1 : 2 : 9 cement–lime–sand mortar makes the walls substantially more expensive to build than thinner concrete block or brick walls. Lower cost mortars such as stabilised soil can be used. Another alternative is to lay the wall in weak soil or soil–lime mortar pointed by a stronger sand–cement mortar at the two faces. This protects the wall from water penetration, adds some stability, and to a limited extent improves the integrity and cohesion of the wall construction.

Vertical reinforcement in the corners of walls and at the sides of window and door openings will contribute to the strengthening of masonry panels already containing horizontal reinforcing bands at the roof and floor or foundation level by enabling shear forces to be resisted by tension in the reinforcement and requiring only compression in the masonry.[29] This reinforcement can be made of the same material as the horizontal bands, either timber or reinforced concrete. Vertical reinforcement is most appropriate for use with squared block masonry, such as dressed stone, brick or concrete block, as it demands considerable construction skill and requires the masonry to act in a composite way. The Quetta bond, shown in Figure 5.8, is a way of constructing brick masonry walls so as to incorporate regularly spaced reinforcement in both vertical and horizontal directions.

Buildings with regular horizontal and vertical bands of reinforcement and stiffened diaphragm floors and roofs are essentially infilled framed structures whether of reinforced concrete or timber and can be expected to have a good earthquake resistance, but they will also be unaffordable by much of the population of rural districts. Given the existing high vulnerability and low incomes of many rural people, it is important to consider a range of possible approaches to upgrading, and to study the additional costs involved, the level of technology associated with them, and the degree of extra protection afforded. Figure 8.12 shows a range of possible alternatives for rubble stone masonry, with approximate costs, based on a study in rural Turkey, and Chapter 10 considers the costs and benefits of alternative upgrading strategies. These techniques are further described in the Building for Safety series of booklets.[30]

---

[29] International Association for Earthquake Engineering (1986).

[30] Coburn *et al.* (1995).

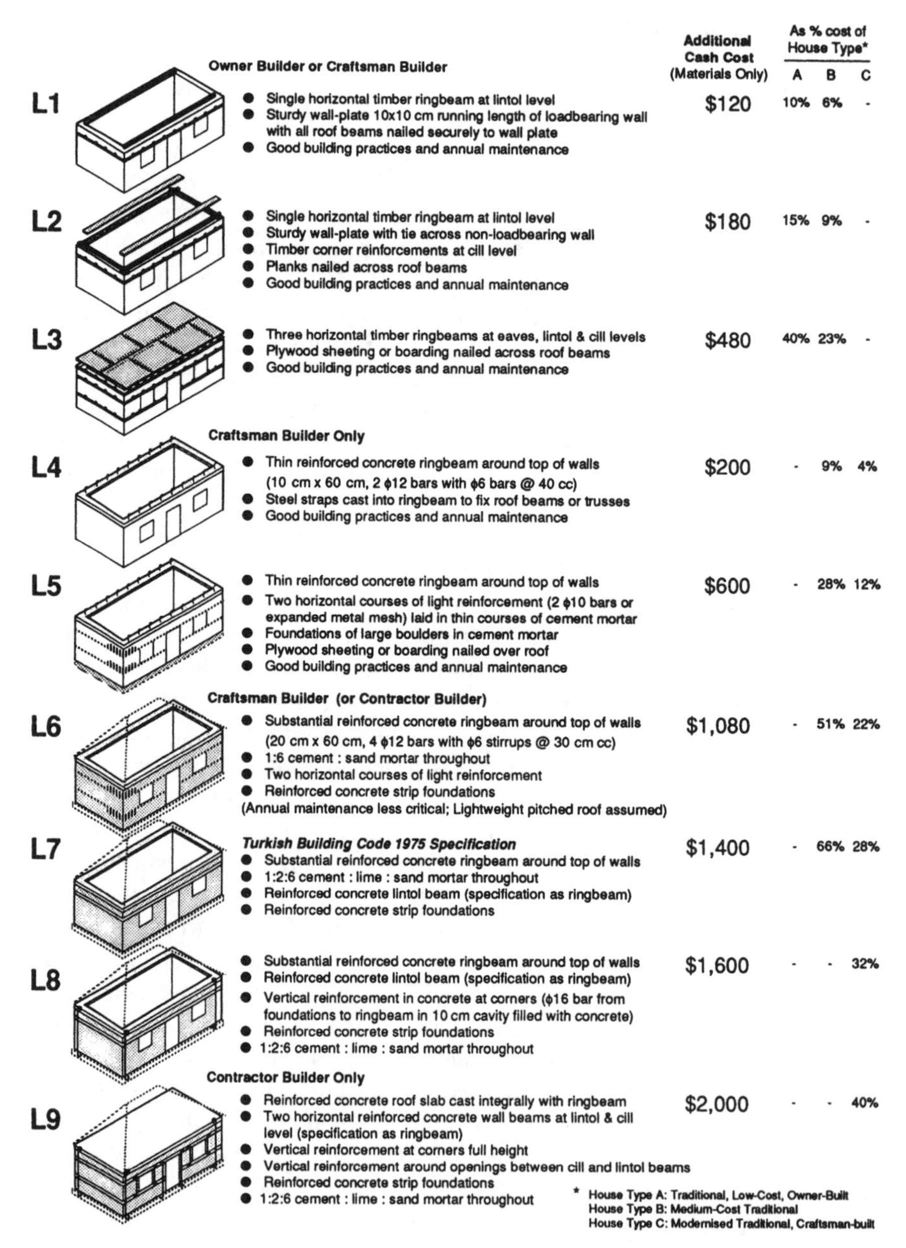

Figure 8.12 Increasing levels of earthquake resistance with increasing levels of cost and building skills required for traditional stone masonry building, Eastern Turkey (after Coburn 1986a)

8.7.3 Manuals for Strengthening Traditional Construction

In an attempt to reach those whose earthquake safety is not influenced by normal codes of practice for construction, a number of guides or manuals for improving the earthquake resistance of traditional construction have been published. These often make use of an illustrated instructive approach, designed to communicate better with rural people with practical skills but little formal education. Some[31] attempt to cover a wide range of traditional construction building types and locations, and give good advice on principles. But their generality and language tend to limit their applicability to specific rural locations. Location-specific manuals, written to communicate with a particular target group of builders, in their local language and using images familiar to them,[32] are more likely to be of direct benefit, though there is always a difficult compromise to make between recommendations which could be economical but unsafe, and on the other hand safe but too costly. An overview of the technical content likely to be covered has been compiled by Coburn *et al.* 1995 in the Building for Safety series of booklets.

Short booklets simply expressed in the local language, clearly illustrated, and incorporating good engineering principles can be useful educational material (Figure 8.13).[33] Often, however, illustrations make assumptions about how people read pictures which are not true for all communities. The design of educational material for rural communities needs to incorporate pre-production testing to get the message across (Figure 8.14). It is also clear that methods proposed in any printed literature would not be widely adopted unless they were accompanied by builder training and other promotional work designed to communicate both the awareness and the skills more directly. Building improvement programmes and training for craftworker builders in rural areas were described in Section 6.6.

8.8 Strengthening Existing Buildings

As the general awareness of the earthquake risk increases, and standards of protection for new buildings become higher, the safety of the older, less earthquake-resistant construction becomes an increasingly important concern. In many earthquakes, damage is concentrated in the older building stock, while recently constructed buildings suffer comparatively lightly. The problem can be expected to diminish over time if improved standards of new construction can be achieved, because the proportion of unsafe buildings in the total building stock will diminish. This is true particularly where there is a high rate of turnover

[31] For example, IAEE (1986), Daldy (1972).

[32] Such as that produced by the Oxfam project in the Yemen (Leslie 1984).

[33] Shortly after the Ecuador earthquake in March 1987, the JNV in Ecuador published a set of five manuals, four of which dealt with the most important types of traditional construction, the fifth with repair and strengthening existing buildings (JNV 1987).

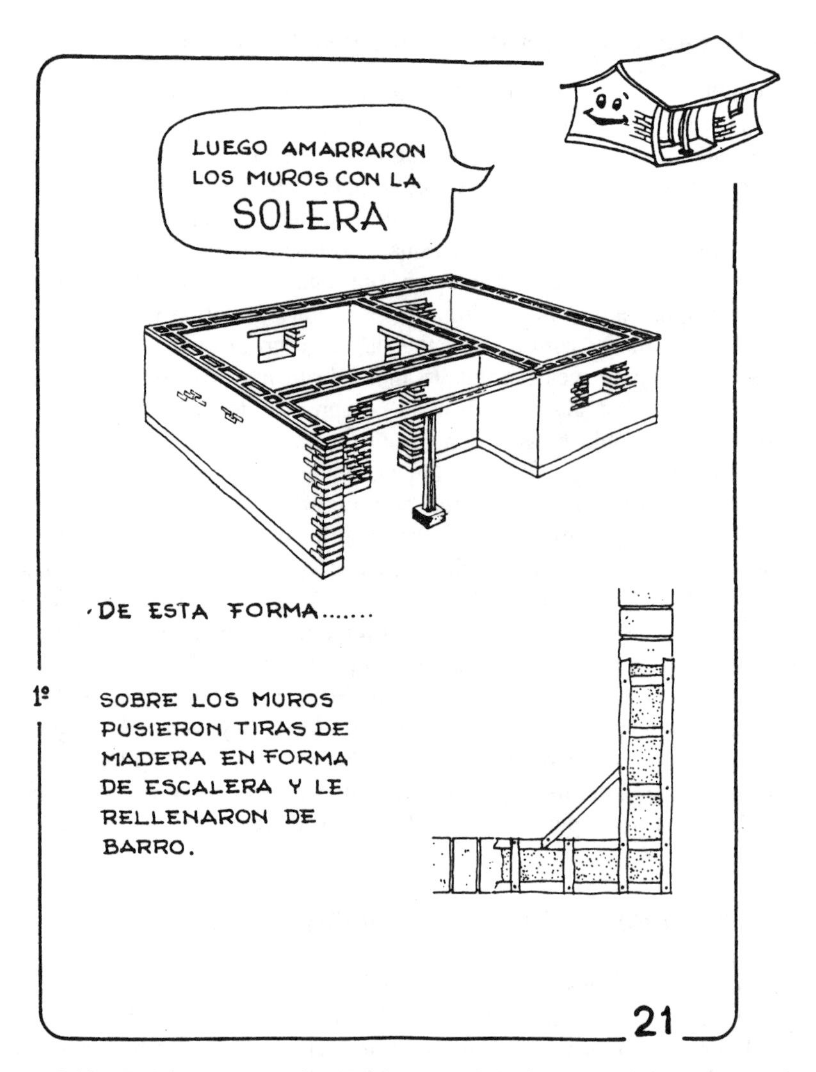

Figure 8.13 Training courses for builders can help improve their understanding of earthquake-resistant construction. The main emphasis of such training has to be through practical demonstration and advice on-site, although printed manuals of this kind can be useful as reference documents for builders. Illustrative page of a manual for builders on adobe construction in seismic areas, Ecuador (after JNV 1987)

of the building stock. But such cases are rare; in many industrialised countries the rate of new building construction is only about 1% per annum, and even in more rapidly developing areas with sustained growth rates of 8% and more, the total number of older buildings does not diminish very rapidly. It has been shown that in the areas affected by the 1980 earthquake in southern Italy the

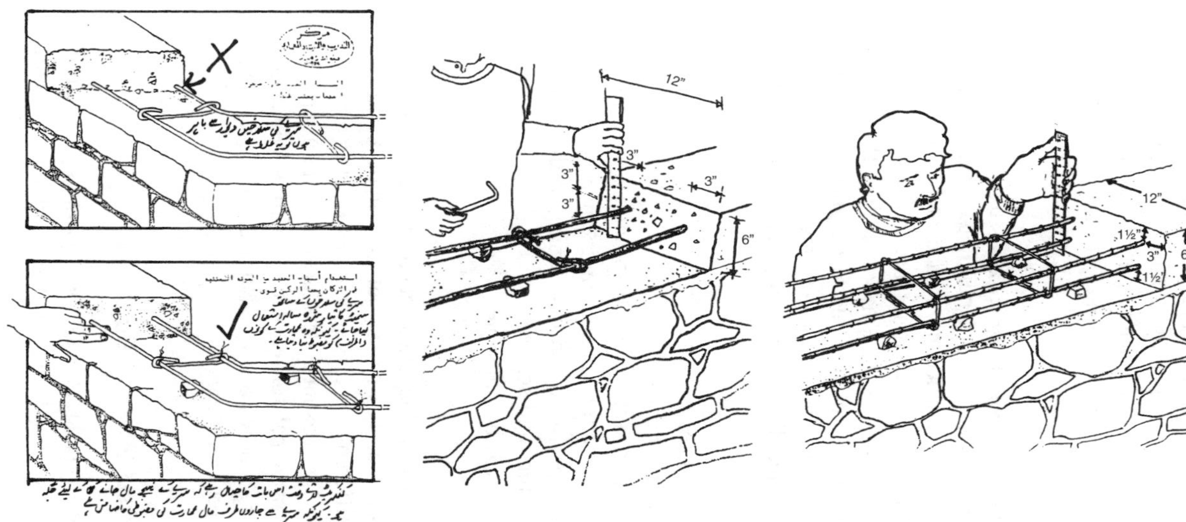

Layout of Reinforcing Steel *Version 1*

The first design for educational material to teach good concrete reinforcement was borrowed from training material prepared for Yemen Building Education Project in 1983, this image contrasts bad and good steel layout in a concrete ringbeam. It was shown to 25 builders, who were asked what they saw in the picture. Although the elements of the drawing were recognised, the use of a tick and a cross to denote good and bad were not understood by the builders of Northern Pakistan; thus the message of the drawing was lost. Further confusion was caused by the use of borders around the two pictures, which were seen as rooms. The fingers were commonly seen as pointing, rather than indicating a dimension.

Layout of Reinforcing Steel *Version 2*

The picture was redrawn to exclude the graphic conventions that were not understood in Northern Pakistan. The 'bad' practice was omitted and only 'good' practice illustrated. picture borders were omitted. The picture emphasised the concept of laying out the steel by showing a man holding a ruler measuring spacing. The actual dimensions were shown. This picture was shown to 15 builders. The drawing was widely recognised to be a man measuring steel layout in a ringbeam. The dimensions were seen as indicating 'correct' sizes, but their relationship to parts of the drawing were not clear. The man without a head and the second hand were distracting and confusing.

Layout of Reinforcing Steel *Version 3*

A further revision of the drawing tried to improve the message of the material. The man's head was included and the scale was reduced to show more context. The man's other hand was removed from view. The dimensions were placed actually in the position of their measurement, rather than using the technical drawing convention of dimension lines outside their location. This picture was shown to another 20 builders. Most said it was a man measuring steel layout in a beam on a stone masonry wall. Nearly all of the respondents could correctly size the overall dimensions of the beam. Many could also answer questions about the spacing of the steel. The overall message of 'good' practice in concrete beam construction was better communicated.

Figure 8.14 Educational material is not universally understood. Literacy levels vary and familiarity with graphic conventions is different from place to place. The design of educational materials for local communities needs to incorporate pre-production testing and redesign to get the message across clearly. Development of builder training literature for earthquake areas of northern Pakistan (after Coburn *et al.* 1991)

existing rate of change of the building stock cannot be expected to bring average risk levels within acceptable levels within 100 years.[34] Indeed, rather than being taken out of use, the older building stock is increasingly being modified and adapted, sometimes in ways which significantly increase the loading or reduce its inherent resistance to earthquake ground shaking. And where old buildings are not being modified, lack of maintenance may lead to decay of already weak and poorly integrated structures, resulting in a continual decline in earthquake resistance possibly made worse by the cumulative effect of low levels of damage in previous earthquakes.

For all these reasons, strengthening existing buildings is assuming increasing importance in earthquake regions. For most types of building, strengthening is a cheaper way of bringing earthquake resistance up to acceptable levels than rebuilding; depending on the situation and construction type, costs of strengthening typically range from 5% to 40% of the cost of a new building. But because strengthening is expensive, careful consideration needs to be given to the type of strengthening best suited to achieve the desired safety level. Factors which need to be taken into consideration in deciding whether to strengthen, and which method to use, will include:

- the required level of structural resistance;
- the general structural form and any changes needed;
- the materials and degree of connection in the existing structure;
- foundation conditions and the effect of strengthening on them;
- the effect of strengthening on the appearance and functioning of the building;
- required strengthening of non-structure and services;
- the time during which the building will be unusable;
- the cost of the work.

The main objective in strengthening is to achieve a structure which satisfies the principles of good earthquake-resistant design as set out in Section 8.4 above. The actual methods used will be different for different types of structure. The following sections discuss methods for masonry structures and for reinforced concrete frame structures, the two principal forms of construction which may require strengthening. Several countries now have regulations, codes of practice or guidelines for strengthening existing buildings.[35]

8.8.1 Strengthening Unreinforced Masonry Buildings

Many cities, towns and villages in earthquake zones consist primarily of unreinforced masonry buildings of a great variety of types and ages which experience

[34] Spence (1988).

[35] FEMA (1997), CEN (1994).

has shown to have poor resistance against earthquakes. The principal sources of the weakness of these buildings are:

- Low-strength masonry units and mortar, inadequately bonded together.
- Insufficient interconnection between inner and outer leaves of external walls.
- Insufficient connection at the junctions of perpendicular walls.
- Insufficient rigidity of floor and roof slabs in their own plane, and inadequate connection between these slabs and the bearing walls.

These weaknesses are frequently compounded by the deterioration of the structure due to weathering and rot, and to extensive structural modification during the lifetime of the building. Figure 8.9 shows the many separate sources of deterioration in stone masonry buildings.

According to the weaknesses identified in particular cases, strengthening may involve any or all of the following interventions:

- modifying the plan form of the building to improve symmetry;
- improving the connections between perpendicular walls;
- strengthening or replacing floor and roof structures, and improving their connection with the load-bearing walls;
- strengthening the walls themselves;
- strengthening the foundations.

The range of techniques available is wide, and details depend on the type of masonry involved. Italy has considerable experience of such work from the repair after the 1976 Friuli, 1980 Irpinia and 1997 Umbria–Marche earthquakes. Details of Italian techniques are given by Benedetti[36] and Croci.[37] A different set of techniques has been used to upgrade the old multi-storey brick apartment blocks found in southern California.[38]

Common strengthening techniques included:

- stiffening existing wooden floors and roofs by covering them with a thin layer of reinforced concrete;
- insertion of reinforced concrete ringbeams into the inner face of external walls at floor and eaves level to tie vertical and horizontal elements together;
- extensive strapping of masonry walls to each other and to slabs using both horizontal and vertical steel straps;
- strengthening of walls (mainly when cracked) by cement injection or by adding a thin layer of cement render reinforced with steel mesh on either side of the wall (Figure 8.15);

[36] Benedetti (1981).
[37] Croci (1998).
[38] Wong (1987).

Figure 8.15 An external and internal skin of wire mesh and cement render is a common method of reinforcing old or damaged masonry. Reconstruction after the 1982 Corinth earthquake, Greece

- adding plywood sheathing;
- strapping parapets.

Where repair and strengthening by these means is not considered feasible, an alternative is to introduce a new independent concrete frame to carry the earthquake loads, and attach the masonry to it. This system was extensively used in China following the 1976 Tangshan earthquake which revealed the inadequacy of much recently built masonry housing.

Methods of evaluating the earthquake resistance of existing unreinforced masonry building have been developed both in California[39] and New Zealand.[40]

[39] Applied Technology Council (1996).

[40] New Zealand National Society for Earthquake Engineering (1985).

These methods can also be used to assess the effectiveness of proposed strengthening interventions, and so to consider the cost-effectiveness of strengthening as against alternative mitigation measures such as reconstruction or change of use.

8.8.2 Strengthening Reinforced Concrete Buildings

The need for strengthening reinforced concrete buildings has become more urgent in recent years in countries where recent earthquake damage has indicated that the resistance requirements in previous codes were inadequate or where buildings have been found to be below code standard.

The principal causes of weakness in reinforced concrete buildings are:

- Insufficient lateral load resistance, as a result of designing for too small a lateral load.
- Inadequate ductility, caused by insufficient confinement of longitudinal reinforcement, especially at beam–column or slab–column junctions.
- A tendency to local overstressing due to complex and irregular geometry in plan and elevation.
- Interaction between structure and non-structural walls resulting in unintended torsional forces and stress concentrations.
- Weak ground floor due to lack of shear walls or asymmetrical arrangement of walls.
- High flexibility combined with insufficient spacing between buildings resulting in risk of neighbouring structures pounding each other during shaking.
- Poor-quality materials or work in the construction.

Unrepaired damage from previous earthquakes may also be a reason for requiring strengthening.[41]

The principal objective in most strengthening interventions is to increase the lateral load resistance of the building; usually increasing the ductility of the structure will be an additional objective. The strengthening may also involve removing or redesigning non-structural walls which may affect the performance of the building (see Section 8.5), and sometimes the strengthening effect may be achieved by removing load from a structure (by, for example, reducing the number of stories). For tall buildings on soft soil deposits, an increase in stiffness may also help to improve a building's performance by reducing its natural period to a value below that of the subsoil. Often the intervention may require simultaneous strengthening of the foundations. The principal technical options for improving the lateral load-carrying ability of existing reinforced concrete structures include:

- Adding concrete shear walls.
- Buttressing.

---

[41] Aguilar *et al.* (1989).

- Jacketing.
- Adding cross-bracing or external frames.

Adding Shear Walls

The most common method of strengthening of reinforced concrete frame structures is the addition of shear walls. These are normally of reinforced concrete, or may exceptionally be of reinforced masonry. In either case, they are reinforced in such a way as to act together with the existing structure, and careful detailing and materials selection are required to ensure that bonding between the new and existing structure is effective.[42] The addition of shear walls substantially alters the force distribution in the structure under lateral load, and thus normally requires strengthening of the foundations.[43] Figure 8.16 shows shear wall addition in progress to strengthen a building slightly damaged by the 1999 earthquake in Bolu, Turkey.

Buttressing

Buttresses are braced frames or shear walls installed perpendicular to an exterior wall of the structure to provide supplemental stiffness and strength.[44] This system is often a convenient one to use when a building must remain occupied during construction, as most of the construction work can be performed on the building exterior. Sometimes a building addition intended to provide additional floor space may be used to buttress the original structure for added seismic resistance. Buttresses typically require the construction of foundations to provide the necessary overturning resistance. Even considering the extra foundation costs, the cost of buttressing an occupied building may be substantially lower than that for interior shear walls or braced frames. The aesthetic impact and the availability of building space adjacent to the existing building are obvious factors affecting choice of this solution. Figure 8.17 shows an example from Naples, Italy.

Jacketing

An alternative technique is to increase the dimensions of the principal frame members by encasing the existing members in new reinforced concrete. The technique is known as *jacketing*. Adequate reinforcement of the new encasing concrete can increase both strength and ductility, and concrete damaged in a previous earthquake can be replaced at the same time. Again careful consideration

[42] Warner (1984).
[43] Jara *et al.* (1989).
[44] Applied Technology Council (1996).

Figure 8.16 Strengthening existing reinforced concrete buildings – adding reinforced concrete shear walls in Bolu, Turkey, following minor damage in the 1999 Kocaeli earthquake

needs to be given to achieving an adequate bond between the existing and new concrete.[45] Jacketing of beams is much harder than columns; jacketing the beams and columns may be ineffective if the beam/column joint is inadequate, and retrofitting joints is also difficult. Jacketing may be a viable option where a significant improvement is available from increasing the strengthening and ductility of some or all of the columns, without substantial intervention to the beams and joints. It may be attractive where there are architectural difficulties in adding shear walls. Jacketing is a valuable technique when complex or deep foundations make the change in the lateral load-bearing system required by a shear wall system impossible or very costly. An example from Mexico City is shown in Figure 8.18.

[45] Jara *et al.* (1989).

Figure 8.17 Retrofitting of reinforced concrete frame buildings: adding shear walls by external buttressing reduces the impact on the internal organisation of the building and may reduce the time during which the building needs to be empty. Residential reinforced concrete block in Naples, Italy

Addition of Cross-bracing

Both of the above techniques involve major interventions to the structure. An alternative technique, involving a less drastic intervention and smaller increase in foundation loads, is the addition of steel cross-bracing to increase lateral load resistance. This generally also involves the strengthening of adjacent columns, which will have to carry increased axial loads, although this is offset by a reduction in column moments and ductility demand. Strengthening of columns

Figure 8.18 Strengthening a reinforced-concrete-framed building can be achieved by encasing the old frame within a new structure. Shear wall structure being cast around a frame to upgrade the earthquake resistance of the Tribunal Law Court building in Mexico City

may be achieved by the addition of an external steel cage surrounding each of these columns. The addition of steel bracing considerably alters the appearance of a building, but is particularly suitable for comparatively low-cost strengthening of buildings which have not been damaged in a previous earthquake. Figure 5.9 shows an example.

Other Methods

Other methods sometimes adopted to improve the performance of reinforced concrete buildings in earthquakes include the addition of separate external frames (see ATC-40),[46] or the removal of one or more storeys to reduce the lateral load. New techniques such as bonding of steel plates to the concrete frame have been proposed but are as yet little tested. In rare instances, base isolation has been used to protect the superstructure from the ground shaking, but this is very expensive. The addition of supplemental damping devices is becoming increasingly used in the United States as a retrofit measure for concrete frame buildings. This is generally suitable for special cases; it would not be recommended unless a high level of the relevant engineering expertise is available. In a few cases,

[46] Applied Technology Council (1996).

strengthening through the use of advanced fibre-reinforced plastic (FRP) has been used in Turkey. While this method has the drawback of cost, it offers the advantage of quick installation, minimum disruption, and no weight increase in the structure.

8.9 Repair and Strengthening of Historical Buildings

Historical buildings constitute a case of special importance. A distinction has to be made between:

(1) historical monuments, and
(2) historical urban centres.

Each is valued for different reasons and the strengthening techniques necessary to retain those values are different, and have different costs and constraints. It is important in planning the repair of older buildings to consider which approach and level of budgeting suits each building. Usually strengthening of historical structures will be done as a result of minor damage from an earthquake, or they may have been damaged in other ways. In either case, the techniques for repair and strengthening are the same.

8.9.1 Historical Monuments

Historical buildings are valued for their cultural associations and interesting physical construction. In restoration and strengthening, the physical fabric of the structure must remain essentially the same as before the earthquake. If the roof is removed, for example, the same roof should be rebuilt using the old elements, replacing as little as possible. Strengthening elements that are added should be unobtrusive and, where possible, reversible, i.e. removable by future renovators.[47] This type of restoration work requires specialist skills and is expensive. There may be only a few buildings for which this sort of expense can be justified. Restoration and strengthening techniques used on historical monuments include:

- Dismantling damaged masonry and reassembling it with improved mortar and concealed reinforcement (e.g. metal cramps, reinforcing bars, mesh, etc.).
- Addition of concealed tension bars, as anchor bolts, ringbeams, corner ties, splay members, arch chords, and other structural connections. These may be drilled through masonry using extended bit drills, capped and grouted into place.
- Internal grout or chemical injection into wall cores where poor-quality rubble has to be stabilised and bonded without altering the external wall finish.

[47] There may often, however, be some conflict between the desire for reversibility and the effectiveness of the intervention (Croci 1998).

Grouting can be gravity fed or pressure injected, but is irreversible and often unpopular with renovators.

- Strengthening or stiffening foundations.

Following the Umbria–Marche earthquake in Italy in 1997, which caused serious damage to the vaulting and external masonry of the Basilica of St Francis (Figures 8.19 and 8.20), a major programme of strengthening and repair was undertaken, to protect the vaults – which carry frescoes of great importance in the history of early Renaissance art – from damage by future earthquakes.[48] The techniques needed for this type of work have been described by Croci.[49]

8.9.2 Historical Urban Centres

The historical centres of many earthquake-prone cities consist of dense residential and commercial districts whose buildings are usually of unreinforced

Figure 8.19 The Basilica of St Francis of Assisi after the 1997 Umbria–Marche earthquake, exterior view. The tower and the tympanum to the south transept sustained repairable damage. The temporary works on the east façade are to provide access to the roof space for repair of the vaulting

[48] Spence (1998b).

[49] Croci (1998).

Figure 8.20 The Basilica of St Francis of Assisi after the 1997 Umbria–Marche earthquake, interior view. One of the two collapsed sections of the vaulting, containing early Renaissance frescoes of great importance to the history of art, showing the twentieth-century reinforced concrete supports for the roofing which may have contributed to the vault failure (Spence 1998). The vaulting was reconstructed with additional external supports (Croci 1998), and the Basilica was reopened in 2000

stone or brick masonry, much altered in unrecorded ways over the centuries and often in poor condition. Although they represent a valuable and irreplaceable part of the urban heritage, they are often under threat from general decay and deterioration and from the pressure for redevelopment in addition to the earthquake risks they face. Figure 8.21 shows buildings typical of the Alfama District

Figure 8.21 Many of Europe's historical city centres are highly vulnerable to destruction in future earthquakes: the Alfama District in Lisbon

of Lisbon today, a city which was destroyed by an earthquake in 1755. To date little has been done to protect any of such buildings from future earthquake damage, and upgrading strategies are needed which will fulfil the sometimes conflicting criteria of life safety for occupants and functional upgrading, limitation of damage from future earthquake, and limitation of alteration to the fabric and appearance of the buildings. Criteria governing the choice of upgrading strategy for historical centres are:[50] first, that interventions should make a significant improvement to the earthquake resistance of the buildings in a way which is both identifiable and measurable; secondly, that interventions should cause only very limited alteration to the external appearance of the building; and thirdly, that interventions should be consistent with existing programmes of upgrading for the buildings in terms of cost, appropriate techniques and the process of design and management.

Repair and strengthening techniques suitable for use in these situations include:

- Use of steel tie rods passing through the floors and external walls with external anchorage plates or bars to connect the walls and the floors together (Figure 8.22).
- Improving the stiffness of floors in their own plane by adding new timber members – for instance, two layers of floorboards laid perpendicular to each

[50] D'Ayala *et al.* (1997).

Figure 8.22 Addition of tie rods is an effective and relatively low-cost way to strengthen existing masonry buildings – buildings in Assisi, Italy, strengthened following the 1997 Umbria–Marche earthquake

other or by cross-bracing with steel straps. The monolithic floors can themselves be made into strengthening diaphragms for their supporting walls by chasing in and casting skirting beams around the edge of the floor.
- Jacketing the walls by application of layers of wire mesh on each face, tied together through the wall at intervals and covered with a layer of dense plaster (Figure 8.15).

Where upper storeys are badly cracked and lower floors are relatively sound, the upper storey may be demolished, a reinforced concrete ringbeam cast on top of the remaining wall, and a new identical upper storey constructed. New masonry should be reinforced and may be in solid brick, high-quality concrete blockwork, or cut stone. Original wall thicknesses should be retained, and all walls should be topped by a reinforced concrete ringbeam at roof level. In reconstruction, the plan of some buildings – for example, 'L'-shaped plans – may be compartmented into interlocking rectangular structural units, by means of ringbeams and cross-ties, for greater seismic rigidity.

In cases of moderately damaged walls with elaborate stucco decoration work, it may be possible to save the wall and its original decoration by using cement injection grout injected into the core of the wall. This should only be used in conjunction with extensive 'stapling', i.e. drilling and grouting steel reinforcing bars as connector reinforcements between walls and from walls to floors.

The skills needed for repair and restoration of the buildings of historical urban centres are general building skills. Techniques of grouting, stapling and mesh reinforcement are relatively straightforward to learn and can be carried out by almost any building professional. Skilled craftwork is needed for repair and for renovation of any original interiors that owners wish to preserve.

Further Reading

Arnold, C. and Reitherman, R., 1982, *Building Configuration and Seismic Design*, John Wiley & Sons, New York.

Booth, E. (ed.), 1994. *Concrete Structures in Earthquake Regions: Design and Analysis*, Longman, Harlom.

Coburn, A., Hughes, R.E., Pomonis, A. and Spence, R., 1995. *Technical Principles of Building for Safety*, Intermediate Technology Publications, London.

Croci, G., 1998. *The Conservation and Structural Restoration of the Architectural Heritage*, Computational Mechanics Publications, Southampton.

Dowrick, D.J., 1987. *Earthquake Resistant Design for Engineers and Architects* (2nd edition), John Wiley & Sons, Chichester.

Fielden, B.M., 1987. *Between Two Earthquakes: Cultural Property in Seismic Zones*, ICCROM, Via di San Michele 13, 00153 Roma, Italy.

Hansen, R.D. and Soong, T.T., 2001. *Seismic Design with Supplemental Energy Dissipation Devices*, Earthquake Engineering Research Institute, Berkeley, CA.

IAEE, 1986. *Guidelines for Earthquake-Resistant Non-Engineered Construction*, International Association for Earthquake Engineering, Kenchiku Kaikan 3rd Floor, 5-26-20, Shiba, Minato-ku Tokyo 108, Japan.

Key, D., 1988. *Earthquake Design Practice for Buildings*, Thomas Telford, London.

Lagorio, H., 1991. *Earthquakes: an Architects Guide to Non-Structural Seismic Hazards*, John Wiley & Sons, Chichester.

Paz, M., 1994. *International Handbook of Earthquake Engineering*, Chapman and Hall, London.

Soong, T.T. and Spencer, B.F., 2000. 'Active, semi-active and hybrid control of structures', Paper 2834, *World Conference on Earthquake Engineering, Auckland*, Elsevier, Amsterdam.

9 Earthquake Risk Modelling

9.1 Loss Estimation

The estimation of probable future losses is of great importance to those concerned with the management of facilities or public administration in earthquake-prone regions. Future loss estimates are of interest to:

- those responsible for physical planning on an urban or regional scale, particularly where planning decisions can have an effect on future losses;
- economic planners on a national or international scale;
- those who own or manage large numbers of buildings or other vulnerable facilities;
- the insurance and reinsurance companies which insure those facilities;
- those responsible for civil protection, relief and emergency services;
- those who draft building regulations or codes of practice for construction, whose task is to ensure that adequate protection is provided by those codes at an acceptable cost.

A variety of different types of loss estimation studies are used depending on the nature of the problem and the purpose of the study. These include:

- *Scenario studies:* Calculation of the effects of a single earthquake on a region. Often a 'maximum probable' or 'maximum credible' magnitude earthquake is assumed, with a best-guess location, based on known geological faults or probabilistic seismic source zones. Historically significant earthquakes, such as the 1906 San Fransisco event, or the 1923 Great Tokyo earthquake are commonly used as scenarios to assess their effects on present-day portfolios. Scenario studies are used to estimate the likely losses from an extreme case, to check the financial resilience of a company or institution to withstand that level of loss, and also to estimate the resources likely to be needed to handle the emergency, i.e. for preparedness planning. The number of people killed, injured, buried by collapsing buildings or made homeless is estimated. From these can be estimated the resources needed to minimise disruption, rescue people buried, accommodate the homeless, and minimise the recovery period.

- *Probabilistic risk analysis:* Calculation of all potential losses and the probability of those losses occurring from each of the different sizes and locations of earthquakes that can occur. For an individual building or for a portfolio of buildings or other assets in a region, this generates a loss exceedance probability (EP) curve, defining the level of loss that would be experienced with different return periods. The EP curve is used to calculate the average annual loss, to use in financial reserving, insurance rate setting or risk benchmarking. The EP curve provides the probability of different levels of loss being achieved, such as the probability of the losses exceeding financial reserves, bankrupting a company, or triggering a reinsurance contract. Probabilistic risk analysis can be used to estimate EP curves for the numbers of buildings destroyed, lives lost and total financial costs over a given period of time. With sufficient detail in the calculation, the likely effect of mitigation policies on reducing earthquake losses can be estimated and costed. The relative effects of different policies to reduce losses can be compared or the change in risk over time can be examined.
- *Potential loss studies:* Mapping the effect of expected hazard levels across a region or country shows the location of communities likely to suffer heavy losses. Usually the maximum historical intensity or a level of peak ground acceleration associated with a long probabilistic return period is mapped across an area. The effect of the intensity on the communities within that area is calculated to identify the communities most at risk. This shows, for example, which towns or villages are likely to suffer highest losses, which should be priorities for loss reduction programmes, and which are likely to need most aid or rescue assistance in the event of a major earthquake.

Table 9.1 summarises the different users of loss estimation and the types of output required.

Because of the importance of loss modelling to so many different groups, and its complexity, the last decade has seen the development of many sophisticated computer models for the computation of likely losses, using scenario studies or on a probabilistic basis. The most advanced of these models have been developed to help the international insurance and reinsurance industry, which has huge financial exposure in earthquake zones, to assess its probable and maximum possible losses. Several specialist companies have developed to supply this demand, and recent earthquakes, particularly the 1994 Northridge earthquake and the 1995 Kobe earthquake, have provided detailed loss data to test and calibrate their

Table 9.1 Users of loss estimation and the information they need.

Who	Why	Information needed
Physical planners	Identify high-risk locations	Risk mapping
Building owners	Identify high-risk buildings Plan mitigation strategies	Building-by-building vulnerability studies
Insurers and reinsurers	Set insurance premium rates Structure risk transfer (reinsurance) deals Identify possible losses Reduce risks	Annualised loss and exceedance probability curves
Civil protection agencies	Plan size and location of emergency services	Estimates of fatalities and injuries, damage, homelessness
Building regulators	Determine optimum resistance levels	Cost–benefit studies

models. This has led in turn to the development of some of the new techniques described later in this chapter for estimating physical and other losses.[1]

Because of the uncertainty of the knowledge available about earthquakes and their recurrence patterns, all loss estimates are necessarily extrapolations into the future of the observed statistical distribution of earthquakes and their effects in the past, and are based on attempts to determine the earthquake risk on a probabilistic basis. The term *risk*, and the associated terms *hazard* and *vulnerability*, have been formally defined by international agreement, and these agreed definitions, which are set out in the next section, will be used in this book.

9.2 Definition of Terms

9.2.1 Risk

The term *earthquake risk* refers to the expected losses to a given *elements at risk*, over a specified future time period.[2] The element at risk may be a building, a group of buildings or a settlement or city, or it may be the human population of that building or settlement, or it may be the economic activities associated with either. According to the way in which the element at risk is defined, the risk may be measured in terms of expected economic loss, or in terms of numbers of lives lost or the extent of physical damage to property, where appropriate measures

[1] The November 1997 issue of the *Journal of Earthquake Spectra* was devoted to loss estimation, and is a useful summary of recent progress in the United States.

[2] According to the international convention agreed by an expert meeting organised by the United Nations Office of the Co-ordinator of Disaster Relief (UNDRO) in 1979 (UNDRO 1979, Fournier d'Albe 1982).

of damage are available. Risk may be expressed in terms of average expected losses, such as:

25 000 lives lost over a 30-year period, or
75 000 stone masonry houses experiencing heavy damage or destruction within 25 years,

or alternatively on a probabilistic basis:

a 75% probability of economic losses to property exceeding $50 million in the City of L within the next 10 years.

The term *specific risk* is used to refer to risks or loss estimations of either type which are expressed as a proportion or percentage of the maximum possible loss. Specific risk is commonly used for financial losses to property, where it usually refers to the ratio of the cost of repair or reinstatement of the property to the cost of total replacement, the *repair cost ratio*.[3]

9.2.2 Hazard

Hazard is the probability of occurrence of an earthquake or earthquake effects of a certain severity, within a specific period of time, at a given location or in a given area. According to the type of analysis that is being made, the earthquake may be specified in terms of either its source characteristics or its effect at a particular site. The source characteristics of earthquakes are most commonly specified in terms of magnitude (see Chapter 1). When considering the hazard of ground shaking, the site characteristics of the earthquake are expressed in terms of an intensity or a parameter for severity of ground motion, such as EMS or modified Mercalli intensity, or in terms of peak ground acceleration (PGA), or some other parameter derived from measured characteristics of the motion. Like risk, hazard may be expressed in terms of average expected rate of occurrence of the specified type of event, or on a probabilistic basis. In either case annual recurrence rates are usually used. The inverse of an annual recurrence rate is an average *return period* . Examples of hazard defined in terms of the earthquake source are:

there is an annual probability of 8% of an earthquake with a magnitude exceeding 7.0 in region E.

[3] The term earthquake risk is still sometimes used to refer not to expected losses but to the expected future occurrence of given levels of earthquake ground shaking. According to the UNDRO definitions, the term 'hazard' is to be used with this meaning, as will be defined below. This is clearly a potential source of confusion, and as both usages are likely to continue for the near future, it is essential to be cautious in reading documents, particularly from US sources, which deal with earthquake risk. The UNDRO definition will be used here.

This is effectively the same thing as saying:

> the average return period for an earthquake of $M \geqslant 7.0$ in region E is 12.5 years, or there is a probability of 85% that an earthquake with a magnitude exceeding 7.0 will occur in region E within the next 25 years.[4]

Examples of earthquake hazard expressed in terms of its site characteristics are:

> an annual probability of 0.04 (or 4%) of an earthquake of EMS intensity VI in the town of N (or expected return period of 25 years for the same event – an equivalent definition),

or:

> an annual probability of 0.20 (or 20%) of a peak ground acceleration exceeding 0.15% g in M City.

The hazard expressed in this way is of course only a partial definition of the ground shaking hazard, related to events of a particular size range. The definition of the hazard for all possible size ranges cannot be done by a single statement of the type given above, but can be presented graphically, as a relationship between the annual probability and the size of the event. An example of a hazard definition in terms of the regional frequency of recurrence of earthquakes of different magnitude is given in Figure 9.1 for several broad regions of the world.

In addition to ground shaking, the potential for other collateral hazards from ground liquefaction, from landslide, dam failure or tsunami, and from direct damage in the fault rupture zone need to be considered at any site; in each case a characteristic hazard parameter needs to be defined, and expressed in a similar way to that for ground shaking hazard.

9.2.3 Vulnerability

Vulnerability is defined as the degree of loss to a given element at risk (or set of elements) resulting from a given level of hazard (i.e. from the occurrence of an earthquake of a given severity). The vulnerability of an element is defined as a ratio of the expected loss to the maximum possible loss, on a scale from 0 to 1 (or 0 to 100%). The measure of loss used depends on the element at risk, and accordingly may be measured as a ratio of numbers killed or injured to total population, as a repair cost ratio or as the degree of physical damage defined on an appropriate scale. In a large population of buildings it may be defined in terms of the proportion of buildings experiencing some particular level of damage.

---

[4]Using the assumed Poisson distribution of earthquake occurrence, explained in Section 9.9.

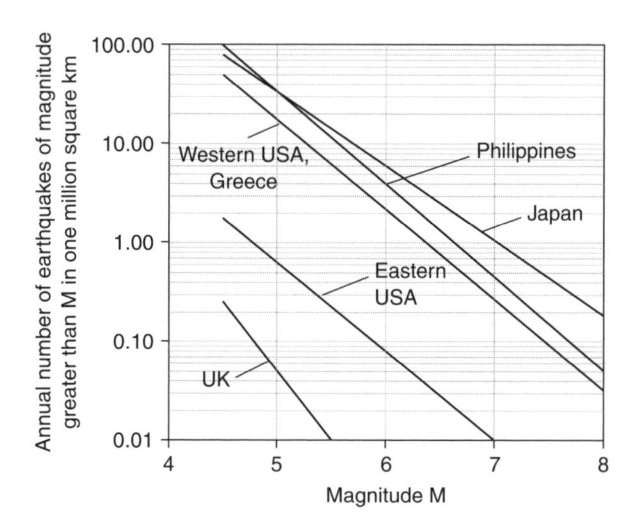

Figure 9.1 Relative seismic activity rates in different parts of the world. In an area of moderate to low seismicity, such as the eastern United States, the probability of an earthquake of magnitude 6.0 or above is little more than one-hundredth of the probability of such an event in Japan

The vulnerability of a set of buildings to an earthquake of intensity VIII may be defined as:

> 70% of buildings suffering heavy damage or worse, at intensity VIII, or average repair cost ratio of 55% at intensity VIII.

Specification of average vulnerability alone is rarely adequate for making loss assessments, however, because the distribution of losses within the set of elements at risk is generally very wide, with some elements sustaining very high degrees of damage, others very little. Thus the vulnerability of elements such as buildings, where a degree of damage may be assessed, is generally expressed by means of a damage distribution which may be expressed as a histogram. The derivation of such distributions is further discussed in Section 9.3.

As in the case of hazard it is clear that the vulnerability to one size of event is only a partial definition of the total vulnerability, which needs to be specified for all possible events which may cause any loss or damage. The complete vulnerability for an element at risk is therefore an assembly of the separate vulnerability distributions for each size of event which may need to be considered. Table 9.5 below shows an example of such a damage probability matrix. Damage probability distributions are defined for events of intensity V or VI to X.

Vulnerability functions such as that shown in Figure 9.2 may be combined with the hazard data defined as shown above in order to estimate the probable distribution of losses for all possible earthquake events in a given time period

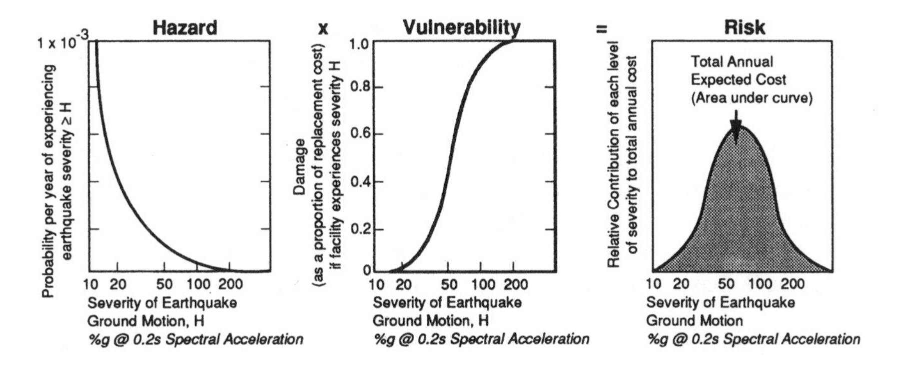

Figure 9.2 Risk is a product of hazard and vulnerability: typical curve shapes

and thus to determine the risk to that element or set of elements at risk. How this is done in particular cases is discussed in Section 9.5 below.

9.2.4 Mathematical Definitions

The definitions given above may also be expressed mathematically, in a way which facilitates the computation of risk. The general equation for the calculation of risk can be given as:

$$[R_{ij}] = [H_j][V_{ij}]$$

where, for an element at risk (e.g. an individual building) i:

$[R_{ij}]$ is the risk, the probability or average rate of loss to element i due to earthquake ground motion of severity j.

$[H_j]$ is the hazard, the probability or average expected rate of experiencing earthquake ground motion (or other earthquake related damaging event) of severity j.

$[V_{ij}]$ is the vulnerability, the level of loss that would be caused to element i as a result of experiencing earthquake ground motion of severity j (where loss is the specific loss; loss as a proportion of the total value of element i).

By summing the risk from all levels of hazard (min $\leqslant j \leqslant$ max), the total risk to any individual element can be derived (see Figure 9.2). The specification of hazard has been discussed in Chapter 7, and that of vulnerability is discussed in more detail in the following sections.

9.3 Vulnerability Assessment

9.3.1 General Approach

Vulnerability is the degree of loss to a given element at risk resulting from the occurrence of a specified earthquake. For assessment of losses due to ground shaking over a population of buildings (or other elements at risk) we need:

1. A means of specifying the earthquake hazard, as discussed in Chapter 7.
2. A classification of the building types or other facilities into distinct types whose performance in earthquakes is likely to be similar both in nature and degree.
3. A method of defining loss so that the extent of loss to a particular building or population of buildings can be quantified.
4. A means of estimating the distribution of losses to each building type for each discrete level of ground shaking (if intensity scales are used), or as a function of ground shaking (if a continuous parameter of ground shaking is used).

A similar approach needs to be used for estimating losses to other collateral hazards.

There are two principal methods of vulnerability assessment, which may be referred to as *predicted* vulnerability and *observed* vulnerability.[5] Predicted vulnerability refers to the assessment of expected performance of buildings based on calculation and design specifications, or, if no other method is available, on judgement based on the assessor's experience. Observed vulnerability refers to assessment based on statistics of past earthquake damage. The former method is suitable for use primarily with engineered structures and facilities, where a reasonable estimate of earthquake resistance may be made, but for which only a limited amount of damage data, if any, is available. The latter method is more suitable for use with non-engineered structures made with low-strength materials such as timber or unreinforced masonry, whose earthquake resistance is more difficult to calculate, but for which substantial statistical damage data may be available. The use of observed vulnerability is increasingly relevant in the case of very common forms of engineered construction, such as reinforced concrete frame structures, as the amount of damage data increases over time.[6] For the most common building types, observed vulnerability methods will continue to be used, but increasingly predicted vulnerability will be needed to assess the performance of newer and better built structures which have not yet been tested in severe earthquake shaking.[7]

9.3.2 Building Type and Facility Classification

The building type and facility type classification to be adopted in any study will depend not only on those characteristics which are expected to influence the earthquake performance of the structure but on the extent of data available. Most studies of earthquake damage have concluded that the form of construction used

---

[5] Sandi (1982).

[6] A relatively new approach is to combine observed and predicted vulnerability measures, as described by Bommer *et al.* 2002.

[7] The effects of changing building stock on vulnerability assessment are discussed by Spence (2000).

for the primary load-bearing structure is the most important factor affecting earthquake damage; in some instances the *vertical structure* ('load-bearing masonry', 'timber' or 'reinforced concrete frame') is a sufficient definition, but in other cases the *horizontal structure* used for floor and roof (timber joists or reinforced concrete slab, for example) may be equally important. In any particular area the definition of the form of construction in this way will embrace the entire building practice associated with it, thus providing a reasonably homogeneous class of buildings in a particular region, but 'load-bearing masonry' buildings located in different regions will not necessarily be similarly well constructed or perform in a similar way in earthquakes.

Because of changing building practices over time, the definition of the *date or period of construction* may be an equally important element of the building classification. Building practices can change radically within a short space of time as a result of economic changes, changes in regulations or building code, rebuilding after earlier disasters or political upheavals. A knowledge of these changes is essential to establish an effective building stock classification. For modern engineered buildings the earthquake performance is likely to be strongly affected by other aspects of the form of construction such as 'moment-resisting frame' or 'shear wall' for reinforced concrete buildings, and by the number of storeys; the building type classification may therefore need to include these factors.

Many other aspects of a building's construction have been shown to have an influence over its performance in earthquakes, some of which are listed in Table 9.2. These have already been discussed in Chapter 8. A classification of common building types in seismic areas of the world is presented in Table 8.1. However, finer and finer subdivision of the building classification would require correspondingly more vulnerability relationships to be defined, and quantitative measures of the separate influence of these factors are difficult to obtain. Their influence is better assumed to be taken account of by the distribution of expected damage within each class, discussed below.

Table 9.3 shows the classification of building types proposed for use with the HAZUS loss estimation methodology (FEMA 1999). This defines 16 model building classes, with further subdivision by numbers of storeys, giving 36 classes in all.[8]

For each country and region, building types will differ, and the classification needed will depend on the range of building uses to be included in the study. For example, where only residential buildings are to be studied, the range of building types considered may not need to include high-rise steel or concrete frames.

9.3.3 Damage Evaluation

Quantification of structural damage presents a number of difficulties. The mechanisms of damage are different for each building type; the cracking and

[8] See FEMA (1999).

Table 9.2 Secondary vulnerability factors (i.e. factors apart from construction type and local subsoil condition) known to influence earthquake damage to structures.

(a) Structural form

Non-symmetrical or irregular plans
Differences in the architectural plans and stiffnesses of different stories (especially framed buildings)
Total number of storeys and stiffness of structure and its effect on natural period and dynamic characteristics of the building
Single directions of strength (e.g. load-bearing walls all in same direction, or frame buildings with a unidirectional structure) and orientation of building with respect to direction of seismic force
Excessive wall openings leaving insufficient wall area to resist lateral shear (masonry)
Heavy roof forms and disposition of loads with height
Foundations; depth, adequacy, protection from frost, etc.
Design faults. Good practice not followed, e.g. vertical load-bearing elements not aligned from one floor to the next

(b) Site planning

Mutual stiffening effects of adjoining buildings
'Pounding' effects of adjacent buildings colliding
Slope effects causing subsidence and weakening buildings before earthquakes
Local ground failure under buildings, triggered or exacerbated by earthquake

(c) Construction quality

Low quality of building materials and failure to comply with specifications (e.g. during wartime construction)
Low quality of work. Good practice not followed or ignorance of the need for details
Deliberate neglect of conforming to design specifications (e.g. misappropriation of concrete reinforcement)
Mixtures of construction materials with different seismic performance (e.g. in load-bearing masonry).

(d) History

Age. Decay and weakening of materials
Pre-existing damage weakening structure, from previous earthquakes, war damage, foreshocks, etc.
Repair, maintenance and strengthening of structure
Modifications to structure (e.g. addition of another storey, extension of plan, alteration of structure to fit services, etc.)

disintegration of load-bearing masonry, for example, is a significantly different process to the deterioration and failure of a reinforced concrete frame. In some cases it is possible to avoid these differences by quantifying and comparing damage in financial terms. The most commonly used economic measure of repair is repair cost ratio (or RCR). This is the ratio of the cost of repair and

Table 9.3 Building structure type classification used in the HAZUS earthquake loss estimation methodology (FEMA 1999).

Label	Building class	Subdivisions
W1	Wood, light frame	
W2	Wood, commercial and industrial	
S1	Steel moment frame	Low, mid- and high rise
S2	Steel, braced frame	Low, mid- and high rise
S3	Steel light frame	
S4	Steel frame with cast-in-place concrete shear walls	Low, mid- and high rise
S5	Steel frame with unreinforced masonry infill walls	Low, mid- and high rise
C1	Concrete moment frame	Low, mid- and high rise
C2	Concrete shear walls	Low, mid- and high rise
C3	Concrete frame with unreinforced masonry infill walls	Low, mid- and high rise
PC1	Precast concrete tilt-up walls	
PC2	Precast concrete frames with concrete shear walls	Low, mid- and high rise
RM1	Reinforced-masonry-bearing walls with wood or metal deck diaphragms	Low and mid-rise
RM2	Reinforced-masonry-bearing walls with precast concrete diaphragms	Low, mid- and high rise
URM	Unreinforced-masonry-bearing walls	Low and mid-rise
MH	Mobile homes	

Low rise = 1–3 storeys
Mid-rise = 4–7 storeys
High rise = more than eight storeys.

reinstatement of the structure (or building) to the cost of replacing the structure (or building). The evaluation of damage in terms of repair cost is unsatisfactory for many purposes, though, because of its dependence on the economy at that time and place. Repair cost ratio varies because there are different ways of repairing and strengthening, and because construction costs vary from place to place and through time – they often rise steeply after an earthquake has occurred. Repair cost ratio is also significantly affected by the type of building, and repair cost for serious damage may be more than replacement cost.

For these reasons, structural damage state is a more reliable measure of damage. If defined with sufficient accuracy, structural damage states can be converted into repair costs in any economic situation. Thresholds of structural damage also correlate with other indirect consequences such as human casualties, homelessness and loss of function, in ways that economic parameters of damage cannot. The definition of structural damage generally used involves a sequence of structural damage states, with broad descriptors such as 'light', 'moderate', 'severe', 'partial collapse', elaborated with more detailed descriptions which may use quantitative

Table 9.4 Definition of damage states for masonry and reinforced concrete frame buildings: brief damage definitions (see also full definitions in Section 1.3).

Damage level	Definition for load-bearing masonry	Definition for RC-framed buildings
D0 Undamaged	No visible damage	No visible damage
D1 Slight damage	Hairline cracks	Infill panels damaged
D2 Moderate damage	Cracks 5–20 mm	Cracks <10 mm in structure
D3 Heavy damage	Cracks >20 mm or wall material dislodged	Heavy damage to structural members, loss of concrete
D4 Partial destruction	Complete collapse of individual wall or individual roof support	Complete collapse of individual structural member or major deflection to frame
D5 Collapse	More than one wall collapsed or more than half of roof	Failure of structural members to allow fall of roof or slab

measures such as crack widths. A commonly used set of damage states is the six-point scale defined in the EMS scale described and illustrated in Section 1.3,[9] since the damage states defined in this scale are relatively easy to assess. A more detailed elaboration appropriate to assessing the performance of particular building types may sometimes be used; damage states, derived from the EMS scale, suitable for assessing the damage to masonry structures and reinforced concrete frame structures, are shown in Table 9.4.

Some damage evaluation methods assess damage levels separately for different parts of the structure and then use either the highest or average values for the overall damage state classification of the structure.

9.3.4 Damage Distribution

In any single location after an earthquake, buildings suffer a range of different types and levels of damage. Surveys record the distributions of structural damage states (numbers of buildings in each damage state) for each building type in each location. The format used for the definition of the probable distribution of damage depends on the method of defining the earthquake hazard parameter. Each of the basic methods of defining the earthquake hazard parameter described in Section 7.3 requires a different format.

Where the hazard is defined from macroseismic site shaking characteristics in terms of intensity, which is a discrete scale, the most widely used form is the damage probability matrix (DPM). The DPM shows the probability distribution of damage among the different damage states, for each level of ground shaking; DPMs are defined for each separate class of building or vulnerable facility.

[9] Grünthal (1998).

Table 9.5 Typical example of a damage probability matrix for Italian weak masonry buildings (based on Zuccaro 1998) % at each damage level.

Damage level		Intensity (European Macroseismic Scale) (%)					
		V	V1	VII	VIII	IX	X
D0	No damage	90.4	18.8	6.4	0.1	0.0	0.0
D1	Slight damage	9.2	37.3	23.4	1.8	0.2	0.0
D2	Moderate damage	0.4	29.6	34.4	10.0	2.0	0.4
D3	Substantial to heavy damage	0.0	11.7	25.2	27.8	12.5	4.7
D4	Very heavy damage	0.0	2.3	9.2	38.7	38.3	27.9
D5	Destruction	0.0	0.2	1.4	21.6	47.0	67.0

Table 9.5 shows an example. In this case, the range of expected damage cost (as a repair cost ratio (RCR)) is sometimes also given for each damage state, along with the estimated mean or central damage factor which may be assumed for each damage state; this makes it possible for the physical damage to be reinterpreted in terms of repair cost ratio.

Where the hazard is defined in terms of an engineering parameter of ground motion such as peak ground acceleration (PGA), similar information may be presented as a continuous relationship, defining, for the particular class in question, the probability that the damage state will exceed a certain level, as a function of the ground motion parameter used. An example of vulnerability defined this way is shown in Figure 9.5. In this case and the above, the damage distribution so defined is assumed to be a unique property of the particular building class, relevant in any earthquake, given the same defined level of ground shaking.

Where the hazard is defined in terms of the spectral displacement of a particular building type, vulnerability is expressed in terms of a set of fragility curves defining the probability of any building being in a given damage state after shaking causing a given spectral displacement. Such fragility curves are based on a standard distribution function, enabling them to be defined by the parameters of the distribution. The approach is discussed in more detail in Section 9.5.

Clearly, to define any such relationships on the basis of observed vulnerability, a substantial quantity of data is required; where data is missing or inadequate, a method is required to enable reasonable assessments to be made. Two such methods will be discussed in this section–the use of standard probability distributions, and the use of expert opinion survey. An alternative approach is described in Section 9.4.

9.3.5 Probability Distributions

In any location affected by destructive levels of earthquake ground motion, buildings will be found in a range of damage states. Surveys of damage, classifying buildings into building type categories and recording damage states for each, can be presented in the form of histograms showing the damage distribution for each building type. This distribution of damage is related to the intensity of ground motion so that, for example, where high intensities have been experienced, the damage distribution shifts towards the higher levels of damage. In the analysis of the damage data from past earthquakes it has been observed that the distributions of damage for well-defined classes of buildings tended to follow a pattern which is close to the binomial distribution.[10] Using this form, the entire distribution of the buildings among the six different damage states D0–D5 could be represented by a single parameter.[11]

The parameter p can take any value between 0 (all buildings in damage state D0, undamaged) and 1 (all buildings in damage state D5, collapsed). The distributions generated for particular values of p are shown in Figure 9.3. Defining damage distributions in terms of p both simplifies these definitions (replacing a six-parameter specification with a single parameter for each building class and level of ground motion) and provides a better basis for the use of limited damage data in generating distributions. The binomial parameter p may be used in the generation of either DPMs or continuous vulnerability functions.[12] Observations suggest that damage distributions of masonry buildings appear to conform quite well to the binomial model. Other building types, such as frame structures, may have a more varied distribution, requiring a more complex description. A similar characterisation of damage distribution in terms of the beta distribution has also been used,[13] which uses two parameters, and hence allows for more flexibility in the shape of the distribution to fit different circumstances.

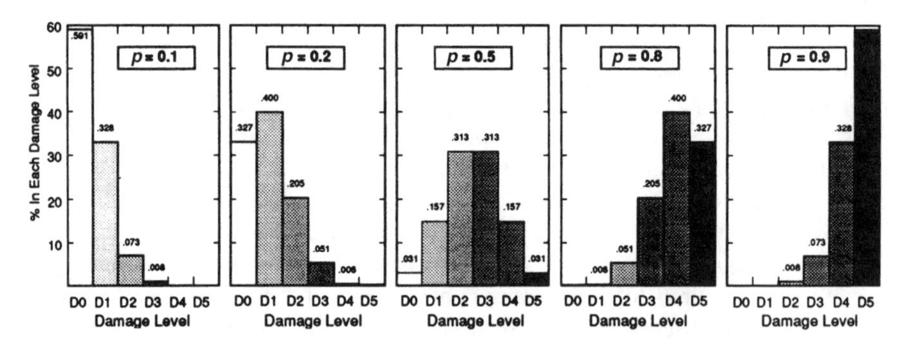

Figure 9.3 Theoretical distributions for each damage level D0–D5 defined by different values of binomial parameter p

---

[10] Braga *et al.*, (1982)

[11] According to this distribution, the proportion of the total building stock falling into damage state Dl is defined by $V_l = \{5!/[l!(5-l)!]\} \times p^l(1-p)^{5-l}$.

[12] Braga *et al.* (1982).

[13] For example, by Spence (1990) and Applied Technology Council (1985).

9.3.6 Expert Opinion Survey

The technique of expert opinion survey may be useful in generating vulnerability functions or DPMs for classes of structures which are reasonably well defined in structural terms, but for which limited damage data is available.

In essence the method is as follows. A number of experts are asked to provide independent estimates of the average damage level (defined in a predetermined way) for each class of building at each level of intensity; the answers are circulated to all the experts, who are then asked to revise their assessment in the light of the responses of others, and by this means a consensus is approached. The average damage levels agreed are then converted into damage probabilities using a standard distribution technique. One use of this method was in developing earthquake damage evaluation data for California.[14]

9.4 The PSI Scale of Earthquake Ground Motion

In many earthquake regions much of the building stock is not built to any code of practice, and there are no instruments available to measure ground motion. Thus, the use of damage data to assess the intensity of shaking at any location is likely to continue to be important both as a measure of the strength of the shaking and as a means to assess likely future losses.

But the use of macroseismic intensity scales as a ground motion parameter for this purpose has a number of difficulties:

- Intensity is a descriptive not a continuous scale, which makes it difficult to use for predictive purposes.
- Significant variations are found to exist between one survey group and another in identifying intensity levels.
- Intensity scales assume a relationship between the performance of different building types which is not found in reality.

The *parameterless scale of seismic intensity* (PSI scale) has been devised to avoid these problems. It is a scale of earthquake strong motion ‘damagingness’, measured by the performance of samples of buildings of standard types. It is based on the observation that, although assigned intensity in different surveys varies widely even with the same level of loss, the relative proportions of a sample of buildings of any one type in different damage states are fairly constant, and so are the relative loss levels of different building classes surveyed at the same location.

[14] Applied Technology Council (1985).

Figure 9.4 shows, for example, the average performance of samples of brick masonry buildings at and above each level of damage D0 to D5, given the proportion of the sample damaged at or above level D3.

The PSI scale is based on the proportion of brick masonry buildings damaged at or above level D3; it is assumed that this proportion is normally distributed with respect to the ground motion scale. The PSI parameter ψ is defined so that 50% of the sample is damaged at level D3 or above when $\psi = 10$, and the standard deviation is $\psi = 2.5$. Thus about 16% of the sample is damaged at D3 or above when $\psi = 7.5$, 84% when $\psi = 12.5$, etc. The curve for D3 thus has the form shown in Figure 9.5(a). Using this curve as a basis, the curves for other damage levels are defined from the relative performance of buildings in a large number of damage surveys. Likewise, vulnerability curves for other building types have been derived from their performance relative to brick buildings in surveys.

Since the vulnerability curves are of cumulative normal or Gaussian form, the proportion of buildings damaged to any particular damage or greater is given by the standard Gaussian distribution function.[15]

Values of the Gaussian distribution parameters M and σ for a range of common building types and damage states have been derived from the damage data in the Martin Centre damage database. These are shown in Table 9.6, with confidence limits on M where appropriate. Some examples are illustrated in Figure 9.6. A fuller description and justification for the PSI methodology is given elsewhere.[16]

9.4.1 Relating PSI to Other Measures of Ground Motion

Figure 9.5(a) shows how the PSI scale relates to the intensity scale defined in the EMS 1998 scale.

---

[15] A normal distribution is defined by a mean, M, and a standard deviation, σ, as:

$$y = \frac{1}{\sqrt{2\pi}\,\sigma} \exp\left[-\frac{1}{2}\left(\frac{x - M}{\sigma}\right)^2\right] \quad (1)$$

The cumulative distribution function, $D = \text{Gauss}[M, \sigma, \psi]$, is then defined by:

$$D = \int_{-\infty}^{\psi} \frac{1}{\sqrt{2\pi}\,\sigma} \exp\left[-\frac{1}{2}\left(\frac{\psi - M}{\sigma}\right)^2\right] \quad (2)$$

where D is the percentage of the building stock damaged (0–1.0) and ψ is the intensity. The inverse function, $\psi = \text{Gauss}^{-1}[M, \sigma, D]$, can also be used to derive an intensity value from a level of damage.

[16] Spence *et al.* (1998).

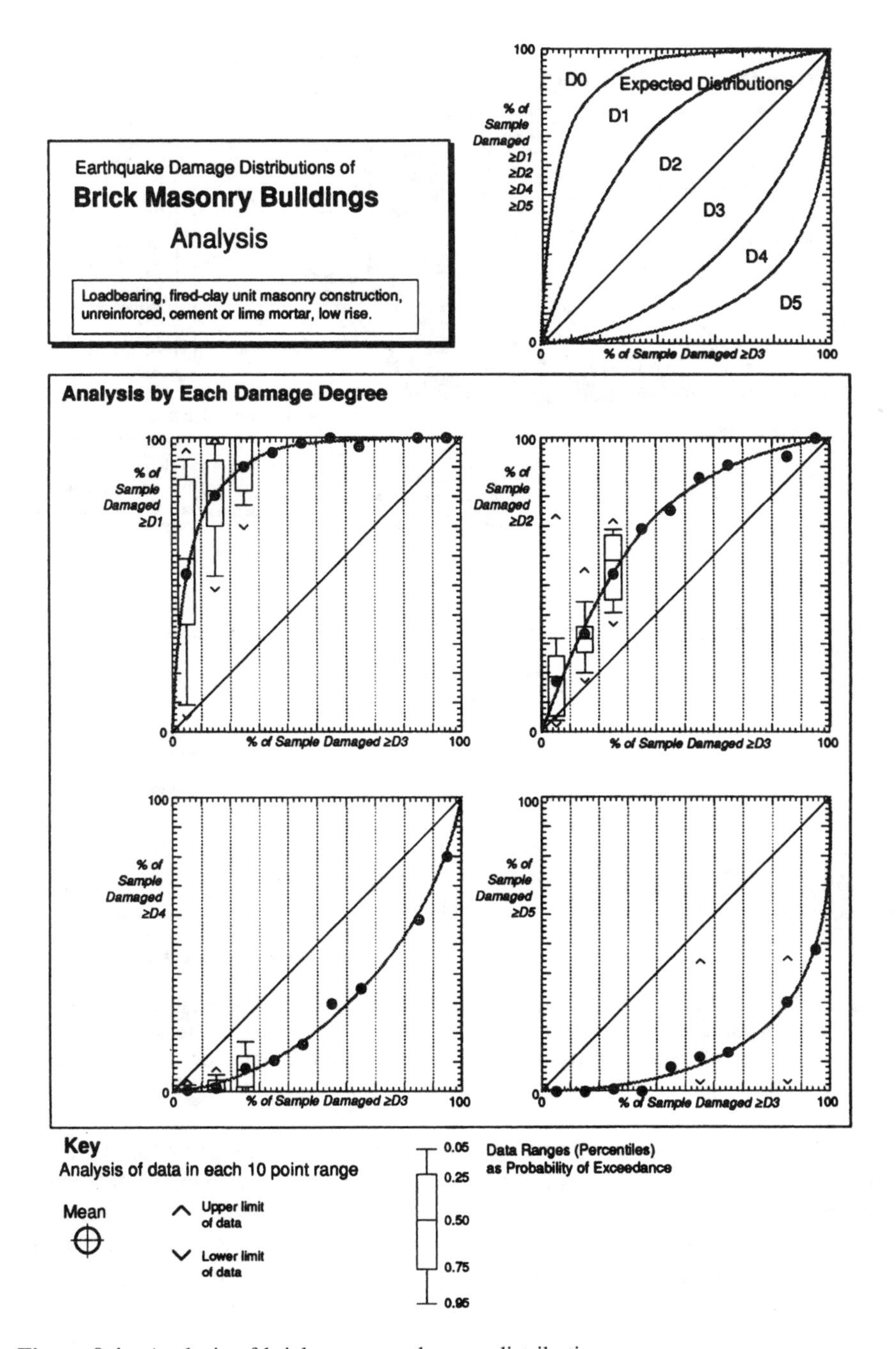

Figure 9.4 Analysis of brick masonry damage distributions

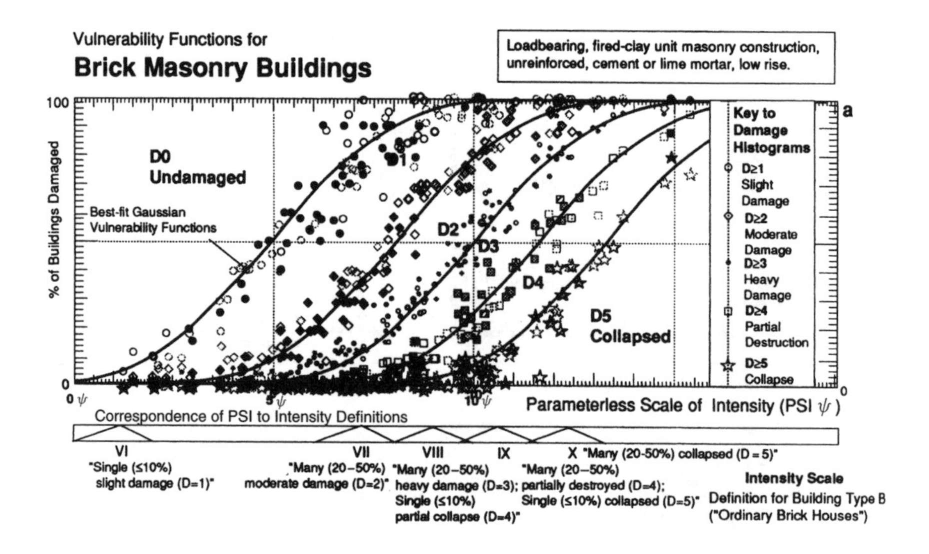

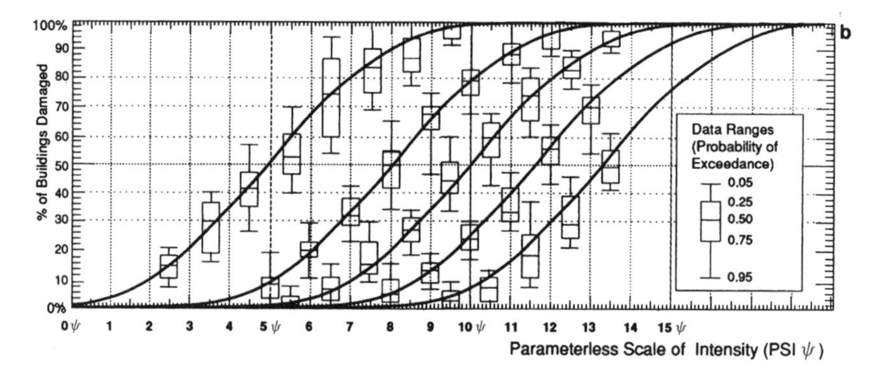

Figure 9.5 (a) Damage distributions of brick masonry buildings arranged as a best fit against Gaussian curves are used to define the parameterless scale of intensity (PSI or ψ). (b) An analysis of the scatter from this gives the confidence limits on predictions using this method

Where it has been possible to carry out statistical damage surveys in the immediate vicinity of recording instruments (within a radius of maximum 400 metres where soil conditions remain constant) it is possible to obtain an approximate correlation between PSI and various ground motion parameters. Figure 9.7 shows data points and linear regression analyses carried out for two particular parameters: peak horizontal ground acceleration (PHGA) and mean response spectral acceleration (MRSA). Peak horizontal ground acceleration is the most commonly used parameter of ground motion, and although the dataset is small, Figure 9.10

Table 9.6 Vulnerability functions for worldwide building types.

			D1	D2	D3	D4	D5
High confidence (20 to 100 damage survey data points)							
BB1	Brick masonry unreinforced	M	4.9	7.8	10.0	11.6	13.3
		σ	2.5	2.5	2.5	2.5	2.5
		Conf. limits (SD)	0.6	0.4	0.4	0.6	0.7
CC1	RC frame, non-seismic	M	7.9	10.3	11.3	12.9	14.1
		σ	2.5	2.5	2.5	2.5	2.5
		Conf. limits (SD)	0.7	0.9	0.5	0.8	1.0
AR1	Rubble stone masonry	M	3.2	5.9	8.2	9.8	11.7
		σ	2.5	2.5	2.5	2.5	2.5
		Conf. limits (SD)	1.0	0.7	0.6	0.8	1.1
Good confidence (up to 20 damage survey data points)							
AA1	Adobe (earthen brick) masonry	M	3.9	6.6	8.9	10.5	12.4
		σ	2.5	2.5	2.5	2.5	2.5
		Conf. limits (SD)					
BB2	Brick with ringbeam or diaphragm	M	6.5	9.4	11.6	13.2	14.9
		σ	2.5	2.5	2.5	2.5	2.5
		Conf. limits (SD)					
BC1	Concrete block masonry	M	5.6	8.5	10.7	12.3	14.0
		σ	2.5	2.5	2.5	2.5	2.5
		Conf. limits (SD)					
BD1	Dressed stone masonry	M	4.0	7.1	9.0	10.5	12.4
		σ	2.5	2.5	2.5	2.5	2.5
		Conf. limits (SD)					
DB1	Reinforced unit masonry	M	7.5	10.6	13.0	15.0	17.0
		σ	2.5	2.5	2.5	2.5	2.5
		Conf. limits (SD)					
There is good evidence from surveys of earthquake damage in Italy (1980) and Turkey (1983) that a reinforced concrete ringbeam or floor diaphragm in load-bearing masonry structures A and B decreases their vulnerability by about 1.6 ψ units (add 1.6 to ψ50 values for these building types).							
Moderate confidence (extrapolated from published estimates by others)							
CT1	Timber frame with heavy infill	M			10.6		
		σ	2.5	2.5	2.5	2.5	2.5
		Conf. limits (SD)					
CT2	Timber frame with timber cladding	M	7.2	9.5	12.0	14.3	15.5
		σ	2.5	2.5	2.5	2.5	2.5
		Conf. limits (SD)					
DC	RC frame seismic design UBC2	M	8.8	10.5	12.5	14.1	15.2
		σ	2.5	2.5	2.5	2.5	2.5
		Conf. limits (SD)					

(*continued overleaf*)

Table 9.6 (*continued*)

			D1	D2	D3	D4	D5
DC	RC frame seismic design UBC3	M	9.4	11.1	13.0	14.7	16.4
		σ	2.5	2.5	2.5	2.5	2.5
		Conf. limits (SD)					
DC	RC frame seismic design UBC4	M	10.6	12.4	14.7	17.0	18.8
		σ	2.5	2.5	2.5	2.5	2.5
		Conf. limits (SD)					

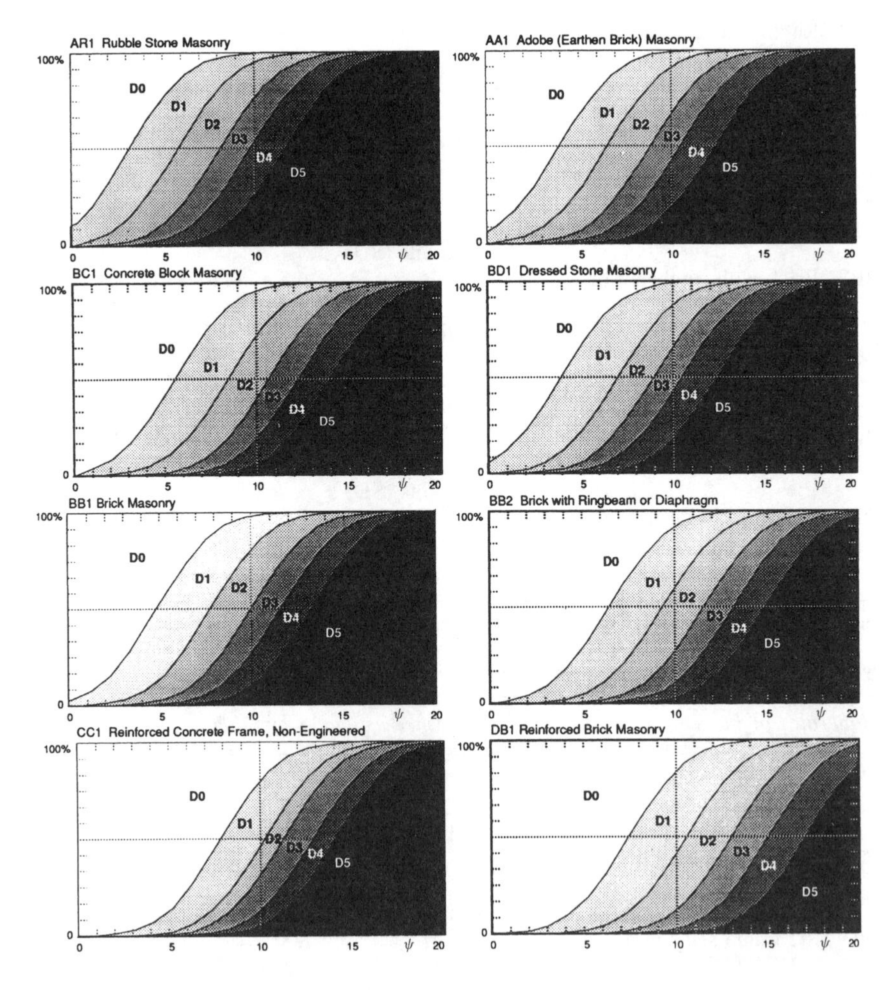

Figure 9.6 Vulnerability functions for some common building types

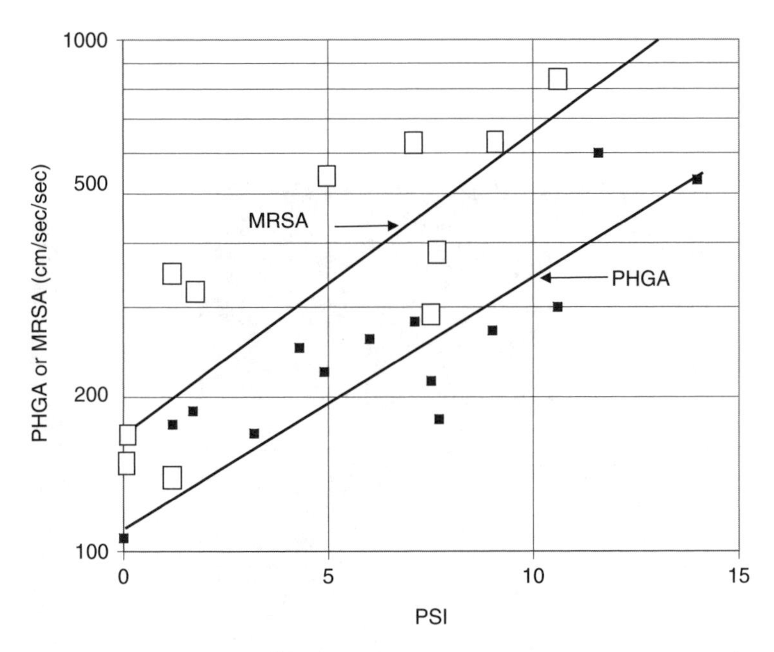

Figure 9.7 The relationship between PSI and instrumental parameters of ground motion, peak horizontal ground acceleration (PHGA) and mean response spectral acceleration (MRSA), over the 0.1–0.3 second period range. Correlation coefficients for the data from the 14 sites are not high, but MRSA is a somewhat better predictor of PSI than PHGA. (Data from Spence *et al.* 1991a)

below shows that it correlates reasonably well with PSI in this dataset: the coefficient of correlation is 0.77. The majority of the masonry buildings in the 14 sites examined here are residential houses one to three storeys high, and it could be expected that a good parameter to describe the 'damagingness' of ground motion to these buildings would be the mean response spectral acceleration over the range of the natural periods of such buildings, i.e. 0.1 to 0.3 seconds. This correlation has also been plotted in Figure 9.7 and was found to give a correlation coefficient of 0.81, slightly better than that for PHGA. Using this relationship and vulnerability functions such as those of Figure 9.6 offers a good basis for estimating losses when ground accelerations or intensities can be predicted.

The PSI scale can also be used to assist in the analysis of post-earthquake damage surveys. Figure 9.8 shows the results of surveys of buildings damaged in a number of locations, as surveyed after the 1999 Kocaeli earthquake.[17] Each of these surveys has been located on the appropriate set of damage curves determining the best-fit value of PSI at that location, and from this an understanding

[17] Johnson *et al.* (2000).

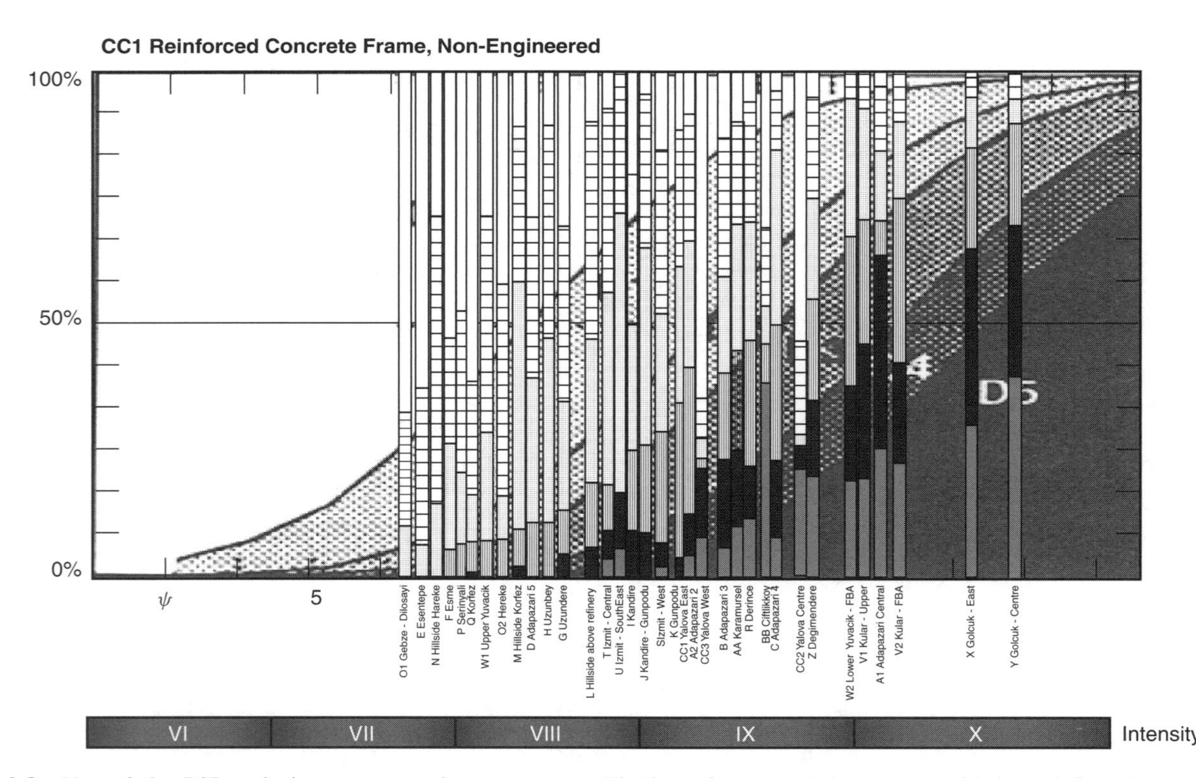

Figure 9.8 Use of the PSI scale in post-event damage survey. Plotting of surveyed damage to mid-rise reinforced concrete frame buildings in the locations of worst damage in the 1999 Kocaeli, Turkey, earthquake. (Risk Management Solutions)

of the geographical distribution of PSI and hence macroseismic intensity was deduced. A similar approach was used for mid-rise reinforced concrete frame buildings damaged in the 2001 Gujarat, India earthquake.[18]

9.5 The HAZUS Methodology

The HAZUS methodology is a predictive method of loss estimation based on recent performance-based procedures for the design of new buildings and for retrofitting existing buildings. For any individual building, these procedures enable levels of earthquake ground motion to be defined which correspond to a range of post-earthquake damage states, from undamaged to complete collapse. The use of such procedures is as applicable to evaluation as it is to design: that is, they can be used for assessing the probable state of an existing building after a given earthquake motion as well as for designing new (or strengthening existing) buildings. The HAZUS methodology has been developed in the United States as part of a FEMA-supported national programme to enable communities or local administrations to assess and thereby reduce the earthquake (and other) hazards they face.

The resulting HAZUS earthquake loss estimation methodology is a systematic approach which combines knowledge of earthquake hazards (from ground shaking, fault rupture, ground failure, landslide, etc.) with building and other facility inventory data and building vulnerability data to estimate losses for a community. One of its strengths is its comprehensiveness: estimation of losses includes losses to lifelines, industrial facilities, etc., and goes beyond direct damage to include estimates of induced damage (fire, hazardous materials release), and to estimates of casualties, shelter requirements and economic losses. But for these modules to be used, there is a large demand for inventory and other data appropriate to each locality. At the heart of the HAZUS loss estimation methodology is a process for developing vulnerability or fragility curves for buildings and other facilities, to estimate the losses from ground shaking, which has been used to define likely losses for a range of different building types found in the United States. Altogether it defines 36 different classes of buildings (Table 9.3) and many other facility classifications, distinguished according to age, height and level of seismic resistance designed for. For each building class a set of parameters defines the expected average earthquake capacity curve for the class. This curve, together with further parameters, then defines the displacement response to any given earthquake ground motion, resulting in an expected loss distribution for a typical population of buildings of any class.

The procedure needed to define the displacement response is rather more complex than that used to develop loss estimates based on MM or EMS intensity as the

---

[18] EEFIT (2002b), Del Re *et al.* (2002).

governing ground motion parameter. However, in many situations its advantages will outweigh the extra computational effort considering that:

- Engineering seismology internationally has for some years been directed towards defining earthquake ground motion in terms of instrumental parameters rather than macroseismic intensity.
- No satisfactory way to incorporate the interaction of earthquake ground motion characteristics (amplitude, frequency, duration) with soil type and building response is possible using intensity or any other single ground motion parameter.
- Intensity-based loss estimation methods are primarily derived from past damage data; this makes it difficult to estimate the losses to newer building types which have not experienced damage.
- The calculated displacement-based procedure can readily be extended to study the effect on losses of strengthening existing buildings in alternative ways, which is not easily achieved using intensity-based procedures.

9.5.1 Damage States

The essence of the HAZUS methodology for estimating losses from ground shaking is that the damage state of a building is taken to be defined by the interstorey drift ratio at the most deformed level of the building. A series of damage states is defined (called slight, moderate, extensive, complete) with detailed descriptors of the state of damage which corresponds with each state for each class. Figure 9.9 shows for example the damage states appropriate for mid-rise reinforced concrete frame buildings. For each separate class of building, each of these damage states is taken to correspond to a threshold level of interstorey drift ratio, at which this damage state would just be triggered.

Performance Point

For a single building, and for any given earthquake ground motion, the interstorey drift is derived from the spectral displacement of the building as a whole in response to the motion. This spectral displacement, at what is described as the 'performance point' for the building, is defined by the interaction of the 'demand' on the building created by the ground motion, and the 'capacity' of the building in terms of a response or capacity curve, which is derived from the elastic response of a single degree-of-freedom system by taking account of the degradation of the building as shaking progresses. Both demand and capacity are defined by curves of *spectral acceleration* S_a against *spectral displacement* S_d, and the performance point (S_a, S_d) is taken to be at the intersection of these two curves. This process is illustrated in Figure 9.10.

Damage state		Description
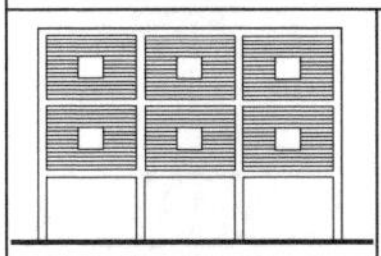	Slight structural damage	Diagonal (sometimes horizontal) hairline cracks on most infill walls; cracks a frame–infill interfaces
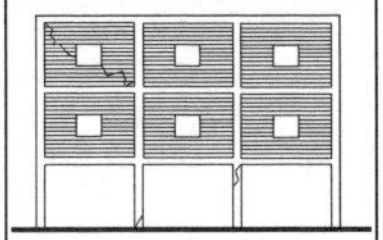	Moderate structural damage	Most infill wall surfaces exhibit larger diagonal or horizontal cracks; some walls exhibit crushing of brick around beam–column connections. Diagonal shear cracks may be observed in concrete beams or columns
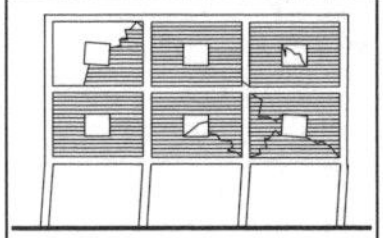	Extensive structural damage	Most infill walls exhibit large cracks; some bricks may dislodge and fall; some infill walls may bulge out-of-plane; few walls may fall partially or fully; few concrete columns or beams may fail in shear resulting in partial collapse. Structure may exhibit permanent lateral deformation
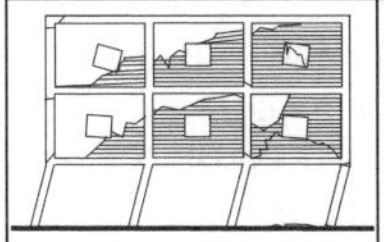	Complete structural damage	Structure has collapsed or is in imminent danger of collapse due to a combination of total failure of the infill walls and non-ductile failure of the concrete beams and columns

Figure 9.9 Damage states for low- and mid-rise reinforced concrete buildings used: the HAZUS loss estimation methodology

9.5.2 Capacity Curve

Detailed rules for the construction of standard capacity curves for each building class are given in the HAZUS manual.[19,20] The capacity curve is derived from static pushover curves using concepts explained in more detail in ATC-40[21] and FEMA 273.[22] For each building type the capacity curve for S_a vs S_d has an initial linear section where the slope depends on the typical natural frequency of the building class, and rises to a plateau level of S_a at which the maximum attainable resistance to static lateral force has been reached (Figure 9.10).

[19] FEMA (1999).

[20] Kircher *et al.* (1997).

[21] Applied Technology Council (1996).

[22] FEMA (1997).

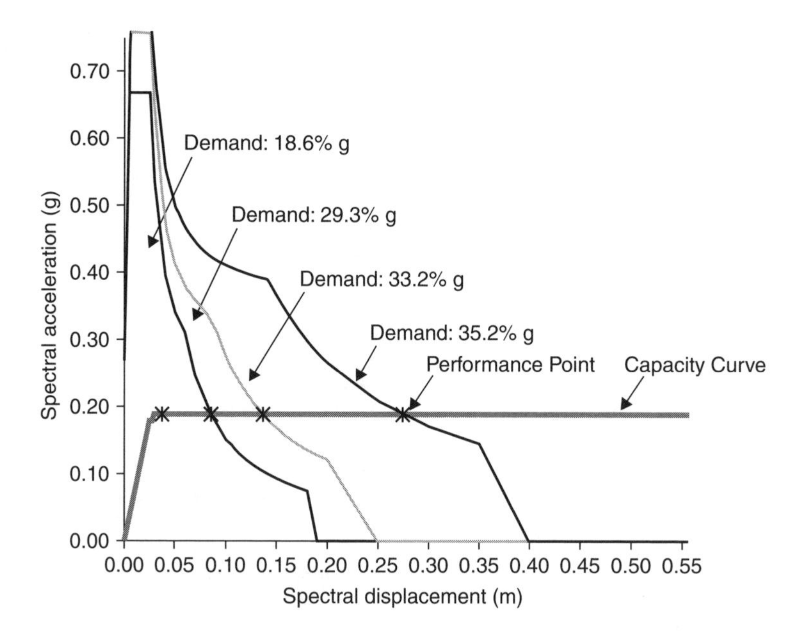

Figure 9.10 HAZUS loss estimation methodology. Definition of damage distribution for earthquakes with different levels of peak ground acceleration for a particular class of building. The 'performance point' at the intersection of the assumed capacity curve for the building class with the demand curve for a given earthquake defines the spectral displacement response, which is used with fragility curves (Figure 9.11) to define the distribution of damage states resulting from that earthquake

9.5.3 Demand Curve

The demand curve derives from a damped elastic spectral response curve built from spectral parameters of the ground motion, as modified according to soil type. This is done by incorporating spectral reduction factors to account for the increased hysteretic damping as the building shifts from elastic into inelastic response (Figure 9.10). Rules for constructing these spectral reduction factors are also given in the HAZUS manual: a different spectral reduction factor is associated with each value of spectral displacement; it depends on the shape of the capacity curve up to that displacement level, and also on a *degradation factor*, to account for the reduction in hysteretic damping occurring in poorly designed buildings, which depends in turn on the duration of shaking and the state of the building.

9.5.4 Damage Distribution

To estimate the performance of a group of buildings of a particular class under given ground shaking, the spectral response of the building at the performance point for the standard building of that class, as defined above, is used in conjunction with a set of four *fragility curves* (Figure 9.11) for that class, which estimate the probability of any particular building being in each of the four damage states after shaking at any given spectral response level. Each of these curves is assumed to be lognormal in form, and is defined by two parameters: a *median value* and a *coefficient of variation*. These curves are used to define the distribution of a set of buildings among the four damage states. The HAZUS manual gives parameters for the construction of these fragility curves for each of the 36 major building classes defined for the US building stock and for zones with different seismic design regulations.

For most building types, the spectral response to be used is the spectral displacement as defined above (such building types are considered 'displacement-sensitive', or 'drift-sensitive'), but some classes of facilities, and some building elements and equipment, are taken to be damaged as a result of the spectral acceleration rather than the spectral displacement (they are 'acceleration-sensitive'), and this is reflected in fragility curves defined in terms of this parameter. For the United States, a set of parameters to construct each of the curves required for each building type has been defined in the HAZUS manual. The method has also been applied for loss estimation studies in Turkey.[23]

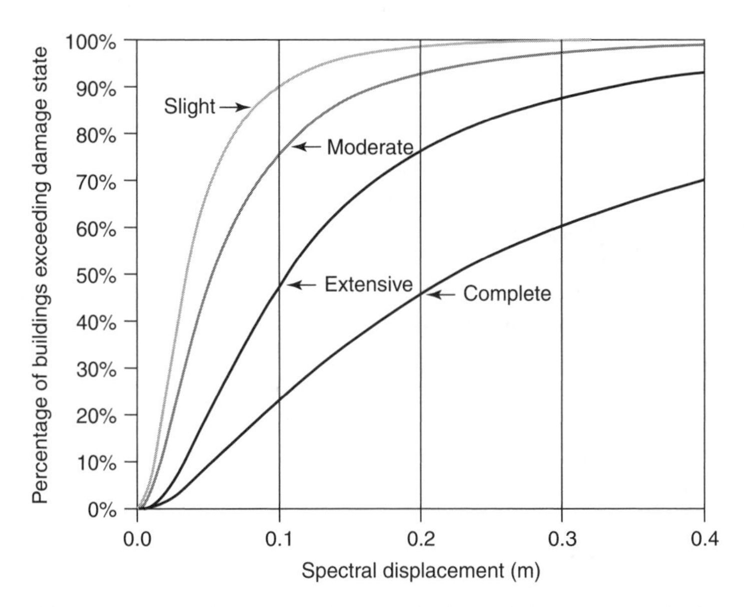

Figure 9.11 Example fragility curves for a particular building type used in the HAZUS loss estimation methodology

[23] Bommer *et al.* (2002).

9.6 Human Casualty Estimation

The purpose of most earthquake protection programmes is to save life. For loss estimation studies to be useful for earthquake protection they need to include an assessment of the probable levels of human casualties, both deaths and injuries, which will be caused by the earthquake.

Casualty estimation is notoriously difficult. Casualty numbers are highly variable from one earthquake to another and data documenting occurrences of life loss in earthquakes is poor. During an earthquake the chaotic disruption and physical damage causes loss of life in many different ways: building collapse, machinery accidents, heart attacks and many other causes. Some earthquakes trigger follow-on secondary hazards which also cause loss of life, like landslides, mudflows and fires.

An approximate classification of earthquake deaths by cause, during the twentieth century, is presented in Figure 1.1. Up to 25% of all deaths are from non-structural causes or follow-on hazards. In some cases, follow-on disasters like urban fires, mudflows, rockfalls and landslides can lead to many more deaths than those caused directly by the earthquake. Follow-on disasters of this type are extremely difficult to predict, but they normally cause only a small proportion of the earthquake casualties. For the large majority of earthquakes, deaths and injury are primarily related to building damage. Over 75% of deaths are caused by building collapse (and if secondary disasters are excluded, building collapse causes almost 90% of earthquake-related deaths). In Figure 9.12, the total number of people killed is plotted against the total number of buildings heavily damaged for earthquakes where both statistics are known with some reliability. Deaths can be seen to be broadly related to the destruction caused by earthquakes. However, casualty totals are much more variable in earthquakes causing low or moderate levels of damage, i.e. those where fewer than 5000 buildings were damaged.

An approach to estimating these casualties is by determining the 'lethality ratio' for each class of building present in a set of buildings damaged by an earthquake.[24] Lethality ratio is defined as the ratio of the number of people killed to the number of occupants present in collapsed buildings of that class. Thus the estimation of casualties derives from an estimate of the number of collapsed buildings of each class, calculated using methods described above, and the lethality ratio for that class.

Lethality ratio has been found from an examination of data from past earthquakes to depend on a number of factors including building type and function, occupancy levels, type of collapse mechanism, ground motion characteristics, occupant behaviour and SAR effectiveness. To obtain overall casualty levels, information on the spatial distribution of earthquake intensity and building

[24] Coburn *et al.* (1992).

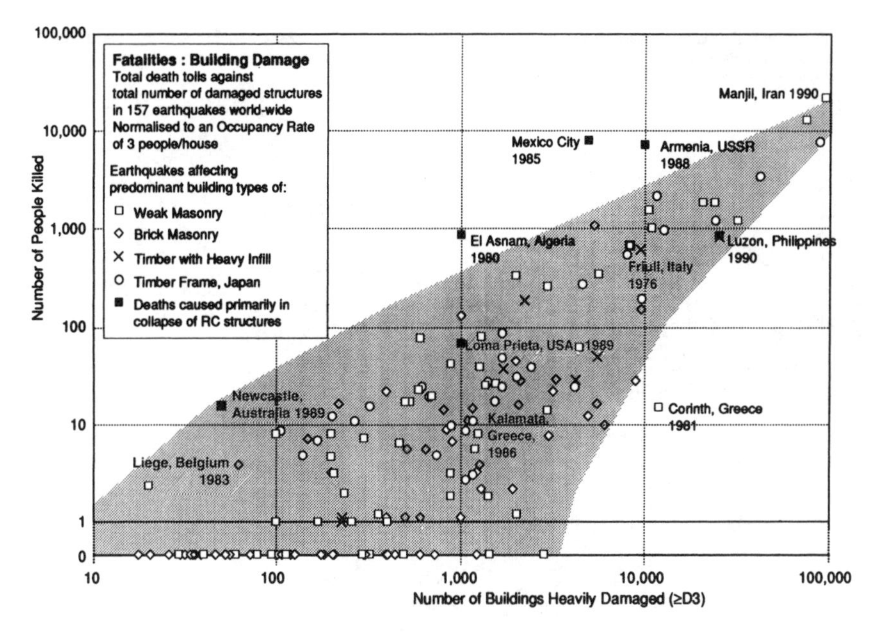

Figure 9.12 Relationship of the number of fatalities to the number of buildings damaged in earthquakes

damage, a suitable building classification, and statistics on the distribution of buildings of each type and their occupancy levels are required.

The lethality ratio for each building class can be estimated using a set of parameters defining the expected proportions of occupants who are trapped, the proportion of those trapped who are subsequently rescued, and the injury distribution in each group. A set of M-parameters is used to estimate the proportions of people rescued and trapped at each stage and the injury distributions among them. Figure 9.13 explains the meaning of these M-parameters. Each building class has its own specific set of M-parameters taking account of the likely collapse characteristics of that class of building and the SAR capability, which are derived from or compared with published casualty data.[25]

The proportion of occupants trapped by collapse (M3) is strongly influenced by building type, and also increases with building height. For tall reinforced concrete or masonry buildings it may reach as high as 50% or 60%. However, for the most numerous one- and two-storey buildings, even if they did collapse, it is unlikely that more than a very small proportion of occupants would be trapped. For collapsed residential timber frame buildings it is estimated that only 3% of occupants would be trapped. The proportion of occupants trapped by collapse is

[25] For example, Murakami (1996), Durkin (1996).

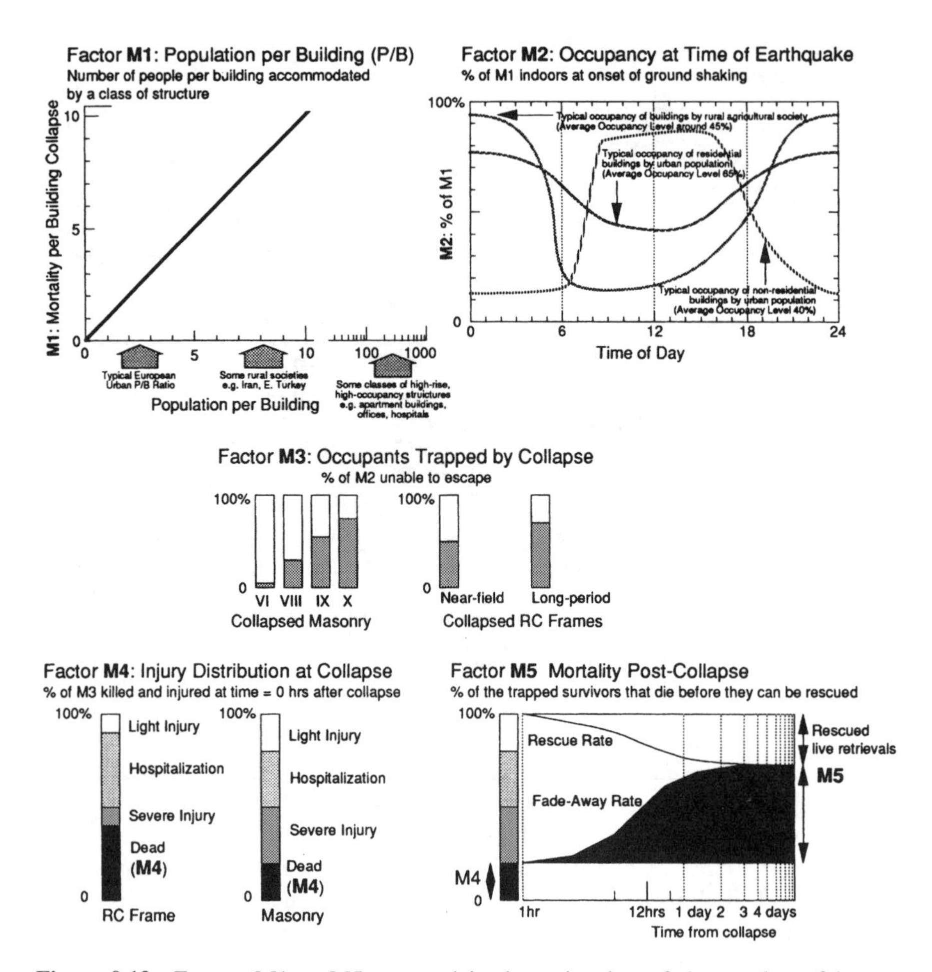

Figure 9.13 Factors M1 to M5 are used in the estimation of the number of human casualties likely to occur in an earthquake

also assumed to be smaller for events of lower intensity. M3 is also affected by the type of ground motion. Table 9.7 indicates the range of available data.

The proportion of occupants killed at collapse (M4) is assumed to depend on building type. For the timber and masonry classes 20% are assumed killed, while for the concrete and steel classes 40% are assumed to be killed at collapse. Table 9.8 shows typical injury distributions for the major building classes.

The mortality post-collapse (M5) depends crucially on the effectiveness of SAR, which will vary considerably between countries and according to the scale of the earthquake and whether the rural or urban population is affected, but the speed of rescue would be slower for concrete and steel buildings for which heavy cutting and lifting equipment would need to be deployed. In timber buildings it

Table 9.7 M3: estimated average percentage of occupants trapped by collapse.

Collapsed masonry buildings (up to three storeys)				
Intensity	VII	VIII	IX	X
	5%	30%	60%	70%
Collapsed RC structures (3–5 storeys)				
Near-field, high-frequency ground motion:				70%
Distant, long-period ground motion:				50%

Table 9.8 M4: estimated injury distributions at collapse (% of trapped occupants).

Triage injury category	Masonry	RC
1. Dead or unsaveable	20	40
2. Life-threatening cases needing immediate medical attention	30	10
3. Injury requiring hospital treatment	30	40
4. Light injury not necessitating hospitalisation	20	10

is assumed that most of those trapped would be quickly rescued; on the other hand, in any reinforced concrete buildings collapsing, rescue might come too late for 50% or more of those trapped. Thus values of M5 range from 10% for residential timber frame to 67% for the tallest pre-code reinforced concrete buildings. Figure 3.3 presents some indicative data on fade-away times for injured trapped victims, and Table 9.9 summarises aggregated data on survival rates from a number of earthquakes.

The injury distribution among those eventually rescued will also depend on the type of building. For steel and concrete buildings it has been assumed that 66% are uninjured, while for masonry and timber only 25% would be uninjured. The injured are roughly equally divided between serious and moderate injuries.

Table 9.9 M5: percentage of trapped survivors in collapsed buildings that subsequently die.

Situation	Masonry	RC
Community incapacitated by high casualty rate	95	–
Community capable of organising rescue activities	60	90
Community + emergency squads after 12 hours	50	80
Community + emergency squads + SAR experts after 36 hours	45	70

Table 9.10 Casualty distributions for collapsed buildings for key building types in the case of Wellington, New Zealand.

Class	Killed (%)	Seriously injured (%)	Moderately injured (%)	Lightly injured or uninjured (%)
Masonry (2–3 floors)	17.5	10	17.5	55
RC frame (2–3 floors)	21	0.8	9.2	70
RC shear wall (2–3 floors)	10	0.7	9.3	80
Steel (2–3 floors)	16	0.6	9.4	75
Timber (1 floor)	0.6	0.2	10.2	89

Table 9.10 shows the resulting distribution of injuries among occupants of collapsed buildings for a few common building types based on the special circumstances of Wellington, New Zealand.[26]

The casualties calculated in this way will constitute most, but not all, of the expected casualties. In addition to the casualties caused by building collapse, other possible causes of casualties need to be considered, including the major secondary catastrophes mentioned above, the collapse of large civil engineering structures, the direct effects of the fault rupture, and miscellaneous other causes.

9.7 Other Losses

The techniques discussed in Sections 9.4 and 9.5 are suitable for the assessment of the physical damage to buildings and other fixed and structured facilities, including the infrastructure of services, roads, power supply networks, resulting from ground shaking. But losses resulting from earthquakes extend well beyond these direct consequences of the ground shaking, and any adequate assessment must take these indirect or secondary effects and their consequences into account, and attempt to evaluate them. In particular it is important to evaluate:

- losses from collateral earthquake hazards such as ground failure, flooding and fire;
- non-structural losses to the buildings and facilities, their equipment and fittings;
- economic loss resulting from loss of function of the facility for the period of time needed to restore its use.

Techniques for the assessment of these losses are much less developed than those for the assessment of structural damage due to ground shaking, since there is much more limited data available, and rather crude assessments must be made,

[26] Spence *et al.* (1998).

generally relying extensively on professional judgement. Methods of assessing these losses are discussed briefly below.

9.7.1 Collateral Hazards

Collateral or secondary hazards are those earthquake-related hazards other than the shaking of the ground itself which threaten life and property. The hazards which usually need to be considered can be grouped under three headings: ground failure, flooding and fire.

Ground Failure

Several types of ground failure can occur in earthquakes. Landslides, rock slides and mudflows are frequently triggered and can be very destructive. The principal factors affecting the occurrence of landslides are the surface geology (including the presence of pre-existing slides), the slope gradient, the water content of the soil, and the intensity of ground shaking. The first two of these factors can be mapped in such a way as to identify different degrees of landslide susceptibility, each associated with a critical level of ground shaking; the level of destructiveness of the potential landslide can also be evaluated by defining damage states ranging from light (insignificant movement) to catastrophic (movement sufficient to carry everything large distances). For each level of landslide susceptibility, a landslide probability matrix can thus be defined, identifying the probability of occurrence of each damage state, for each intensity of ground shaking. In practice the data for the construction of such matrices is insufficient, and those so far produced rely heavily on professional judgement.

A second type of ground failure is earthquake-induced liquefaction. Loose fine sands which are in a saturated state are most susceptible, and these can generally be identified from existing subsoil maps. The probability of liquefaction for any susceptible deposits is greater the greater the level of ground shaking, and liquefaction probabilities can be estimated for particular known deposits based on *in situ* soil testing and professional judgement.[27]

A third type of ground failure is that which occurs as a result of ground disturbance at or close to a fault break. The disturbance can take the form of a local deformation, often a clean linear break in the ground surface, with the two sides moving relative to each other. The relative movement at a fault break can be either horizontally along the fault or vertically, or a combination of the two. Alternatively the ground disturbance may take the form of a more general regional deformation. Local deformations may be severe up to a few hundred metres at most from the fault, and are potentially highly destructive to any facilities in this area. If the location of a potential fault is known, the approximate

[27] FEMA (1999).

length of fault break and the extent of movement on the fault as a function of magnitude can be estimated, and this can be used to judge the damage potential.[28] This is of minor importance for buildings but of great significance in the case of subsurface lifelines and roads. In most cases a knowledge of the precise location of the potential fault is not available, but this may not be important for assessing its effects on lifelines. Another local hazard commonly associated with fault breaks is the amplification of ground motion in a region near the end of the fault break in the direction of propagation of movement along the fault.

Flooding

Flooding in earthquakes can result from tsunamis and seiches, or from dam or reservoir failure. The damage potential from tsunamis and seiches in low-lying coastal areas may be considerable, and can result from large undersea earthquakes with very distant epicentres. Damage assessment requires a knowledge of the potential height of the waves, velocity of the water, the topography of the coastal areas, and the damageability of the facilities in these areas to saturation and to water at various velocity rates. The assessment of damage potential from dam failure requires a knowledge of the vulnerability of the dam to earthquake ground shaking, the area susceptible to flooding in the event of the failure, and the vulnerability of the facilities in these areas to flooding.[29] If a reservoir is at a low level at the time of the earthquake, dam failure may occur several months later as seasonal rainfall refills it.

Fire

Fire following earthquakes is a common occurrence, and can be a major cause of damage as described in Section 3.5. Some fires are started in almost every damaging earthquake, but losses become significant only in cases where fire spreads in an uncontrolled manner. There are many factors influencing the probability of such a 'conflagration'. The first is the number of fires started initially, which will depend on the type of heating and cooking equipment in use and fuel storage and distribution methods; the second is the density of combustible material available, and the rate of spread which will also depend on the weather and climatic conditions; finally, the action of the firefighting services in suppressing fires will be a key factor. This will in turn depend on the capability of those services, the availability of water, accessibility of the fires, and the extent of involvement of the firefighting services in parallel activities such as SAR. Because of this large range of variables, it has proved exceptionally difficult to develop useful

---

[28] FEMA (1997), p. 4–40.

[29] FEMA (1997), Chapter 9.

quantitative procedures for predicting fire losses. Models have been developed for low- and mid-rise buildings in Japan, where available data is sufficient to justify the use of empirical relationships between the numbers of collapsed timber frame buildings.[30] A method has also been proposed for use in the western United States where timber frame single-family dwellings and apartment blocks are common, though its application depends on many assumptions about density of development, windspeed, temperature and so on.[31] Prediction of fire losses in earthquakes is likely to be highly uncertain.

9.7.2 Non-structural and Economic Losses

Non-structural losses are of two types: first, losses to the non-structural elements and components of the actual building, such as cladding, partitions, windows and services; and, secondly, losses to the building contents, such as furniture and equipment. The losses to non-structural components are usually included with those to the structure itself, since they are often indistinguishable, and these may be measured either in terms of direct physical damage (damage level D0–D5) or in terms of repair cost or damage factor. The damage probability matrix (DPM) or other vulnerability functions generally include such non-structural damage.

Losses to contents can usually be measured only in terms of value; the degree of loss can be expected to relate not only to the extent of physical damage to the building, but also to the use to which the building is put. Thus to determine damage probability distributions for contents a social function classification of buildings is required in addition to their structural vulnerability classification. For residential buildings, contents losses are relatively predictable as a function of damage state. The HAZUS manual, for example, gives mean estimates of 1% for slight damage, 5% for moderate damage, 25% for extensive damage, and 50% for complete damage.[32] For other uses a detailed understanding of the nature of the contents, and for industries their inventories or stocks of raw materials or unsold products, is needed to assess probable contents losses.

Economic losses arising from an earthquake are not limited to the monetary value of the physical damage, but must also include losses of industrial production, commercial and other economic activities consequent on the physical damage, which have been discussed in Chapter 2. These economic losses are associated with (or the consequences of) the loss of function of the buildings or the unavailability of employees, and to assess probable economic losses it is important to try to assess the degree of loss of function of each building and facility, and the length of time needed for partial and complete restoration of function.[33] This will depend not only on the degree of physical damage (to

---

[30] Scawthorn *et al.* (1981).

[31] FEMA (1997), Chapter 10.

[32] FEMA (1997), Chapter 15.

[33] FEMA (1997), Chapter 15.

building and contents), but also on the degree of damage to lifelines (roads, power networks and other infrastructure) on which the building depends, and other external factors such as casualties among the workforce. The loss of production of a large number of individual enterprises can have a significant effect on the economy of a whole region through broken chains of backward linkages (to suppliers) and forward linkages (to buyers or consumers).[34] The scale and nature of these losses can be investigated by using input–output modelling, and standard procedures have been developed.[35]

9.8 Applications of Loss Estimation

This section will discuss methods of combining hazard and vulnerability to undertake loss estimation applicable to different situations. Location, the type of buildings and facilities involved, the extent of data available and the purpose for which the loss estimation is being made all have an influence over the choice of the method used. The problems involved in estimating losses in rural and urban areas can be quite different, as the following examples of studies carried out by the authors will show.

9.8.1 Loss Estimation in Rural Areas

Estimates of probable future losses in rural areas may be needed in order to plan relief and emergency preparedness at a regional level, and in order to support and evaluate plans for upgrading traditional housing. Often traditional low-strength rural housing is the principal cause of earthquake loss.

To evaluate losses over a large, predominantly rural area an approximate first estimate of losses may be adequate, and it may be possible to assume:

- a single homogeneous seismic source zone
- a uniform population distribution
- a single predominant type of dwelling applicable to the majority of the population.

Further, if the form of construction used is not changing rapidly, it may be possible to develop damage–attenuation relationships for the predominant type of construction based on the distribution of damage from past earthquakes in the region. Using such assumptions relationships can be used either to estimate the total losses which can be expected in the event of earthquakes of different magnitudes, or, in conjunction with a magnitude–recurrence relationship such as

[34] A major cause of economic loss following the Kobe earthquake was the closure of the key port of Kobe.

[35] Brookshire *et al.* (1997).

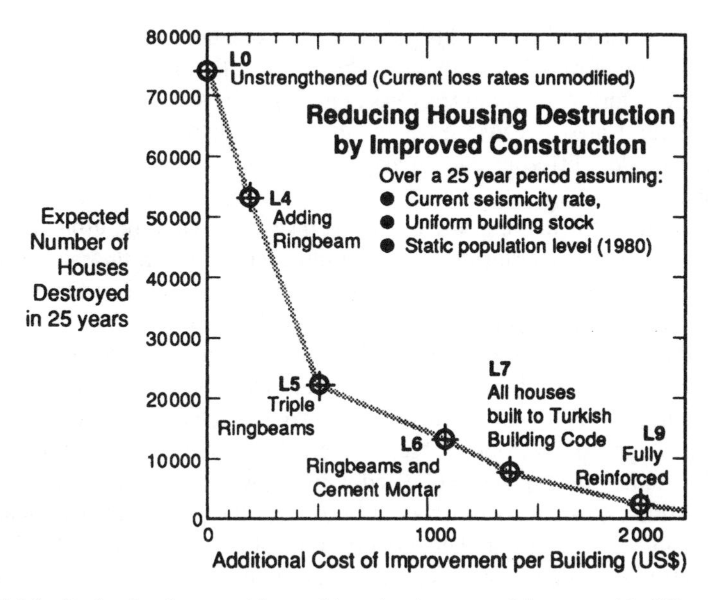

Figure 9.14 Reduction in casualties and housing loss over 25 years with different levels of strengthening (after Spence and Coburn 1987a)

that shown in Figure 7.3, to estimate the total losses which can be expected from all earthquakes over a given period of time.

The number of people killed and injured in earthquakes depends on many variables, but within a particular rural region with unchanging building technology, it is primarily related to the number of buildings which totally collapse (D5). Estimates of numbers of people killed and injured can be derived using an empirical relationship derived from past experience in the area (as discussed in Section 9.6).

One of the most important uses of loss estimates of this sort is that they can be used to assess the impact of a building improvement programme of upgrading the traditional houses, and to compare the effectiveness of different levels of technology in upgrading, if the relative vulnerabilities are known or can be estimated. Figure 9.14 shows the impact, over 25 years, in the expected numbers of deaths and houses destroyed in eastern Turkey, if different levels of strengthening corresponding to some of those shown in Figure 8.12 were generally introduced. Data of this sort can be used in a cost–benefit or cost-effectiveness evaluation of alternative possible government intervention programmes. This is discussed in Chapter 10.

9.8.2 Loss Estimation in Urban Areas

For urban loss estimation it may be reasonable to assume that with the occurrence of an earthquake some distance away, the attenuation of ground shaking across the breadth of the city will be insignificant. Thus standard methods of making a hazard assessment may be used, with whichever is the most appropriate ground motion parameter. Vulnerability assessment is likely to be much more complex than for a rural area because most urban settlements contain a wide range of building types of differing earthquake vulnerabilities and a variety of ground conditions.

One technique for making a loss estimation is to divide the urban area into a number of distinct vulnerability zones, within each of which the mix of building types may be assumed uniform, the ground conditions may be assumed uniform, and the total population (or number of dwellings) is known.

This subdivision into zones can be done using whatever large-scale mapping or aerial survey of the city is available, coupled with the use of subsoil maps and field investigation. Frequently administrative zones such as districts or subdistricts will be most appropriate, since these are the units within which building stock or population data will have been collected. Often it will be found that the zoning so far as building types is concerned closely follows the pattern of historical development of the city, with a higher proportion of older, more vulnerable buildings in the centre, and predominantly newer, less vulnerable buildings towards the periphery. There is often a close coincidence between the pattern of historical development and subsoil ground conditions, with the earliest settlement located on firm ground conditions and later development occupying progressively less satisfactory ground.[36] The mix of building types in each zone can be established either using census data or by sample field survey if needed. The building types defined should correspond to those for which vulnerability data already exists in the form of damage distributions from previous earthquakes. The development and availability of damage distributions in the form of the DPM and vulnerability functions has been discussed in Section 9.3. The total number of buildings in each vulnerability zone can be estimated from maps and aerial photographs, from field survey or from census data depending on the size of the zone and the availability of mapping.

The effect of soil conditions can be dealt with either by using modified damage probability distributions for poor ground conditions, or by assigning one or more increments of MM or EMS intensity, or even an adjustment of the PSI for these sites to derive an appropriate damage distribution. Where damage distributions are based on spectral parameters of ground motion, the effect of soil conditions is incorporated as a site-specific or zonal modifier of the ground motion parameter used.

A useful technique for dealing with the variation of building types and soil conditions within a city is to divide it into a grid, and assume that the soil type, building type distribution and population density appropriate to the centre of each grid square apply to the whole of that grid square. The accuracy of such estimates

[36] Coburn *et al.* (1986).

can be improved by increasing the fineness of the mesh, but it has been found that for a medium-sized city (0.5 million population) a grid square of 0.5 km side gives sufficiently good results.[37]

These sorts of estimates are useful for regional planning of emergency services and also for investigating the impact of mitigation policies, but given the uncertainties a close correlation with experience cannot be expected. Uncertainty is discussed further below.

9.9 Uncertainty in Loss Estimation

The uncertainties involved in all loss estimation methods are large, combining uncertainties in both hazard and vulnerability assessment.

9.9.1 Hazard

The uncertainties involved in hazard assessment include those in:

- the definition of seismic source zones
- the recurrence rates and the actual time of occurrence
- the ground motion – attenuation relationships
- the effect of ground conditions on ground motion.

Most of these uncertainties are difficult to quantify because the amount of available data is limited, but where they can be it is usually in the form of a standard deviation of the error between the data and the proposed relationship.[38] Typical variations of $\pm 10\%$ of the mean value within different parts of a seismotectonic region are reported, implying variations in the recurrence rates of earthquakes of $\pm 25\%$.

The actual earthquake recurrence pattern can be estimated assuming earthquakes are independent of one another, and that their recurrence pattern follows a Poisson process. This seems to fit reasonably well with observed earthquake behaviour in many large regions, if aftershocks are excluded. This allows the probability of occurrence in any time interval to be evaluated if the average recurrence rate is known.[39] For example, if the annual probability of occurrence of an earthquake of a particular intensity is 0.25 (average recurrence interval 4 years), the probability of one or more event in any particular year can be

[37] See Department of the Environment (1993).

[38] Variations in the values of the constants A and b used in the Gutenberg linear regression relationships have been discussed by Kaila and Madhava (1975).

[39] The Poisson distribution gives the probability of just k events in time interval s as being $p(k) = (e^{-Ls}(L \cdot s)^k)/k!$, where L is the average rate of ocurrence of events (Ang and Tang 1976).

shown to be 22%, and the probabilities of one or more event occurring in a period of 2, 5 or 10 years are 39%, 71% and 87% respectively.

The Poisson model, by assuming independence of events. It does not allow for the inclusion of aftershocks, or the clustering of events. It also assumes a stationary process, with a constant average rate of occurrence of events, and therefore does not allow for the possibility of periodic changes in the seismicity of a region, or time-dependent changes in seismicity caused by strain energy build-up and release, which are known to occur (see Chapter 3).

In a study of the uncertainty in ground motion attenuation relationships,[40] standard deviations on the logarithm of peak ground acceleration (PGA) and peak velocity were both around 0.25, implying a 66% probability of actual values between 0.55 and 1.8 of the mean value. Because the uncertainties in the different aspects of hazard estimation interact, the uncertainty in the final hazard assessment is best approached by studying its sensitivity to likely errors in the various assumptions made. Experience suggests that the uncertainty in the effect of subsoil ground conditions on likely ground motion levels is likely to be particularly significant.

9.9.2 Vulnerability

Vulnerability relationships also involve a high degree of uncertainty. The uncertainties involved here are in the 'damagingness' of an event of a particular severity, the definition of the building stock, and the appropriateness of the chosen vulnerability functions to the particular building stock or other facilities involved. The uncertainty is even greater when indirect losses derived from the primary losses of building stock, such as human casualty and economic losses, are made.

It is possible to examine the effect of cumulative uncertainties in loss estimates using discrete event simulation (or Monte Carlo) techniques if it is assumed that the hazard is known and that the probability distribution of each of the constituent relationships is known. This was done for losses in eastern Turkey as a part of the study discussed above.[41] The results are shown in Figure 9.15. Estimated total losses have a 90% probability of being within ±50% of predicted losses, when a damage–magnitude model is used. However, when losses are calculated for traditional construction using a magnitude–distance damage model, the probability of the actual losses being within 50% of the predicted losses falls to 73%, and when losses to other building types are inferred through relative vulnerability functions, the probability that the actual losses will be within 50% of predicted values drops further to 40%. Estimates of human casualties are derived by uncertain relationships from already uncertain building loss estimates,

---

[40] Joyner and Boore (1981).

[41] Coburn (1986a).

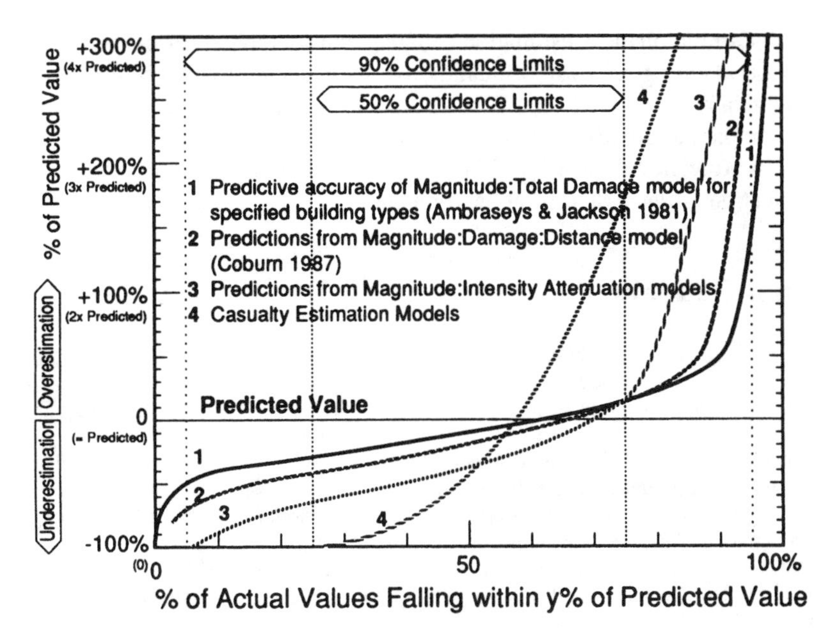

Figure 9.15 Estimated confidence limits on earthquake damage estimations for eastern Turkey (after Coburn 1986a)

so the uncertainties in these estimates are compounded. The study concluded that casualty estimates have only 10% to 20% probability of being within ±50% of predicted values. Where further losses such as loss of function and economic losses are to be inferred from human casualty and building losses, the uncertainty of the prediction increases still further.

To date, there have been very few cases in which loss estimates have been tested by the subsequent occurrence of an earthquake. However, one study in Italy was able to compare predicted vulnerability with observed earthquake damage in two earthquakes.[42] It was found that the correlation improved as the intensity of ground shaking increased with acceptable correlation for areas of intensity IX and X; however, for areas with intensity levels of about VII or less the correlation was too low to be satisfactory for use in loss prediction.

To date, it appears that earthquake loss estimation is a somewhat inexact science depending to a considerable extent on professional judgements. However,

[42] Vulnerability was measured using a vulnerability index which took into account 10 different contributing factors to the vulnerability of a building, whose contribution was determined by a weighting factor. This was calculated for a sample of over 1500 masonry buildings previously damaged by the 1976 Friuli earthquakes, and a separate survey was carried out in the small town of Gubbio in 1983, which subsequently experienced a moderately damaging earthquake in 1984. The damage level was thus in both cases able to be compared with the vulnerability index (Benedetti and Benzoni 1985).

within its limitations loss estimation can give considerable information for use in protection planning. The use of quantitative methods such as those described for assessing risk or the likely outcome of various scenario studies makes it possible to compare alternative protection strategies and to obtain maximum value for money in protection investment. The use of these techniques in making decisions on protection is discussed in the next chapter.

Further Reading

Bommer, J., Spence, R., Tabuchi, S., Aydinoglu, N., Booth, E., del Re, D., Erdik, M., and Peterken, O., 2002. 'Development of an earthquake loss model for Turkish catastrophe insurance', Special Issue of the *Journal of Seismology* on the Turkish Earthquakes of 1999, Vol. 6, No. 2.

Earthquake Spectra, 1997. The whole of the November 1997 issue is devoted to the subject of Earthquake Loss Estimation.

FEMA, 1999. *HAZUS Earthquakes Loss Estimation Methodology*, US Federal Emergency Management Agency, Washington.

Kircher, C.A., Nassar, A.A., Kustu, O. and Holmes, W.T, 1997. 'Development of building damage functions for earthquake loss estimation', *Earthquake Spectra*, **13**(4), 663–682.

Scawthorn, C. (ed.), 1986. *Techniques for Rapid Assessment of Seismic Vulnerability*, American Society of Civil Engineers, New York.

Spence, R., 2000. 'Recent earthquake damage in Europe and its implications for loss estimation methodologies', Chapter 7, pp. 77–90, in *Implications of Recent Earthquakes for Seismic Risk*, Imperial College Press, London.

Spence, R.J.S., Coburn, A.W., Sakai, S. and Pomonis, A., 1991. 'A parameterless scale of seismic intensity for use in seismic risk analysis and vulnerability assessment', in SECED (ed.), *Earthquake, Blast and Impact: Measurement and Effects of Vibration*, Elsevier Applied Science, London.

UNDRO, 1979. *Natural Disasters and Vulnerability Analysis: Report of Expert Group Meeting*, Office of United Nations Disaster Relief Co-ordinator (UNDRO), Palais des Nations, CH-1211 Geneva 10, Switzerland.

Woo, G. 1999. The Mathematics of Natural Catastrophes, Imperial College Press, London.

10 Risk Mitigation in Action

10.1 Introduction

The previous two chapters have shown how we can design buildings better equipped to deal with the impact of earthquakes, and how we can estimate the losses which will occur when earthquakes strike, both for buildings that have been improved and for those which have not. These are vital tools in the creation of a risk mitigation strategy, but they must be formulated into action programmes clearly understood by building owners, government legislators and their primary beneficiaries, the building's day-to-day occupants, if they are to be implemented. This chapter will first of all review three different aspects of risk mitigation programmes which have been identified in the preceeding chapters:

1. Improving standards of construction for new buildings and infrastructure.
2. Strengthening existing buildings.
3. Upgrading rural construction.

How the costs and benefits of mitigation programmes can be evaluated in such a way as to strengthen the economic case for mitigation is then discussed; and the chapter then looks at the question of public perception of earthquake risk and its impact on the formulation of public policy for earthquake protection. The book concludes with a discussion of what has been achieved and remains to be done to bring about a global culture of action for disaster mitigation in the years ahead.

10.2 Improving Standards of Construction for New Buildings

As urban populations grow, an unprecedented boom in the construction of new buildings is taking place in many earthquake-prone areas. The life of most of these buildings is likely to be 50 years or more, which means that there is a high probability that they will experience a damaging earthquake at least once. It is the responsibility of the owners, designers and builders of these buildings to provide them with the best possible protection from earthquake hazards at the time of their construction. The cost of providing protection at this stage is relatively small, while the cost of any subsequent strengthening is very large. Unfortunately, as recent earthquakes have shown, too much of today's construction is well below the standard needed for future safety. Three important aspects of the problem are:

1. Improving the codes of practice which define safe design and construction practice.
2. Building and development control.
3. Research into better design and construction techniques.

10.2.1 Improving Codes of Practice

Earthquake protection – especially in the cities – depends greatly on the application of appropriate regulations for earthquake-resistant construction, generally in the form of a code of practice for earthquake-resistant design. There has been substantial progress in recent years in both the extent of coverage and the quality of these codes. In 1973 the International Association for Earthquake Engineering (IAEE) was able to identify only 27 countries with an earthquake design code out of at least 60 earthquake-prone countries. The most recent IAEE list[1] shows that the number has increased to 36 and that this number continues to grow.

The quality of these earthquake design codes is also improving. It was reported in 1977[2] that all codes were either inadequate or misleading with respect to one or more of the essential components of an earthquake code, in giving guidance on loading and risk, overall structural performance criteria, or detailing. Many gave lateral force coefficients for design which were too low for the structure concerned. Partly as a result of the experience of some disastrous earthquakes, and partly due to the efforts of many committed scientists and building professionals, regulations in many countries have been substantially improved over the last decades. The lateral force coefficients have also, in many cases, been increased, or seismic zoning modified to enlarge the areas subject to existing regulations. In many cases the requirements for detailing to achieve earthquake resistance have been improved, particularly with respect to the requirements for ductility, and the means to incorporate ductility into the design of building structures.

The development of codes of practice is a vital area of earthquake protection, but codes are only effective if they are enforceable, and there is an ever-present danger that builders will ignore a strict code – particularly as the elapsed time

[1] Paz (1994).
[2] Dowrick (1977).

since the last damaging earthquake grows. An interesting example is Quetta in Pakistan (Box, Chapter 5), which developed one of the first earthquake codes after the disastrous 1935 earthquake, but has in more recent years been unable to enforce it in the face of rapid urban growth. It also has to be borne in mind that many established cities change their building stock by no more than a few per cent per year, so it takes a long time before the introduction of new building regulations can have a significant impact on overall vulnerability, even where they are enforced.

10.2.2 Building Control

The experience of recent earthquakes, especially those in Turkey and Taiwan, 1999 and Gujarat, India, 2001, has demonstrated that even when carefully formulated codes of practice for construction exist, widespread failure of apparently engineered buildings often occurs. Usually the press and the public attack the builders as the guilty party,[3] with some justification, but in reality the inadequate standard of construction is the result of a more extensive inadequacy of building control involving not just the builders, but government, the building design professions, the property developers, the client and eventual owners, the builders and also the eventual occupants.

A study of the causes of poor-quality construction in Turkey[4] pointed to deficiencies in both the nature and implementation of laws and regulations concerning the planning system, the project supervision at the design stage, and the system of supervision on site.

The principal deficiencies in the planning system are:

- a lack of basic mapping of areas especially prone to high earthquake ground shaking or other associated hazards such as landslides;
- a lack of any process to identify 'risk areas' within municipalities in which development should be prevented or controlled;
- a lack of integration and communication between the various government agencies involved;
- failure to implement those development controls which do exist.

The failures in project supervision include:

- a lack of properly qualified staff in the municipalities to undertake design checks;
- a lack of simplified procedures for carrying out design checks;
- no system of continued responsibility for the quality of the design by either the designer or the checking authority.

---

[3] For example, *India Today*, 2001.

[4] Gülkan *et al.* (1999).

Even more serious are the following deficiencies in building construction supervision:

- no requirement for adequate expertise on the part of the supervising engineer;
- supervising engineer has little contact with the process on-site;
- a lack of personal liability insurance on supervisors;
- no mechanisms for municipalities to become aware of, to refuse utility connection to, or to demolish unpermitted buildings;
- no adequate system for prosecuting negligent builders;
- no requirement for registration of builders or contractors.

Such inadequacies as these are commonly found in developing countries, particularly those undergoing rapid urbanisation; the consequences in human lives when the buildings concerned are multi-storey apartment blocks were shown all too clearly in Izmit and Golcük in Turkey, in Ahmedabad and Bhuj in India, and in Taipei in Taiwan.

Since 1999, serious efforts have been made to overcome these deficiencies in Turkey through new legislation and through setting up new training programmes. One particular innovation proposed, which has international significance, is the establishment of a new role of building supervision specialist. Private building supervision firms take on, in return for a fee, the responsibility for supervision of building projects, in both the design and construction phases; that responsibility carries with it the liability for offsetting any losses which might occur to the owner, during 10 years, resulting from poor construction. This liability is backed by indemnity insurance on the part of the supervising firm. This measure in effect removes from the municipalities to the private sector the task of building control which they have failed (or been unable) to undertake adequately.

Other aspects of the recommended new provisions for building control in Turkey include:

- the requirement for a resident site engineer for all substantial projects;
- proper registration of contractors as well as engineers and architects taking responsibility for all buildings;
- compulsory testing of materials used in all construction projects;
- establishment of a compulsory national earthquake insurance system.

The principal purpose of the compulsory insurance scheme is to create a financial pool (backed by international reinsurance) which can be used to support repair and reconstruction following future damaging earthquakes, replacing the increasingly unsustainable burden on the government for compensation payments under the current system (discussed in Chapter 2). But this system has potentially huge implications for building control. By requiring householders to purchase insurance, premiums for which depend on the quality and location of construction, they are forced to consider and to some extent pay for the risks they face. This will in turn bring pressure to bear on the builders and designers to demonstrate that construction standards are being maintained. Similar compulsory earthquake insurance schemes, already practised in New Zealand,[5] are currently being considered in other countries.

Improvement of building control is not of course simply a matter of introducing the right laws and procedures. In a society where most people are poor and unaware of the nature of earthquake risk, builders are driven to build as cheaply as possible and evade regulations they see as unnecessary and cut safety margins to reduce costs. Improvement in building control inevitably goes hand in hand with the development of the general 'safety culture' in society as a whole, a subject discussed further in Section 10.8.

10.2.3 Earthquake Engineering Research

The growth in the expenditure and output of earthquake engineering research in recent years is extraordinary. As one measure of this growth, the First World Conference on Earthquake Engineering in California in 1956 was attended by about 40 participants; 45 years later in 2001, the Twelfth World Conference on Earthquake Engineering, held in New Zealand, was attended by over 3000 delegates from about 60 countries, and over 1500 papers were presented. During this time, very considerable progress in understanding has been achieved in the aspects which have traditionally been the concern of these conferences, especially in understanding the source mechanisms of earthquakes and their effect on ground motions, and the analysis and design of structures and their foundations to resist those effects. Some of the milestone developments over these years are as follows:

- The theory of plate tectonics has become established, providing a firm framework for understanding the source mechanism of most damaging earthquakes.
- The World Wide Standard Seismograph Network recording instrument has been unified on a worldwide basis, providing the opportunity for the international exchange of data both in teleseismic records and near-field records, and better definition of regional seismicity and of ground attenuation characteristics.
- The relationship between ground motion and the dynamic analysis response of buildings has been clarified and safe design rules devised.
- The influence of the near-failure behaviour of structures on their earthquake performance has been studied using new testing techniques and earthquake simulators, and rules for providing structures with improved ductility and resistance to failure in earthquakes have been developed.

But valuable though this work is, its overall effect on reducing earthquake losses globally is surprisingly limited, partly because of the excessive concentration of this

[5] Earthquake Commission, New Zealand.

research activity on increasingly sophisticated engineering. Most losses are suffered in developing countries and in non-engineered or low-engineered structures. It has been claimed that only 2% of all the world's R&D effort is directed towards the problems of developing countries, 98% being concerned with the needs of the already rich. A similar bias is observable in earthquake engineering, with only a handful of the 1500 papers in the 2001 World Conference being concerned with the problems of earthquake risk in the rural areas of developing countries, or with materials such as unreinforced masonry in which the most vulnerable people, in rural and urban areas alike, still live.

There is also a focus on procedures for the design and construction of new buildings, when, throughout the world, it is the older existing building stock which constitutes the greatest risk. About 5% of the research effort appears to be directed towards this problem, which constitutes perhaps 80% of the risk for the immediate future.

In spite of its prodigious output, earthquake engineering research has not in the past been sufficiently directed to these problems; an urgent reorientation of research will have to take place in the years ahead if research is to improve its impact on future earthquake losses.

10.3 Strengthening Existing Buildings and Infrastructure

Most large-scale strengthening programmes have taken place immediately after a major damaging earthquake, at a time when public concern and awareness of the risk is at its highest, when substantial building repair work is in progress, and when there may have been changes in the code of practice. After the 1985 Mexico City earthquake, the lateral resistance requirements for buildings in the worst-hit parts of the city were substantially increased, and the new building code also specified that the increases should apply not only to all new buildings and all buildings damaged in the earthquake, but also to all existing buildings whose failure would put the public or essential services at risk, even where they had not been seriously damaged by the 1985 earthquake. This resulted in a very substantial programme of strengthening, affecting many thousands of existing buildings.

After the 1999 Kocaeli, Turkey earthquake, a number of building owners decided to strengthen buildings which were either lightly damaged or undamaged, recognising that their resistance was below that specified in the current code. Such strengthening involved in many cases the addition of concrete shear walls over the whole height of the building, a substantial intervention requiring total evacuation (Figure 8.16), and often costing as much as 40% of the cost of rebuilding.

A substantial programme of repair and strengthening of stone masonry buildings also followed both the 1976 Friuli earthquake and the 1980 Irpinia earthquake in Italy and a number of other post-earthquake projects have begun to tackle the problem of the existing building stock at risk.

Strengthening of buildings in areas threatened by (but not recently damaged by) earthquakes is not yet happening very widely, although there is some progress in the United States. According to a Hazardous Buildings Ordinance, issued in 1981, the City of Los Angeles requires all buildings of unreinforced masonry constructed before 1934 to be brought up to a minimum standard of structural resistance. The standard required is somewhat lower than that required for new building, but sufficient to reduce the risk of loss of life or injury to acceptable levels. The rules were introduced as a recognition of the extent to which building damage and casualties in earthquakes in southern California during this century were concentrated in these older buildings. A method of assessing the resistance of existing and strengthened buildings is specified, and all building owners are expected to have surveys carried out within nine months, and have completed any necessary strengthening within a specified period depending on the extent of the risk involved.[6] However, compliance with this ordinance has been slow. A study among owners of unreinforced masonry buildings in Los Angeles in 1989, by which time all buildings should have started strengthening works, showed that only 24% had completed, and another 5% were in progress.[7]

Similar rules have now been introduced in other building authority areas in California and in New Zealand. New legislation in Italy provides for public funds to assist local authorities in high-risk areas to carry out a selective programme of strengthening key vulnerable buildings. But in Europe too, progress in strengthening existing weak unreinforced masonry buildings in earthquake risk areas is very slow, and often hampered by local planning restrictions and rent control zones.

Strengthening programmes of this sort are likely to result in substantial reductions in future earthquake losses in those countries which are able to implement them, but they are costly (see Section 10.6), and require a high degree of official control over building to be effective. There is now a need for concentrated research into the development of lower cost, affordable measures for achieving life safety in older buildings in order to enable such strengthening to start happening.

10.4 Upgrading Rural Construction: Building for Safety

Given the limited resources of the rural population in most developing countries, and the scarcity and high price of manufactured building materials, upgrading schemes need to respond to local priorities, and be designed as far as possible around existing skills and crafts. The Building for Safety project 1990–1993 produced a series of source documents designed to support such upgrading schemes.[8]

[6] Los Angeles Hazardous Buildings Ordinance.

[7] Comerio (1989).

[8] Aysan *et al.*, 1995, Clayton *et al.*, 1995, Dudley and Healand, 1993, Coburn *et al.*, 1995.

The techniques adopted in the reconstruction programmes following four different earthquakes give an indication of the range of possibilities.

ASAG Seismic Safety Confidence Building Programme in Latur, 1993–1996

After the Latur earthquake in rural India in 1993, the Indian voluntary agency Ahmedabad Study and Action Group (ASAG) worked intensively with victims of the earthquake to help them develop confidence to repair and rebuild their houses using small modifications of the existing technology rather than the alien concrete block technology introduced in the larger scale rebuilding programmes. Traditional technology for rural houses used thick crudely bonded stone walls and a mud roof on timber beams supported on a timber structure set within the walls. The modifications introduced by ASAG involved stitching elements in stone masonry walls, integral reinforced concrete bands below roof level, introducing a plastic water barrier below the mud roof, and knee bracing in the timber supporting structure[9] (Figure 10.1). A programme of demonstration and learning, working with several individual families, was followed by the preparation of a repair and retrofitting manual. This led to a training programme for newly appointed engineers working in the reconstruction programme, and to a programme of collaboration with government and NGOs, during which the techniques developed were disseminated and widely adopted.

Builder Training Project in Ecuador

The Ecuador earthquakes of March 1987 severely damaged rural housing over a wide area in the remote and sparsely populated Andean highlands. The principal form of construction in the area is to use rammed earth walls, with a clay

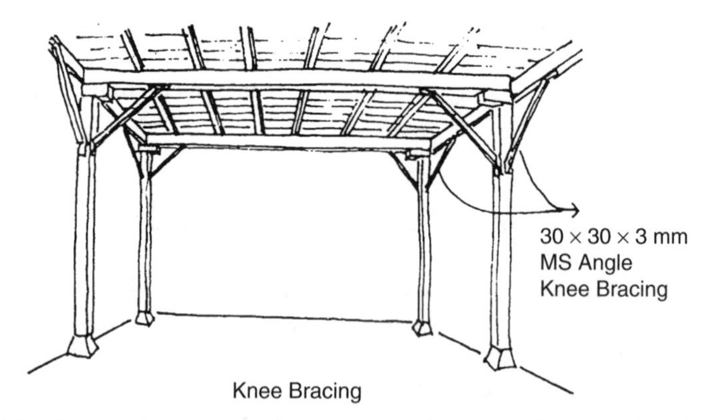

Figure 10.1 Interventions in traditional construction technology for the Latur District of India proposed by (ASAG 1996) the Ahmedabad Study and Action Group (ASAG) following the Killari earthquake of 1993 (Reproduced by permission of ASAG.)

[9] ASAG (1996).

tile roof. A common structural weakness of these dwellings was that the corners of the walls were inadequately bonded and fell out under the earthquake shaking, leaving the rest of the wall unrestrained; this resulted from the vertical joints being inadequately staggered, and being too close to the corners. An additional weakness was the lack of rigidity of the roof structure and a poor junction between roof and wall. A local NGO, Centro Andino de Accion Popular (CAAP), devised a training programme for reconstruction based on a technology essentially the same as is traditionally used, but with simple modifications to strengthen the weak corner junctions[10] (Figure 10.2). Builders from the affected community projects were trained in the construction and use of the mould, and also shown how a roof can be built so that it is stronger, and how a ringbeam can be provided set in a channel cut in the top of the earth wall. Using these techniques thousands of strengthened houses were rebuilt after the earthquake, with limited financial assistance and technical assistance from voluntary agencies.

Yemen Rural Building Education Project

The earthquake which occurred in Dhamar Province of the Yemen Arab Republic on 13 December 1982 caused widespread damage and destruction in an area where the traditional form of construction is of rubble stone. Rural as well as urban houses are often two or more storeys high, with walls of rubble or dressed stone and timber floors and heavy, flat timber roofs. Oxfam established a builder training programme for reconstruction, which, like that in Ecuador, was aimed at local builders, with the intention of introducing some simple techniques for strengthening houses, using locally available materials and skills.[11] The principal causes of weakness in traditionally constructed dwellings were found to be at the wall-to-wall junctions, where separation occurred, and at the junctions of walls and roofs, where the timber joists separated from their supporting walls, and in the separation and disintegration of the masonry walls themselves, due to inadequate bonding. The training programme emphasised single-storey building, and demonstrated techniques (such as better mortar, stone dressing and through-bonding) for constructing a wall with better integrity and earthquake resistance. It also offered a range of techniques for both strengthening the corners and providing a ringbeam to connect the tops of the walls and the roof (Figure 10.3).

Over a period of four years over 1000 builders were trained, about 25% of the total number of builders in the area, and most were found (in a subsequent study) to have changed their practices as a result of the course.[12]

---

[10] Dudley (1987).

[11] Leslie (1984).

[12] Coburn and Leslie (1985).

Figure 10.2 Corner mould developed for improving earthquake resistance of rammed earth walls after the Ecuador earthquake (after Dudley 1987. Reproduced by permission of Eric Dudley.)

Guatemala Subsidised Materials Programme

After the massive earthquake of 1976 in Guatemala, Oxfam and World Neighbours took a different approach. It was observed that the great majority of the casualties in the area were caused by the collapse of heavy tiled roofs, inadequately supported on weak masonry walls. Galvanised steel sheets were

Figure 10.3 Improved construction techniques demonstrated in a builder training programme after the 1982 Yemen earthquake (after Leslie 1984. Reproduced by permission of Jolyon Leslie.)

a feasible alternative climatically, and were already used by those who could afford them, and they would evidently reduce loss of life in any future earthquake. Thus the voluntary agencies subsidised the sale of these sheets at a very low cost to all those who had lost houses; the sheets were initially used for roofing the temporary shelter, but could subsequently be incorporated into permanent reconstruction.[13]

In none of these cases has the effectiveness of the techniques used been fully tested in a subsequent earthquake, and no claim was made that the improved techniques would be entirely adequate against future earthquake damage. Yet in each case the techniques adopted were in accordance with sound engineering understanding of the causes of earthquake vulnerability and how to reduce it; they made use of materials and skills available to local builders at little or no extra cost; and their implementation depended on action very soon after the event, while reconstruction was taking place. In any rural upgrading the extent of reduction in vulnerability will inevitably be limited by the degree of change which is possible, and there is a continuous need for innovation in techniques to upgrade rural construction practices worldwide.

[13] Cuny (1983).

10.5 Evaluating Alternative Protection Strategies

10.5.1 Cost–Benefit Analysis

Earthquake protection measures such as those described are in many cases costly to apply, and frequently involve public funds. As a general rule, it can be expected that the higher the level of protection, the higher the cost of protection will be. But the protection will result in lower levels of damage and fewer lives lost in any future earthquake, so the costs of future losses will be reduced. Clearly it is important to have some means of deciding on the right level of protection, and of choosing between alternative ways in which limited resources might be spent to improve protection. Questions to which answers are needed include the following. What is the appropriate level of earthquake force for which new buildings should be designed? Which existing buildings should be strengthened, and to what level? Should certain types of building development be prohibited in certain areas? How much is it worth investing in earthquake prediction or emergency planning measures?

The most widely used method for choosing between alternative investments designed to achieve some socially desirable outcome is *cost–benefit analysis* (CBA). At its simplest the idea is that all the benefits of the project are computed in financial terms, the costs are then deducted, and the difference is the value of the project. All projects with a positive value are worthwhile, but in a situation with a number of possible alternative projects and with limited resources available for investment, the projects with the highest value are chosen. Where some of the costs and benefits occur in the future, their value is discounted by a discount rate designed to reflect society's, or the funding agency's, preference for present benefits over future benefits.

Where earthquake protection strategies involving improved building design are to be considered, the cost of the project is the additional cost of providing earthquake resistance over the cost of construction in which no special provision for earthquake resistance is made, while the benefits are the reductions in future losses. Costs of any particular protection strategy can be calculated conventionally, summing the additional labour, materials and other investments needed in the earthquake-resistant components of the project. If the alternative strategies being considered are alternative sets of building code requirements, for example, it is a relatively simple matter to carry out designs according to alternative sets of requirements, and calculate the cost difference based on current building costs. Where the alternative strategy may involve strengthening existing buildings, the initial cost has to include not only the complete structural cost, but also the cost associated with the loss of function of the building while the project is carried out. In either case, the additional cost is calculated as a proportion of the cost of a new building.

Future losses may usefully be divided between physical losses (building and infrastructure damage) and incidental losses (loss of life, injury, relief and rescue costs, economic losses and all other consequences of future earthquakes).[14] Direct losses are computed for an expected sequence of future earthquakes each large enough to cause damage over a chosen strategy lifetime, perhaps 30 years; all costs have to be computed in monetary units, including costs of human life and injury. Estimation of future losses for any particular level of earthquake may be carried out by the methods discussed in Chapter 9. Obviously if the building stock changes in any way which affects the vulnerability, this changes the loss estimation and can be incorporated into the analysis. Determination of the annual probability of an earthquake of a particular severity, i.e. the seismic hazard, has been discussed in Chapter 7, and is a complex procedure, with its own uncertainties. Often it is worth making a range of assumptions, from the most pessimistic to the most optimistic, in order to determine the sensitivity of the evaluation to this factor.

Indirect losses may be assumed to include human casualties and loss of life, loss of building contents, and economic losses due to loss of function of the building. Separate computations need to be made of the expected level of each of these losses, and procedures for making such estimates are discussed in Chapter 8. To be consistent with the method of expressing physical losses, assessment of indirect losses, including loss of life, needs to be expressed as a cost. This requires a knowledge of the expected occupancy level per unit building cost, and requires further assumptions to be made about occupancy levels and value of building contents and a value attributable to the function for the class of building involved. Equally, a cost needs to be attributed to the loss of human life. This is the most difficult and controversial element of a cost–benefit analysis, and because of the high value placed on human life, it has a major influence on the outcome. Two ways of avoiding this difficulty are:

1. The use of a cost-effectiveness criterion in relation to human life saving.
2. The development of an 'acceptable risk' approach.

10.5.2 Cost-effectiveness Criterion

It is not necessary to equate the value of a human life and the costs of earthquake damage if the two are calculated separately. The cost-effectiveness of spending to save life can be evaluated as a separate measure of the value of alternative strategies. For a range of possible strategies, the financial costs and benefits are

[14] Mathematically, the problem can be restated as follows. Choose the strategy which has the least value of net present cost NPC, where:

$$\mathrm{NPC} = \mathrm{PCC} + \mathrm{PPL} + \mathrm{PIL}$$

and where PCC is the present value of the additional construction costs, PPL is the present value of the future physical losses, and PIL is the present value of the future incidental losses.

assembled, but without including a valuation of human life. The expected benefits in terms of saved human lives, and saved injuries, are computed, allowing the financial *cost per saved life* to be calculated. Decision-makers are then faced with choosing between the projects on the basis of these separate attributes, and may use the cost per saved life and per saved injury as indicators of the cost-effectiveness of a particular strategy.

This approach can be useful in setting appropriate levels of seismic resistance in building codes: a graph of the marginal cost per saved life, as a function of the level of seismic protection, can be expected to be of a form similar those shown in Figure 10.4.[15] For areas of low seismicity, the marginal cost per saved life will always be positive (i.e. it will always cost additional money to save additional lives) and this cost will also increase as the level of protection increases. For areas of high seismicity (and more vulnerable building types), the marginal cost per saved life for low levels of protection may actually be negative, implying that strengthening may save more in financial terms (by reducing physical damage) than it costs; as the level of protection rises, more lives will be saved, but at a higher cost per saved life. So the marginal cost increases, and at higher levels of protection will become positive as in an area of low seismicity.

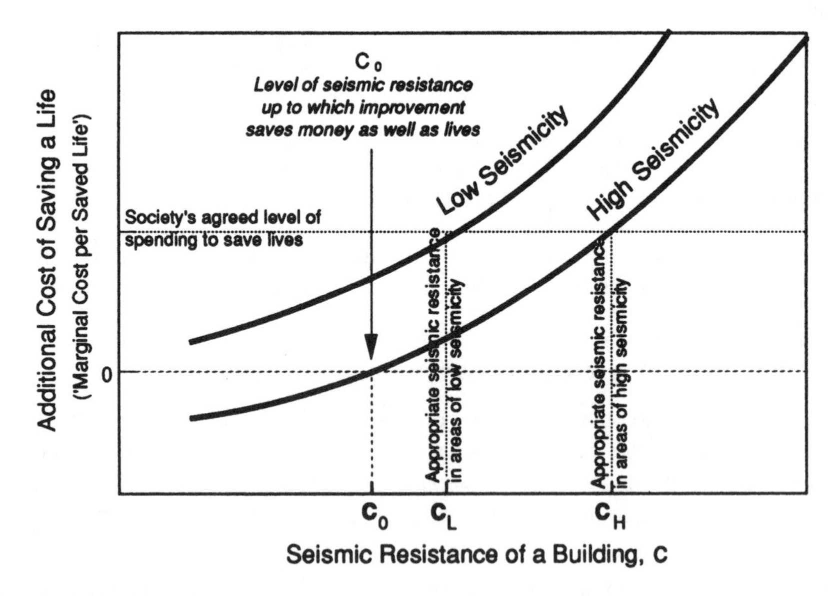

Figure 10.4 Cost-effectiveness criterion in deciding alternative protection strategies: the marginal cost per saved life in areas of high and low seismicity (after Grandori and Benedetti 1973)

[15] Grandori and Benedetti (1973).

The point at which marginal cost per saved life is zero (point C_0 in Figure 10.4) has a special importance. A level of protection lower than this will cost society more in economic terms than it would to provide this level of protection, even if human lives are not considered. Put another way, if current protection levels are lower than this, a project to bring protection up to this level will save money as well as saving lives. But this does not imply that C_0 defines the *best* level of protection. Presumably society will be willing to pay something for protecting human lives, and if this amount can be decided, then the appropriate level of protection can be determined from a graph like that of Figure 10.4. One way of deciding could be to look at the amount society is willing to pay for life saving by other means.

The cost-effectiveness criterion is useful for another reason. One important consideration in earthquake protection programmes is how best to provide protection over an area where the seismicity varies. If a given amount of money is to be spent in such a way as to minimise the total number of victims, it has been shown[16] that this can be achieved by making the marginal cost per saved life the same at all sites. It can be seen that this implies a lower level of protection for a site of lower seismicity. Thus, if the cost-effectiveness curve can be defined, this helps define also the appropriate level of protection.

The cost-effectiveness approach is very appropriate for comparison of earthquake protection strategies, and has been used in a variety of situations described in more detail in Section 10.6.

10.5.3 Acceptable Risk

In effect the purpose behind the cost–benefit analysis is to define the acceptable level of risk using economic criteria. There are essentially two elements to this risk: the risk to human life, and the risk to property. If acceptable levels of either or both of these types of risk can be defined, these levels can be used to define the appropriate level of protection. This approach is implicitly adopted in the formulation of many codes of practice for building design. The background documents for the seismic regulations in California state explicitly that the level of resistance aimed for is based on the concept of an *acceptable risk*, and what is taken to be acceptable is that buildings designed according to the code should resist minor earthquakes without damage, resist moderate earthquakes without significant structural damage but with some non-structural damage, and resist major or severe earthquakes without major failure of the structural framework of the building or its equipment, and maintain life safety.[17]

Once the meaning of 'minor', 'moderate' and 'major' earthquakes has been more precisely established in terms of earthquake severity or intensity levels, the

---

[16] Grandori (1982).

[17] Applied Technology Council, ATC3-06 (1978), see also Section 8.6, Codes of Practice.

above criteria can become the basis for defining suitable levels of protection. But the procedure implies that the acceptable level of risk has already been defined. How is it possible to decide whether this level of risk is right or too high or too low? One method proposed is to use the concept of balanced risk, using as a decision criterion the level of risk which is acceptable in other risky human activities. The intention of this approach is to discover the levels of risk which are acceptable in society by examining risk levels in a range of comparable activities.

Table 10.1 shows a range of risk of death for different regions and different causes. It shows that risks of death from disasters tend to be considerably lower than the risks from more everyday causes such as disease and road accidents. As expected, disaster risks vary widely according to the community affected. For example, it can be seen that the risk of being killed in an earthquake is nearly 100 times higher for an average Iranian than for an average California resident. However, such general comparisons are not really very helpful, because risks vary widely in any community between those most at risk and those least at risk. Moreover, the fact that they exist does not imply that these risks are acceptable. How can we assess what level of risk is acceptable?

Research suggests that the public is willing to accept voluntary risks around 1000 times greater than involuntary risks, and on this basis it has been suggested that an annual level of 1 death per 10 million persons exposed might be used

Table 10.1 Probability of an individual dying in any one year from various causes.*

Smoking 10 cigarettes a day	1 in 200
All natural causes, age 40	1 in 850
Any kind of violence or poisoning	1 in 3300
Influenza	1 in 5000
Accident on the road (driving in Europe)	1 in 8000
Leukaemia	1 in 12 500
Earthquake, living in Iran	**1 in 23 000**
Playing field sports	1 in 25 000
Accident at home	1 in 26 000
Accident at work	1 in 43 500
Floods, living in Bangladesh	**1 in 50 000**
Radiation working in radiation industry	1 in 57 000
Homicide living in Europe	1 in 100 000
Floods, living in northern China	**1 in 100 000**
Accident on railway (travelling in Europe)	1 in 500 000
Earthquake, living in California	**1 in 2 000 000**
Hit by lightning	1 in 10 000 000
Wind storm, northern Europe	**1 in 10 000 000**

*From BMA (1987), with statistics in bold type added from the Cambridge University Human Casualty database and other sources.

as a target for seismic protection levels which are to be determined by public authorities.[18]

The actual level of exposure to seismic risk varies widely, even within industrialised countries. For the City of Boston, it was estimated in 1975[19] that the annual risk of death to inhabitants of concrete frame apartment buildings not designed with any specific design requirements varied from 0.2 to 80 per million inhabitants according to the assumed level of seismic hazard. The estimated annual risk of death to the inhabitants of buildings in southern California designed to zone 3 seismic regulations was then estimated as 50 per million. In parts of southern Italy, the authors have estimated that the annual risk of death to inhabitants of existing stone masonry buildings may vary from 90 to as high as 2000 per million for lower and higher local levels of seismicity. In Europe, the current rate of fatalities from motor accidents varies little from country to country, all countries reporting annual risk levels between 100 and 200 per million per year. This range of values indicates that it is very difficult to pin down an acceptable level of risk of loss of life due to earthquakes even if we could calculate accurately the rate resulting from any policy. However, it is clear that there are few if any societies which can afford the cost of building to standards which reduce the level of risk as low as 1 death per year per 10 million population.

10.6 Evaluation of Alternative Strategies: Some Examples

10.6.1 Evaluating Programmes for Upgrading Rural Buildings

In countries where a high proportion of the population live in dwellings of traditional construction in the rural areas, changes in the design coefficients in codes of practice for new construction are largely irrelevant as a means of improving earthquake protection. The very high vulnerability of such buildings has prompted governments to look for means of intervening in rural construction which would be effective in reducing future losses. In these situations strengthening of existing buildings is seldom likely to be cost-effective, but since rebuilding is comparatively frequent the option of introducing subsidised improvements at the time of rebuilding is worth considering. But before embarking on any such upgrading programme a government will need to look at a range of alternative technical options and study their costs and benefits as well as other aspects.

The authors conducted such a study in collaboration with the Turkish government for the high-seismicity regions of eastern Turkey.[20] A range of different upgrading options were considered, some of which were already used to a limited extent, and which were of gradually increasing cost and providing an increasing

[18] Wiggins and Moran (1970).

[19] Whitman *et al.* (1980).

[20] Spence and Coburn (1987a).

level of protection. The additional cost (above standard traditional technology) for each house was calculated, and the expected future losses in terms of physical losses and casualties were estimated for each option. The method for making these estimates is discussed in Section 9.6, and the resulting reductions in losses are shown in Figure 8.12. These figures were used to calculate the cost-effectiveness of each upgrading option both in terms of the cost per saved house and the cost per saved life (Figure 10.5).

Upgrading weak masonry buildings would make a considerable impact on expected casualty levels over, say, a 25-year period. Figure 10.5 shows that if

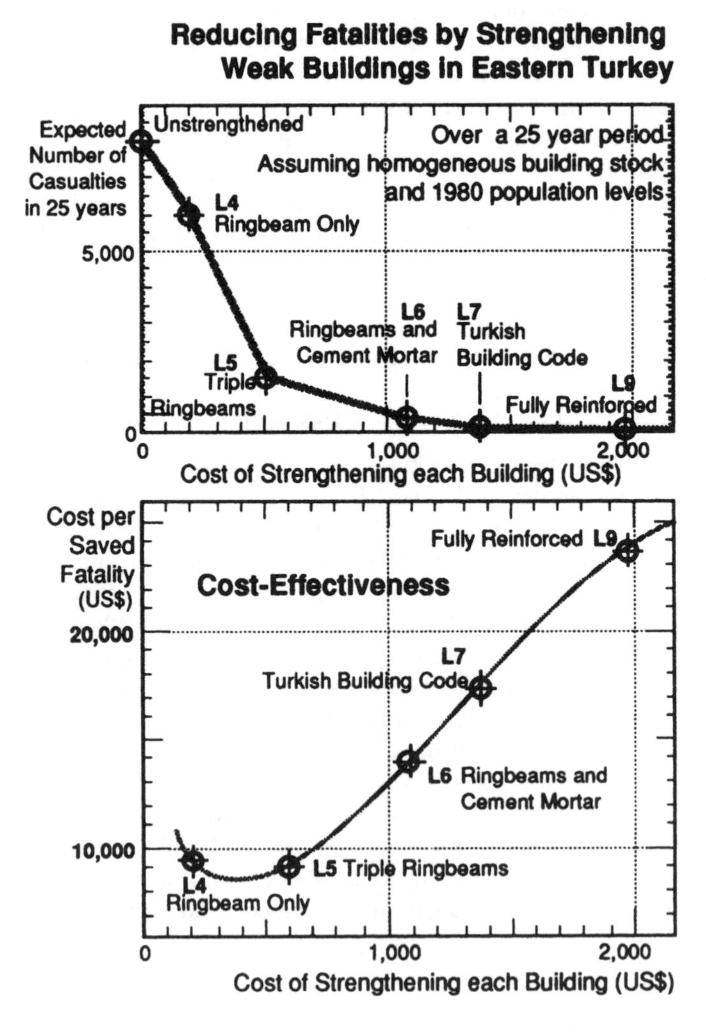

Figure 10.5 Cost-effectiveness of building safer traditional buildings in rural areas

all the traditional weak masonry buildings of eastern Turkey had a single timber reinforcement ringbeam the expected fatalities would be reduced from 8000 to somewhere in the region of 6000. The 1 million houses in the region would all need to be built with strengthening during their initial construction and the additional total capital investment would be $200 million. But 18 000 fewer buildings would be destroyed, which would reduce reconstruction costs by $252 million during the same period. Thus the programme actually saves money in the long term, and the net cost per saved life is negative.

This benefit of the programme could be made even better if the targeting of strengthening was improved. If only half the number of houses, or, say, the 10% in the most seismic areas, were improved and still achieved similar life and damage savings, the net benefit would increase from $26 million to $76 million over the same period.

More savings of life could be made if each building was stronger still. Three timber ringbeams at every 1 metre height up the wall (upgrading option L5) would tie the structure together even more effectively and reduce expected fatalities from 8000 to around 1500 and the cost-effectiveness of such strengthening would be even better than the single ringbeam, although the total capital investment needed is almost three times as large.

As the level of strengthening goes up, the buildings become safer and the cost increases. The most cost-effective strengthening measures, measured in terms of the expenditure necessary to save a life, are the lower cost ones, and the better the targeting of structures at risk, the more cost-effective the strengthening measures become. The cost per saved house for the two lowest cost options (options L4 and L5) is actually lower than the cost of reconstruction, thus the use of either of these options would be expected to save money in addition to saving lives.

Introducing upgrading programmes along these lines is technically feasible and appears economically worthwhile for national governments, but difficult social and political problems nevertheless arise in devising effective programmes. These questions are explored in Section 10.7.

10.6.2 Evaluating Strategies for Strengthening Existing Urban Buildings

Reducing the vulnerability of existing buildings is a key aspect of any earthquake protection programme. Maintaining the existing building stock is essential for social and cultural as well as economic reasons, yet in many areas of high seismicity older, weaker buildings, often inhabited by the poorer people, are the main source of expected future losses. Studies of nineteenth-century unreinforced masonry buildings in Boston[21] suggested that their vulnerability to collapse was about 50 times that of a building designed to the current code of practice.

---

[21] Whitman *et al.* (1980).

Options exist for strengthening, but they are often expensive compared with the incorporation of additional strengthening into new building, and thus careful analysis of costs and benefits is needed in deciding whether to strengthen or demolish, and what level of protection society should demand, or an individual owner should provide. Moreover, it is difficult to assess accurately the effect of any particular strengthening measure in reducing vulnerability.

A study of the upgrading of both nineteenth-century unreinforced masonry and early twentieth-century reinforced concrete warehouse buildings in Boston, an area of moderate seismicity, indicated that the cheapest of the upgrading options was substantially the most cost-effective in terms of life saving, and that the upgrading cost-effectiveness was much greater for the weaker masonry building. The minimum cost per saved life was estimated as $240 000 for a masonry building and $1.25 million for a concrete building.[22]

In areas of higher seismicity, upgrading is likely to be more cost-effective, i.e. the cost per saved life is likely to be lower. A cost–benefit analysis of the upgrading of unreinforced masonry apartment buildings in Los Angeles[23] indicated again that the lowest cost upgrading options were most cost-effective, showing a cost per saved life below $200 000, while further upgrading cost more than $800 000 for each extra saved life.

In Turkey, the authors made a study of the cost-effectiveness of strengthening existing reinforced concrete frame apartment buildings, a very common class of structure which was found to be especially vulnerable in the 1999 earthquakes.[24] Such upgrading requires heavy intervention into the existing structure and is typically found to cost about 40% of the cost of reconstruction. Without considering the cost of human lives saved, the payback period was between 200 and 300 years depending on the level of seismicity assumed – insufficient to act as an incentive to investment. Measuring the cost-effectiveness in terms of life saving, however, the cost per life saved was between $250 000 and $750 000.

Studies of this sort give a useful comparative picture of overall social costs and benefits: it is interesting for example to compare the cost of saving lives in these US examples, ranging from $200 000 to $1 million, with the cost of saving lives in upgrading traditional Turkish houses, ranging from $20 000 to $120 000, and between $250 000 and $750 000 for modern Turkish apartments. But such studies are inadequate if they do not identify the costs and benefits to the different parties involved – the various risk stakeholders.

The owner of a property is normally required to pay for any upgrading, but has no liability for death or injury caused by earthquakes. Hence the only benefit to the owner to set against the cost of upgrading will be in the higher rents chargeable, and any tax reduction associated with upgrading. The Los Angeles

[22] Pate-Cornell (1985).

[23] Sarin (1983).

[24] Spence *et al.* (2002b).

study established that occupants were willing to pay a rented accommodation increase of $20 per month for increased safety, enough to pay for the cheapest upgrading option, but that any higher cost would not be acceptable. The much higher standard of safety and significantly higher cost introduced in the Los Angeles regulations met strong resistance from building owners.

10.6.3 Evaluating Targeted Strengthening

The cost-effectiveness of strengthening can be further enhanced by targeting only the most vulnerable structures. In a pilot study in Mexico City, screening of the building stock to target residents most at risk was carried out in a small area of the city to identify the worst 1%, 5% and 10% of the building stock in terms of its potential contribution to future casualty levels.[25] The concentration of population in high-occupancy buildings in Mexico City makes this approach relatively effective.

In the event of a severe earthquake, expected fatalities in the pilot study area would be reduced by almost 50% if the worst 5% of buildings had been strengthened, giving a cost per saved life (assuming the earthquake occurs) of around $5000. The benefits of this strengthening programme (referred to as programme I), in terms of fatalities, are shown in Figure 10.6. The calculated cost per saved life is also highly dependent on the severity of the assumed earthquake. There is a low probability of such a severe earthquake occurring within the lifetime of the strengthened building: a less severe earthquake is more likely, but an earthquake with a return period of only 25 years would cause fewer casualties, and the strengthening would save the lives of fewer people and so the cost of per saved life would be of the order of $50 000. A larger scale programme (programme II), involving the strengthening of over 10% of the pilot area building stock, would reduce expected fatalities by over 80%, at a cost per saved life of $7000, again in the event of the most severe earthquake, and would further reduce the numbers of homeless and repair costs as a function of the severity of the earthquake as shown in Figures 10.6. The total cost of such a programme is high because it involves strengthening all of the worst 5% or 10% of buildings in the city to be effective, but it is substantially cheaper and more cost-effective than one involving the strengthening of all buildings.

10.6.4 Evaluating Strategies for Protection of Historical Centres

A special application of cost–benefit analysis is in the evaluation of programmes for upgrading historical city centres. In these cases the upgrading must be designed to fulfil a range of potentially conflicting criteria: the limitation of future damage to the buildings and protecting the lives of occupants must be

[25] Aysan *et al.* (1989).

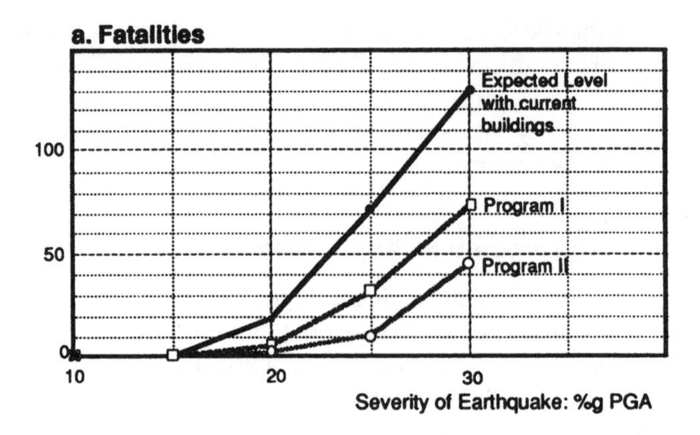

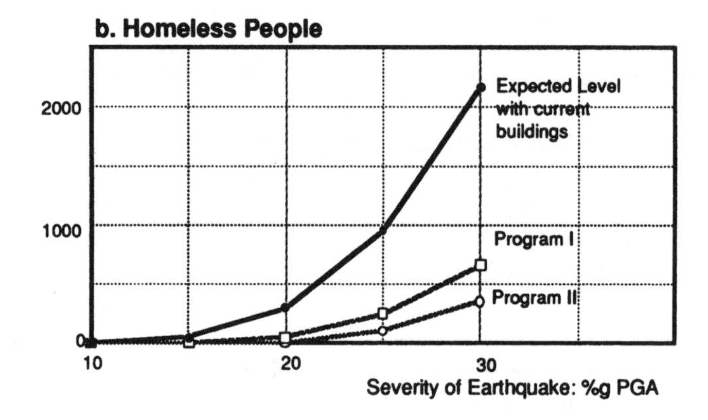

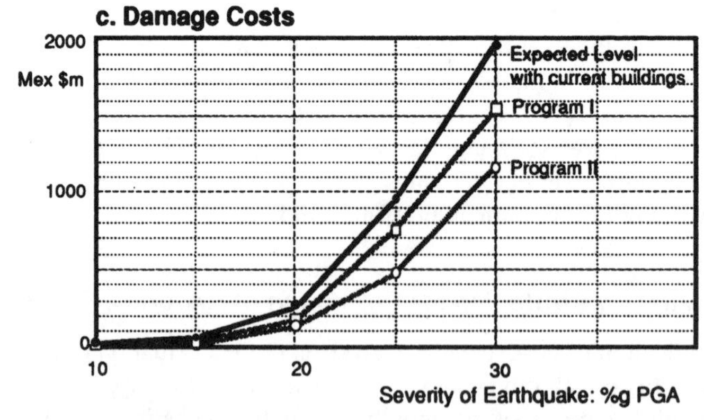

Program I : Strengthening of targeted worst 8% of building stock in high-risk areas
Program II: Strengthening of targeted worst 18% of building stock in high-risk areas

Figure 10.6 The effects of building strengthening programmes on reducing earthquake consequences in a sample area of Mexico City

considered alongside the limitation of alteration to the appearance of the buildings and neighbourhoods, and their future economic viability. But strengthening programmes designed to achieve the last of these criteria can be evaluated by considering their cost-effectiveness in terms of saved future reconstruction costs and saved lives. In such a study for part of the Alfama District of the historical centre of Lisbon, in Portugal, using low-cost wall ties for strengthening (see Chapter 7), the authors found a payback period between 5 and 15 years compared with no action, indicating that intervention would be highly beneficial financially. Life safety was an additional benefit. In an earthquake of intensity VII (with an annual probability of 1–3% of occurrence), 40 lives would be lost per 10 000 inhabitants in the unstrengthened buildings, and only 2 in the strengthened buildings.[26]

10.7 Social and Public Policy Aspects of Earthquake Protection Strategies

In the formulation of public policy for earthquake protection, proposed risk mitigation strategies have to win support politically, and from the public at large, if they are to succeed. The effectiveness claimed for these strategies by their proponents has to be weighed against the degree of confidence which individual householders place in them and the willingness of society and individuals to pay for them. This in turn is affected by the perception of the risk by individuals. Alternative approaches to public policy formation are broadly of three types:[27]

1. The do-nothing-until-it-happens approach
2. The market approach
3. The planning approach.

The *do-nothing-until-it-happens approach* recognises that protective strategies of the sort described earlier are not going to happen, either because of lack of understanding of the risk or because they are too expensive. After the earthquake, it is left to local government backed by national government and international aid agencies to provide recovery assistance not only for public property (roads, buildings) damaged, but also for house owners to repair or rebuild their property and for small businesses to re-establish themselves. Eventually such payments are reimbursed by general taxation or by increasing national debt burdens. It is the model used in the past for recovery and reconstruction after most earthquake disasters; it has some merit in that the community at large is asked to pay for assistance to the comparatively few disaster victims (and may be willing to do

[26] D'Ayala *et al.* (1997).

[27] Comerio (1998).

so). But its consequences are poor standards of construction in earthquake-prone areas, with no incentive to improve; long delays in funds becoming available and political conflict about their disbursement; and a debt burden which can be crippling to weak economies. The repetition of failed reconstruction programmes in the past is proof that this is not the answer, and it is the contention of this book that there are better ways.

The basis of the *market approach* is that there are no government programmes of assisting recovery for private homes and businesses, and it is left to individual owners to protect themselves by insurance and reinsurance, or by improving their own property to provide a level of safety they find acceptable. Adopting this model in its simplest form requires that people are extremely well informed about the risks and building safety issues, and are prepared to give the payment of very substantial earthquake insurance premiums a high priority among other day-to-day expenditures. Comerio[28] believes that even in California, perhaps the world's most earthquake-aware society, few people would do this, with the result that earthquakes would cause many thousands to lose their homes and livelihoods without hope of compensation, and the only winners would be property speculators cashing in on the acquisition and refurbishment of damaged property.

A *semi-market approach* has been adopted in some countries in which earthquake insurance is made available to all householders through a compulsory national scheme, operated through normal household insurance contracts and backed by international reinsurance, thus privatising the risk, but ensuring through the system of planning and development control that everyone is covered. Such a scheme has operated in New Zealand for some years through the Earthquake Commission, and has been introduced in Turkey.[29] A scheme has also been proposed for the United States whereby federally backed home loans would be available, tied to the requirement for earthquake (and other disaster) insurance cover.[30]

The alternative *planning approach* involves the adoption, by the local community as a whole, of policies of protection such as improved building codes and building control measures, the strengthening of high-risk buildings and emergency planning, and the creation of a climate of opinion in which such measures can be supported and funded. These are the measures which have been described and advocated in this book, but they have to date been applied only to a limited extent, because they are costly; where they are applied only to a few buildings, their impact on future losses will not be very great.

The three approaches are not mutually exclusive and can be combined in various ways. For the United States, Comerio,[31] for instance, suggests a combination of limiting government assistance to the replacement of public sector

[28] Comerio (1998).

[29] Bommer *et al.* (2002). Two million policies have been sold by mid 2002.

[30] Comerio (1998).

[31] Comerio (1998).

losses, using government backing to stimulate a more active involvement by insurance companies in carrying the earthquake risk, and tax incentives to stimulate mitigation actions (such as strengthening) by individuals. For the support of the newly formed Turkish national insurance pool, the authors have suggested that the insurance pool could support retrofitting action by a combination of reduced insurance premiums, tax advantages and easing of planning restrictions.[32]

Which approach to earthquake protection is adopted in any situation will depend on national and local circumstances, and will also be affected by who pays, who benefits and on how the risk is perceived.

Perception of Risk

The public perception of risk is playing an increasingly important role in the formation of public policies for risk mitigation. Perception of risk can differ from one group to another. Experts like to use statistics, but most other people are less comfortable with statistical concepts and prefer to base perceptions of risk on a range of other values, philosophies, concepts and calculations.

In general, research into perception[33] shows that people evaluate risks though a number of subjective concepts and beliefs in a multi-dimensional way. The calculated risk is less important to most people than some of the qualitative attributes of the risk – the image of that risk and conjecture associated with it. Four factors appear to be important in perception of risk:

1. Actual quantitative risk level ('exposure').
2. Personal experience of the hazardous events ('familiarity').
3. The degree to which the hazard is perceived as controllable or its effects preventable ('preventability').
4. The concept of the hazard that some researchers term 'dread' – the horror of the hazard, its scale and consequences.

It is clear that earthquake disasters score very highly on the dread factor, and are widely perceived as unpreventable. Disasters that cause large numbers of deaths are more dreadful than low-fatality catastrophes. Perception of risk appears closely related to the dread factor, and only generally related to exposure levels or to personal familiarity.

For most people, personal contact with hazards is fairly rare and so knowledge of them is acquired more through the news media than from first-hand experience. The way the media report hazards is therefore extremely influential in risk perception. Research has shown that, partly because of intensive media interest, there is a general tendency in well-informed subject groups to overestimate the

---

[32] Bommer *et al.* (2002).

[33] Lichtenstein *et al.*, 1978.

incidence of rare causes of death and underestimate the frequency of the more common ones. A summary of some tests in Oregon in the United States[33] is given in Figure 10.7. Conversely populations without regular exposure to news media may underestimate the environmental risks they face: the limited studies of less informed communities facing high risks have concluded that the individuals are probably more at risk from hazards than they realise.[34]

An important ingredient of disaster mitigation programmes is therefore a campaign of public education to increase disaster awareness. This is not only to increase perception of risk where it is judged too low, but also to educate the public that disasters are preventable and to encourage them to participate in protecting themselves.

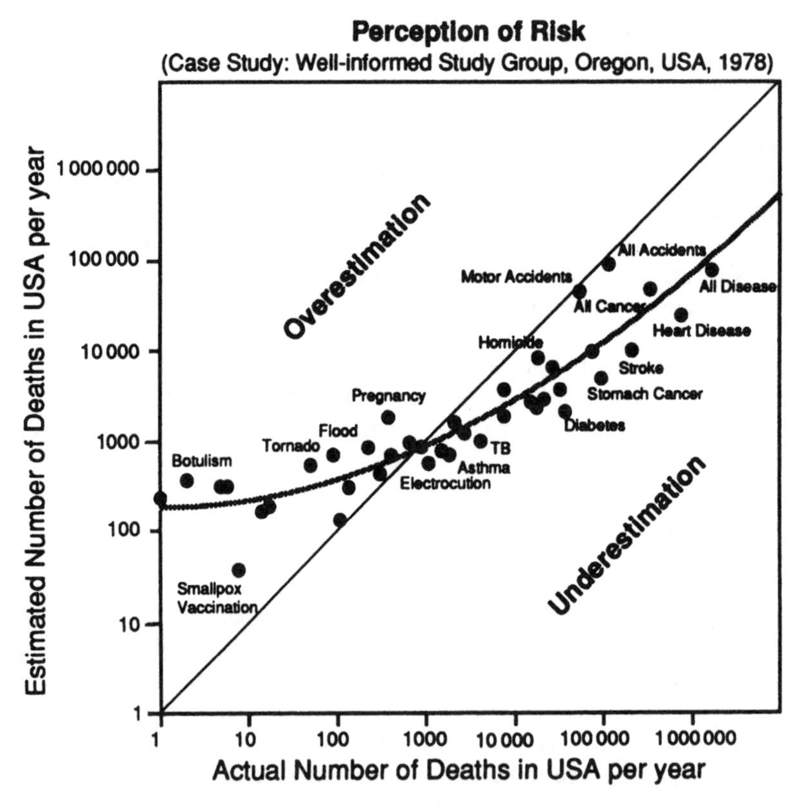

Figure 10.7 Perception of risk by a well-informed population (after Lichtenstein *et al.* 1978)

[34] Villagers living in areas of high seismic risk were interviewed by social scientists in eastern Anatolia, Turkey, as part of a study of risk reduction programmes by the Turkish National Committee for Earthquake Engineering (Coburn 1982b).

Decision-making and Evaluation of Risk

Studies such as those described above have tended to emphasise the decision-making potential of the information generated, but little attention has been paid to the range of uncertainty involved. Where it has been possible to do this, a wide variation in the 'optimum' strategy may become apparent.[35] A further difficulty is that studies rarely distinguish the viewpoints of the different interest groups involved; it is clear from the experience of Los Angeles[36] that what is in society's interest may not be acceptable to those who are required to pay. And there is often a significant discrepancy between the measured or estimated risk and the public perception of it. Public perception can to a degree be modified by better information, but always has to be taken into account in policy formation.

Clearly in any evaluation of alternative earthquake protection strategies, as much attention needs to be devoted to the formulation of an acceptable policy for implementation as to the technical details and the economic evaluation.

A valuable study[37] of the earthquake mitigation legislation programmes of three communities in southern California aimed to discover the reasons why such a long period had elapsed from the time at which the need for such programmes was widely recognised (after the 1933 Long Beach earthquake) until a mitigation ordinance was finally passed. This took place in 1972 in Long Beach, but only in 1981 in Los Angeles. The main conclusion of the study was that hazard mitigation is not primarily a technical exercise: it is 'inherently and often intensely political because mitigation usually involves placing some cost burdens on some stakeholders, and may involve a redistribution of resources'. Its author argues that advocates for risk mitigation strategies must develop political as well as technical solutions. The author concludes that a decision on a policy for earthquake protection has four prerequisites:

1. A recognised and well-defined problem, and a belief that something can be done about it which will be politically acceptable.
2. A possible technical solution to the problem that non-technical policy-makers view as practical and effective.
3. A group of policy advocates who believe in the policy, are seen by the policy-makers as credible, and are persistent in their pursuit of the policy.
4. A window of opportunity for the policy to be enacted such as that which appears when an earthquake has occurred that affects the community directly or indirectly.

The precise nature of the political solution will vary from place to place but one essential ingredient will be to find some way to offer some incentives or

[35] Whitman *et al.* (1980).

[36] Sarin (1983).

[37] Alesch and Petak (1986).

compensation to those who are required to pay the cost of the programme. This should include at least some benefits in addition to the rather intangible benefit of reducing earthquake risk, and a means of complying that will not put an unacceptable financial burden on them.

In the case of Los Angeles, perhaps the largest and best-documented programme of retrofit strengthening to existing buildings so far, the building owners were offered certain tax concessions and phased requirements for compliance so that the cost could be spread over a number of years. Even then progress in compliance has been slow.

In proposed upgrading programmes for the historical centre of Mexico City the use of the transference of development potential has been considered, imposing a tax on the developers as a means to finance the upgrading of the historical buildings of the city.[38]

Policies for Developing Countries

The earthquake risk is unlikely to be considered important in a community that faces much greater everyday threats of disease and food shortages. Even if the risk is quite significant it is unlikely to compare with the risk of child mortality in a society with minimal primary health care. For example, villagers in the hazardous mountain valleys of northern Pakistan, regularly afflicted by floods, earthquakes and landslides, do not perceive disaster mitigation to be one of their priorities[39] – their priorities are protection against the greater risks of disease and irrigation failures. Thus systematic community-wide strengthening programmes are unlikely to be effective, and post-earthquake disaster relief is likely to continue to be needed. However, the demonstration effect of building for safety programmes such as those described earlier, should not be underestimated, if only to help people understand the value of their local building traditions better, and thereby avoid the catastrophic impact of inappropriate urban building techniques following a major disaster.

10.8 The Way Ahead

10.8.1 The Sanitary Revolution: a Precedent for Disaster Mitigation

A useful analogy can be made between the recently developing science of disaster mitigation and the implementation of public health measures that began in the mid-nineteenth century. Before then tuberculosis, typhoid, cholera, dysentery, smallpox and many other diseases were major causes of death and tended

---

[38] Aysan *et al.* (1989).

[39] Davis (1984), D'Souza (1984).

to assume epidemic proportions as the industrial development of cities fuelled increasing concentrations of population. These diseases had a major effect on life expectancy at the time and yet were regarded as unavoidable everyday risks. The apparent randomness with which the diseases struck and the unpredictability of epidemics meant that superstition, mythology and a certain amount of fatalism were the only public responses to the hazards; the high risk of disease was generally accepted because there was little alternative.

As the understanding of what caused diseases increased, chiefly through the efforts of scientists and epidemiologists in the nineteenth century, so the incidence of epidemics and illnesses generally became demystified. It became evident that disease was preventable and gradually the concept of public protection against disease became accepted.

It became evident that sanitation, purification of water supply, garbage disposal and public hygiene were key issues for public health. The measures necessary to reduce the risk of disease were expensive – massive infrastructural investment was needed to build sewers and clean water supply networks – and required a major change in public practices and attitudes of individuals. Social historians refer to this as the 'Sanitary Revolution'. Garbage collection and disposal had to be organised. It became socially unacceptable to throw garbage or to dispose of sewage in the streets. Personal hygiene, washing and individual sanitation practices emerged as important, initially encouraged by public awareness campaigns and gradually becoming part of the social norms and taught by parents to their children. Attitudes changed from the previous fatalism about disease to a public health 'safety culture', where everyone participated in reducing the risk of communal disease.

Public health advances went hand in hand with public medicine, medical care, vaccination, primary health care and a health industry that in most developed countries today consumes a very significant proportion of national economic production. Today public epidemics are unacceptable. High levels of risk from disease are not tolerated and outbreaks of disease are followed by outbursts of public opinion demanding medical and government response to protect them. Most people now consider it normal to participate in their own protection against health hazards and accept the high levels of cost involved in society's battle against disease. The level of risk from public health hazards that is judged acceptable by modern society is far lower than it was three or four generations ago.

Earthquake risks today are seen in much the same way as disease was in the early nineteenth century: unpredictable, unlucky and part of the everyday risk of living. Concentrations of people and rising population levels across the globe are increasing the risk of disasters and multiplying the consequences of earthquakes when they occur. However, the 'epidemiology' of earthquakes – the systematic science of what happens to turn an earthquake into a disaster – shows that disasters are largely preventable. There are many ways to reduce the impact of an earthquake and to mitigate the effects of secondary hazards and damage consequences.

Just like the fight against disease, the fight against earthquakes has to be fought by everyone together and involves public and private sector investment, changes in social attitudes and improvements in the practices of individuals. Just as the Sanitary Revolution occurred with the development of a 'safety culture' for public health, so earthquake protection has to develop through the evolution of an equivalent 'safety culture' for public safety. Governments can use public investment to make a stronger infrastructure and a physical environment where a disaster is less likely to occur, but individuals also have to act to protect themselves. Just as public health depends on personal hygiene, so public protection depends on personal safety. The type of cooking stove each family uses, and its awareness that a sudden earthquake could tip the stove over, is more important in reducing the risk of conflagration than the community maintaining a large fire brigade. The type of house each family builds and where it considers a suitable place to live affect the potential for disaster in a community even more than sophisticated earthquake warning systems or large engineering projects to prevent rockfalls or stabilise landslides.

The science of earthquakes is in a similar state of development to epidemiology in the latter half of the nineteenth century: the causes, mechanisms and processes of disasters are becoming understood. As a result of this understanding, the more developed countries have begun to implement individual measures to reduce the risk of future disasters. A catalogue of techniques have been described here for earthquake protection, and their relevance to the countries that need it most is now clear.

Disasters are very largely a developmental issue. The great majority of casualties and disaster effects are suffered in developing countries. Development achievements can be wiped out by a major disaster and economic growth reversed. The promotion of earthquake mitigation in the projects and planning activities of development protects development achievement and assists populations in protecting themselves against needless injury.

Returning to the precedent of the Sanitary Revolution, the situation at the present time is in many ways not unlike the period in the mid-nineteenth century, when the great public health programmes – piped water and sewage disposal – began to be implemented in the large cities. These depended on:

- a scientific understanding of the causes of disaster (in this case water-borne diseases);
- the availability of technical means to eliminate or mitigate the disaster, and a knowledge of the costs of protective measures;
- a widespread public belief that disasters are not random, and that mitigation is possible;
- the political will and opportunity to act.

It can be expected that when these same four conditions are achieved, disaster protection schemes will begin to be implemented on a very much increased scale. At the present time the first two of these conditions are already met, and the third – public belief in the possibility of mitigation – is steadily increasing as each successive disaster is shown to have been caused not by the natural phenomena, but by an avoidable failure of protection planning. As it already has in a number of communities, well-organised pressure by those at risk, backed by the support of building professionals, can generate the political lobbying will to implement protection programmes. Indeed, as the issues become more widely understood we can hope to see earthquake and other disaster protection programmes as a significant part of all urban and national development programmes.

10.8.2 The Twenty-first Century

The world's population growth, which topped 6 billion in 2000, is still increasing rapidly. The United Nations 'medium' growth assumptions foresee a global population of 7.15 billion by the year 2015 and 8.1 billion by 2030. Gradually declining rates of growth suggest a possible stabilisation of the world's population at around 10 billion perhaps by 2075.

What is certain is that the bulk of this growth will take place in the developing countries, and in countries that already suffer some of the world's worst earthquake disasters. Of the top 20 earthquake countries listed in Table 1.1, 15 of them had population growth rates in the last decade above the world's average (1.5%) and 8 had growth rates of over 2% annually.

Within all these countries, urbanisation continues apace and with it the potential for massive earthquake disasters. In 1990, there were 90 'super-cities' in the world with populations of over 2 million. At the turn of the new century, the world had over 160 such 'super-cities', of which 70% are in seismic zones (likely to experience damaging intensities of VII or more), 20% are in zones likely to experience destructive intensities (VIII or more) and at least 9% are in severe seismic zones, capable of experiencing highly destructive intensities (IX). In Chapter 1, we identified 29 cities with high loss potential, based on their earthquake hazard, their population size and the quality of their building stock. Most of these cities are growing rapidly, increasing the potential for catastrophic disasters in the future. The main seismic belts that rim the Pacific and cross southern Asia and the Middle East are areas of some of the fastest urbanisation in the world. The growth of cities is outstripping population growth in nearly every country, as the magnet of economic earning power draws people into the urban areas.

The greatest challenge for earthquake protection today is to establish policies and strategies that can be sustained into the next century and beyond to offset the inexorable increases in disaster potential. Long-term administrative and social structures need to be established now so that protection becomes ingrained. Social attitudes have to be shaped so that each new generation build ever stronger and safer within their hazardous environment.

The process of building structures and facilities, planning them, locating them and investing in them needs to incorporate earthquake protection considerations as part of their normal ingredients. The investment needed to build an earthquake-resistant environment must be established as part of the baseline costs, not seen as some additional cost or optional extra. This investment is a basic element of social protection and is the necessary price of living in safety from the elemental forces of nature.

The doubling of the world's urban population over the next half century entails a massive increase in the physical infrastructure and building stock of the globe. Much of this investment will be in the world's great earthquake zones. This massive investment and construction activity on an unprecedented scale must be carried out with earthquake safety as an integral part of it. The attitudes and procedures that will shape this future construction need to be established now. The laws, codes, investment standards and procedures established now will form the foundation of a safer future.

People will continue to live on our restless planet. Over our history the other elements – the cold, heat, wind and rain – have gradually been tamed. The earth will continue to unleash its elemental forces but increasingly our buildings and creations are capable of withstanding their worst effects. Through systematic measures of earthquake protection, we can expect that the earthquakes that once destroyed our cities will one day roll beneath our buildings as impotently as the weather today rolls by above them.

Bibliography

Abolafia, M. and Kafka, A.L., 1978. 'Towards a measure of socio-seismicity', *Proceedings Second International Conference on Microzonation*, 1489–1497.

Aguilar, J., Juarez, H., Ortega, R. and Iglesisas, J., 1989. 'Statistics of building damage and of retrofitting techniques in reinforced concrete buildings', *Earthquake Spectra*, **5**, (1), 145–152.

AIJ, 1993. *Earthquake Motion and Ground Conditions*, Architectural Institute of Japan, 5-26-20 Shiba, Minato-ku, Tokyo 108, Japan.

AIJ, 2001. *Report on the Damage Investigation of the 1999 Kocaeli Earthquake in Turkey*, Architectural Institute of Japan, 5-26-20 Shiba, Minato-ku, Tokyo 108, Japan.

Akbar, S., 1989. 'Urban housing in seismic areas: a computerised methodology for evaluating planning strategies for risk mitigation', PhD Dissertation, University of Cambridge.

Alesch, D.J. and Petak, W.J., 1986. *The Politics and Economics of Earthquake Hazard Mitigation*, Institute of Behavioral Science, University of Colorado.

Alex, W., 1963. *Japanese Architecture*, Prentice Hall, Englewood Cliffs, NJ.

Alexander, D.E., 1984. 'Death and injury in earthquakes', *Disasters*, **9**, (1), 57–60.

Ambraseys, N.N., 1971. 'Value of historical records of earthquakes', *Nature*, **232**, 375–379.

Ambraseys, N.N., 1978. 'Middle East – a Reappraisal of the Seismicity', *Quarterly Journal of Engineering Geology*, **11**, 19–32.

Ambraseys, N.N., 1985. 'Intensity-attenuation and magnitude-intensity relationships for northwest European earthquakes', *Earthquake Engineering and Structural Dynamics*, **13**, 733–778.

Ambraseys, N.N., Anderson, G., Bubnov, S., Crampin, S., Shadidi, M., Tassias, T.P. and Tchalenko, J.S., 1969. *Dasht-e-Bayaz Earthquake of 31 August 1968*, UNESCO 1214/BMS.RD/SCE, Paris.

Ambraseys, N.N. and Jackson, J.A., 1981. 'Earthquake hazard and vulnerability in the northeastern Mediterranean: the Corinth earthquake sequence of February–March 1981', *Disasters*, **5**, (4), 355–368.

Ambraseys, N., Lensen, G. and Moinfar, A., 1975. *The Patan Earthquake of 28 December, 1974*, UNESCO FMR/SC/GEO/75/134, Paris.

Ambraseys, N.N. and Melville, C.P., 1982. *A History of Persian Earthquakes*, Cambridge University Press, Cambridge.

Anderson, M.B., 1990. 'Analyzing the costs and benefits of natural disaster responses in the context of development', Environment Working Paper No. 29, Policy Planning and Research Staff, The World Bank.

Ang, A.H.S. and Tang, W.H., 1975. *Probability Concepts in Engineering Planning and Design*, Vol. 1, John Wiley & Sons, New York.

Ansal, A. and Marcellini, A., 1999. 'Variability of source and site factors in seismic zonation: state of the art report', *Proceedings 11th European Conference on Earthquake Engineering*, CD-ROM Balkema, Rotterdam.

Applied Technology Council, ATC3-06, 1978. *Tentative Provisions for the Development of Seismic Regulations for Buildings*, Applied Technology Council, Redwood City, CA.

Applied Technology Council, ATC-13, 1985. *Earthquake Damage Evaluation Data for California*, Applied Technology Council, Redwood City, CA.

Applied Technology Council, ATC-14, 1985. *Evaluating the Seismic Resistance of Existing Building*, Applied Technology Council, Redwood City, CA.

Applied Technology Council, ATC 20-1, *Field Manual: Post-Earthquake Safety Evaluation of Buildings.* 1989. Applied Technology Council, Redwood City, CA.

Applied Technology Council, ATC-40, 1996. *Seismic Evaluation and Retrofit of Concrete Buildings*, 2 vols, Applied Technology Council, Redwood City, CA.

Armillas, I., Petrovski, J., Coburn, A.W., Corpuz, A. and Lewis, D., 1990. *Technical Report on Luzon Earthquake of 16 July 1990, Republic of Philippines with Recommendations for Reconstruction and Development*, Expert Mission to Assist in Reconstruction and Development After the Philippines Earthquake of 16 July 1990, United Nations Centre for Human Settlements (Habitat), PO Box 30030, Nairobi, Kenya.

Arnold, C. and Reitherman, R., 1982. *Building Configuration and Seismic Design*, John Wiley & Sons, New York.

Arya, A.S. and Chandra, B., 1977. 'Earthquake-resistant construction and disaster prevention', Souvenir Volume, *Sixth World Conference on Earthquake Engineering*, New Delhi, India.

Ashimi, T., 1985. 'Urban planning of earthquakes: Japanese perspective', *Proceedings of US-Japan Workshop on Urban Earthquake Hazards Reduction*, EERI Pub. No. 85–03, Earthquake Engineering Research Institution, 2620 Telegraph Avenue, Berkeley, CA 94704, USA.

Assar, M., 1971. *Guide to Sanitation in Natural Disasters*, World Health Organization, Geneva.

ASAG, 1996. *ASAG's Intervention Through Technology Upgrading for Seismic Safety*, ASAG, Ahmedabad, India.

Atkinson, G.M. and Boore, D.M., 1990. 'Recent trends in ground motion and spectral response relations for North America', *Earthquake Spectra*, **6**, (1), 15–34.

Aysan, Y., 1983. *Report on the Immediate Local and Governmental Response after the Erzurum-Kars Earthquake of 30 October 1983*, Report No. 4 on Cultural Aspects of Housing in Seismic Areas (Turkey), Oxford Polytechnic Department of Architecture, Oxford.

Aysan, Y., Clayton, A., Cory, A., Davis, I., Sanderson, D., 1995. *Developing Building for Safety Programmes*, Intermediate Technology Publications, London.

Aysan, Y.F., Coburn, A.W., Davis, I.R. and Spence, R.J.S., 1989. *Mitigation of Urban Seismic Risk: Actions to Reduce the Impact of Earthquakes on Highly Vulnerable Areas of Mexico City*, Report of Bilateral Technical Cooperation Agreement between the Governments of Mexico and United Kingdom, April.

Aysan, Y. and Davis, I. (eds), 1992. *Disaster and the Small Dwelling*, James and James Science Publishers, London.

Aysan, Y. and Oliver, P., 1987. *Culture and Housing after Earthquakes: A Guide for Future Policy Making on Housing in Seismic Areas*, Disaster Management Centre, Oxford Polytechnic, Headington, Oxford OX3 OBP, UK.

Bachman, R.E. and Bonneville, D.R., 2000. 'The seismic provisions of the 1997 Uniform Building Code', *Earthquake Spectra*, **16**, (1), 85–100.

Barbano, M.S., Cosentino, M., Lombardo, G. and Patane, G., 1980. *Isoseismal Maps of Calabria and Sicily Earthquakes (Southern Italy)*, Consiglio Nazionale Delle Ricerche Progetto Finalizzato Geodinamica, Gruppo di Lavoro 'Catalogo dei Terremati', Pubblicazione n. 341, Istituto di Scienza della Terra, Università di Catania, Italy.

Båth, M., 1979. *Introduction to Seismology*, Birkhäuser Verlag, Basle.

Bayulke, N., 1985. *Earthquake Behaviour of Buildings in Turkey*, Address to the Society for Earthquake and Civil Engineering Dynamics at the Institution of Civil Engineers, London.

Baur, M., Bayraktrali, Y., Fiedrich, F., Lungu, D. and Markus, M., 2001. 'EQSIM – a GIS-based damage estimation tool for Bucharest', *Earthquake hazard and countermeasures for existing fragile buildings* (ed. Dan Lungu and Saito Taiki), Bucharest, Independent Film, 245–254.

Bayülke, N., 1985. *Earthquake Behaviour of Buildings in Turkey*, Address to the Society for Earthquake and Civil Engineering Dynamics at the Institution of Civil Engineers, London.

Bayülke, N., 1989. *The State of Masonry Buildings in Turkey*, Report to Working Group 6, European Association of Earthquake Engineering, Earthquake Research Department, Ministry of Public Works and Settlements, Republic of Turkey.

Beinin, L., 1985. *Medical Consequences of Natural Disasters*, Springer Verlag, Berlin.

Benedetti, D., 1981. 'Riparazione e Consolidamento degli Edifici in Muratura', Chapter XI in *Costruzione in Zona Sismica*, Masson Italia Editori, Milan.

Benedetti, D. and Benzoni, G.M., 1985. 'Seismic vulnerability index vs damage for unreinforced masonry buildings', *US – Italy Workshop on Seismic Hazard and Risk Analysis*, Varenna, NSF, CNR.

Berger, G.M., 1985. 'Perspectives on political and social features of disaster mitigation in Japan', *Proceedings of US – Japan Workshop on Urban Earthquake Hazards Reduction*, EERI Pub. No. 85–03, Earthquake Engineering Research Institution, 2620 Telegraph Avenue, Berkeley, CA 94704, USA.

Bernstein, P.L., 1996. *Against the Gods: The Remarkable Story of Risk*, John Wiley & Sons, New York.

BMA (British Medical Association), 1987. *Living with Risk*, John Wiley & Sons, Chichester.

Bolt, B.A., 1999. *Earthquakes* (4th edition), Freeman, New York.

Bommer, J., Spence, R., Tabuchi, S., Aydinoglu, N., Booth, E., del Re, D., Erdik, M., Peterken, O., 2002. 'Development of an earthquake loss model for Turkish catastrophe insurance', Special Issue of the *Journal of Seismology* on the Turkish Earthquakes of 1999, **6**, (2).

Boore, D.M., Joyner, W.B. and Fumal, T.E., 1997. 'Equations for estimating horizontal response spectra and peak accelerations from western North American earthquakes: a summary of recent work', *Seismological Research Letters*, **68**, (1), 128–153.

Booth, E. (ed.), 1994. *Concrete Structures in Earthquake Regions: Design and Analysis*, Longman, Harlow.

Booth, E.D., Chandler, A.M., Wong, P.K.C. and Coburn, A.W., 1991. *The Luzon, Philippines Earthquake of 16 July 1990, A Field Report by EEFIT*, Institution of Structural Engineers, 11 Upper Belgrave Street, London SW1X 8BH, UK.

Booth, E., Spence, R., Bommer, J., Peterken, O., Aydinoglu, N. and Gülkan, P., 2002. 'Earthquake risk mitigation: lessons from recent Turkish earthquakes', *12th European Conference on Earthquake Engineering, Paper 743*, Elsevier Science, Amsterdam.

Booth, E. and Vasavada, R., 2001. 'Effect of the Bhuj, India earthquake of 26 January 2001 on heritage buildings' (unpublished manuscript).

Braga, F., Dolce, M. and Liberatore, D., 1982. 'A statistical study on damaged buildings and ensuing review of the MSK-76 scale', in *The Southern Italy Nov 23 1980 Earthquake*, Chapter 5, Geodynamics Project, CNR Publ. 503, Rome.

Bressan, G., Poli, G., Sani, G. and Stucchi, M., 1986. 'Preliminary low-cost urban planning-oriented investigations in seismic areas: a methodology and some applications', *Proceedings, International Symposium on Engineering Geology Problems in Seismic Areas*, Bari, 397–406.

Bronson, W., 1986. *The Earth Shook, The Sky Burned: A Photographic Record of the 1906 San Francisco Earthquake and Fire*, Chronicle Books, San Francisco.

Brookshire, D.S., Chang, S.E., Cochrane, H., Olson, R.A., Rose, A. and Steenson, J., 1997. 'Direct and indirect economic losses from earthquake damage', *Earthquake Spectra*, **13**, (4), 683–702.

Buckle, I.G. and Mayes, R.L., 1990. 'Seismic isolation: history, application and performance – a world view', *Earthquake Spectra*, **6**, (1), 161–202.

Burton, I., Kates, R. and White, G., 1978. *The Environment as Hazard*, Oxford University Press, New York.

Burton, P.W., 1979. 'Seismic risk in southern Europe through India examined using Gumbel's third distribution of extreme values', *Geophysical Journal, Royal Astronomical Society*, **59**, 249–280.

Burton, P.W., McGonigle, R., Makropoulos, K.C. and Uçer, S.B., 1984. 'Seismic risk in Turkey, the Aegean and the Eastern Mediterranean: the occurrence of large magnitude earthquakes', *Geophysical Journal, Royal Astronomical Society*, **78**, (2), 475–506.

California Earthquake Authority, 1998 (www.earthquakeauthority.com).

Campbell, K.W., 1981. 'Near-source attenuation of peak horizontal acceleration', *Bulletin of the Seismological Society of America*, **71**, (6), 2039–2070.

Carter, R.L., Lucas, L.D. and Ralph, N., 2000. *Reinsurance* (4th edition), Reactions Publishing Group in association with Guy Carpenter & Company.

CDI, 1998. *California Earthquake Zoning and Probable Maximum Loss Evaluation Program*, California Department of Insurance, Los Angeles.

Celibi, M., 1990. 'Topographical amplification – a reality?', *Proceedings of the Ninth World Conference on Earthquake Engineering*, Moscow, Vol. II, 459–464.

CEN, 1994. *Eurocode 8: Design of Structures for Earthquake Resistance*, EN-1998, Comité Européen du Normalisation, Brussels.

City of Los Angeles, 1985. *Los Angeles Building Code, Division 88: Earthquake Hazard Reduction in Existing Buildings.*

Clayton, A. and Davis, I., 1994. *Building for Safety Compendium*, Intermediate Technology Publications, London.

Coburn, A.W., 1982a. 'Models of vulnerability', *Ninth Regional Conference on Earthquake Engineering*, Istanbul, Turkey.

Coburn, A.W. (ed.), 1982b. *Bingol Province Field Study, 2–24 August 1982*, Turkish National Committee for Earthquake Engineering and The Martin Centre for Architectural and Urban Studies.

Coburn, A.W., 1984, 'Data on structural vulnerability derived from post-earthquake surveys', Appendix II, Report to the European Association on Earthquake Engineering by Working Group 10, *Vulnerability and Seismic Risk Analysis* in Vol. IX, *Proceedings of Eighth European Conference on Earthquake Engineering*, Lisbon, Portugal.

Coburn, A.W., 1986a. 'Seismic vulnerability and risk reduction strategies for housing in Eastern Turkey', PhD Dissertation, University of Cambridge.

Coburn, A.W., 1986b. 'Relative vulnerability assessment', *Eighth European Conference on Earthquake Engineering*, Lisbon, Portugal.

Coburn, A.W., 1987. *The Use of Intensity in Earthquake Vulnerability Assessment*, Appendix to Report to the European Association of Earthquake Engineering by

Working Group 3, Vulnerability and Risk Analysis, *Proceedings of Ninth European Conference on Earthquake Engineering*, Moscow, USSR, 1990.

Coburn, A.W. and Hughes, R.E., 1984. *Report on Damage to Rural Building Types in the Erzurum-Kars Earthquake 30.10.1983*, Martin Centre, Department of Architecture, Cambridge University.

Coburn, A.W., Hughes, R.E., Illi, D., Nash, D.F.T. and Spence, R.J.S., 1981. 'The construction and vulnerability to earthquakes of some building types in the Northern areas of Pakistan', in *The International Karakoram Project* (ed. K.J. Miller), Vol. 2, Cambridge University Press, Cambridge.

Coburn, A.W., Hughes, R.E., Pomonis, A. and Spence, R., 1995. *Technical Principles of Building for Safety*, Intermediate Technology Publications, London.

Coburn, A.W. and Kuran, U. (eds), 1985. *Emergency Planning and Earthquake Damage Reduction for Bursa Province: A Preliminary Evaluation of Earthquake Risk*, Project on Regional Planning for Disasters, Earthquake Research Department, Ministry of Public Works and Housing, Republic of Turkey, and The Martin Centre for Architectural and Urban Studies, University of Cambridge.

Coburn, A.W., Leone, L.P.G., Spence, R.J.S. and Wilhelm, R., 1984b. 'An approach to urban seismic vulnerability assessment: the case study of Noto, Sicily', *Proceedings of International Seminar Vulnerabilita ai Terremoti e Metodi per la Riduzione del Rischio Sismico* (ed. C. Latina), Noto.

Coburn, A.W. and Leslie, J.D.L., 1985. *Dhamar Building Education Project, Project Assessment*, Oxfam Publication for the Executive Office of the Supreme Council for Reconstruction of the Earthquake Affected Areas, Yemen Arab Republic.

Coburn, A.W., Leslie, J.D.L. and Tabban, A., 1984a. 'Reconstruction and resettlement 11 years later: a case study of Bingöl Province, Eastern Turkey', *International Symposium on Earthquake Relief in Less-Industrialised Areas*, Zurich.

Coburn, A.W., Ohashi Murakami, H. and Ohta, Y., 1987. *Factors Affecting Fatalities and Injuries in Earthquakes*, Chair for Engineering Seismology and Earthquake Disaster Prevention Planning, Department of Architectural Engineering, Hokkaido University, Japan.

Coburn, A.W., Petrovski, J., Ristic, D. and Armillas, I, 1990a. *Technical Review of the Impact of the Earthquake of 21 June 1990 in the Provinces of Gilan and Zanjan, Iran*, Earthquake Reconstruction Programme Formulation Mission to the Islamic Republic of Iran, Office of the United Nations Disaster Relief Coordinator (UNDRO), Palais des Nations, CH-1211 Geneva 10, Switzerland.

Coburn, A.W., Pomonis, A. and Sakai, S., 1989. 'Assessing strategies to reduce fatalities in earthquakes', *International Workshop on Earthquake Injury Epidemiology*, Baltimore, MD.

Coburn, A.W., Pomonis, A., Sakai, S. and Spence, R.J.S., 1990b. *The Parameterless Scale of Seismic Intensity*, Martin Centre Working Paper, Department of Architecture, University of Cambridge.

Coburn, A.W., Pomonis, A. and Spence, R.J.S., 1992. 'Factors determining human casualty levels in earthquakes: mortality prediction in building collapse', *10th World Conference on Earthquake Engineering*, Madrid.

Coburn, A.W., Spence, R.J.S. and Zuccaro, G., 1988. *Seismic Risk to Populations in Campania, Italy: The Preparation of SISMA, a Seismic Impact Simulation Model for Regional Planning*, Laboratorio di Urbanistice e Pianificazione Territoriale Centro di Richerche Interdipartimentale, Università degli Studi di Napoli, Italy.

Cohen, M. and Noll, R., 1981. 'The economics of building codes to resist seismic shock', *Public Policy*, **29**, (1), 1–29.

Comerio, M., 1998. *Disaster Hits Home: New Policy for Urban Housing Recovery*, University of California Press, Berkeley, CA.

Comerio, M.C., 1989. *Seismic Costs and Policy Implications; Report for Los Angeles City Council Community Development Department*, George Miers Associates, San Francisco.

Commission of the European Communities, 1989. *Eurocode 8: Structures in Seismic Regions – Design, Part 1 General and Building*, EUR 12266 EN.

Cornell, C.A., 1968. 'Engineering seismic risk analysis', *Bulletin of the Seismological Society of America*, **58**, (5), 1583–1610.

Crampin, S., Evans, R. and Atkinson, B.K., 1984. 'Earthquake prediction: a new physical basis', *Geophysical Journal, Royal Astronomical Society*, **76**, 147–156.

Crampin, S. and Zatsepin, S.V., 1997. 'Changes of strain before earthquakes: the possibility of routine monitoring of both long-term and short-term precursors', *Journal of the Physics of the Earth*, **45**, 41–66.

CRESTA, 1998. CRESTA Manual, CRESTA, Munich Reinsurance, D-80791, Munich (www.cresta.org).

Croci, G., 1998. *The Conservation and Structural Restoration of the Architectural Heritage*, Computational Mechanics Publications, Southampton.

Cuny, F., 1983. *Disasters and Development*, Oxford University Press, Oxford.

Dahle, A., Bungum, H. and Kranne, L.B., 1990. 'Attenuation modelling based on intraplate earthquake recordings', *Proceedings of the Ninth World Conference on Earthquake Engineering*, Moscow, Vol. 4a, 121–129.

Daldy, A.F., 1972. *Small Buildings in Earthquake Areas*, Building Research Establishment, Watford, UK.

Davis, I., 1978. *Shelter after Disaster*, Oxford Polytechnic Press, Headington, Oxford OX3 0BP, UK.

Davis, I., 1981. 'The proof of the pudding...', *International Workshop on Earthen Buildings in Seismic Areas*, Albuquerque, NM.

Davis, I., 1983. 'Disasters as agents of change? or form follows failure', *Habitat International*, **7**, (5/6), 277–310.

Davis, I., 1984. 'A critical view of the work method and findings of the housing and natural hazards group', in *The International Karakoram Project* (ed. K.J. Miller), Vol. 2, Cambridge University Press, Cambridge.

Davis, I. and Spence, R., 2000. 'The impact of the Building for Safety Project on earthquake risk reduction in developing countries', *12th World Conference on Earthquake Engineering, Paper 2163*, Auckland.

Davis, I. and Wilches-Chaux, G., 1989. *The Effective Management of Disaster Situations*, Disaster Management Centre Guidelines No. 1, Oxford Polytechnic, Headington, Oxford OX3 0BP, UK.

Davis, J. and Lambert, R., 1995. *Engineering in Emergencies*, for RedR, Intermediate Technology Publications, London.

D'Ayala, D. and Spence, R., 2000. 'Earthquake damage and vulnerability assessment of historic masonry structures', *Reducing Earthquake Risk to Structures and Monuments in the EU, Conference Proceedings*, EU Environment and Climate Programme.

D'Ayala, D., Spence, R., Oliveira, C. and Pomonis, A., 1997. 'Earthquake protection for historic town centres', *Earthquake Spectra*, **13**, (4), 773–794.

DDF, 1987. *Manual de Analisis Sismico de Edificios*, General Secretariat of Public Works, Department of the Federal District, Mexico, in collaboration with United Nations Development Programme (UNDP), and United Nations Centre for Human Settlements (UNCHS Habitat).

del Re, D., Spence, R. and Patel, D., 2002. 'Damage, repair and strengthening of residential buildings following the Bhuj earthquake of 26.1.01, *12th European Conference on Earthquake Engineering, Paper 744*, Elsevier Science, Amsterdam.

Department of the Environment, 1991. *UK Seismic Hazard and Risk: A Preliminary Study*, Ove Arup and Partners, Consulting Engineers.

Department of the Environment, 1993. *Earthquake Hazard and Risk in the UK*, HMSO, London.

Dowrick, D.J., 1977. *Earthquake Resistant Design: A Manual for Engineers and Architects*, John Wiley & Sons, Chichester.

Dowrick, D.J., 1987. *Earthquake Resistant Design for Engineers and Architects* (2nd edition), John Wiley & Sons, Chichester.

D'Souza, F., 1984. 'The socio-economic cost of planning for hazards: an analysis of Barkulti Village, Yasin, N. Pakistan', in *The International Karakoram Project* (ed. K.J. Miller), Vol. 2, Cambridge University Press, Cambridge.

D'Souza, F., 1986. 'Recovery following the Gediz earthquake', *Disasters*, **10**, (1).

D'Souza, F., 1989. *Coping with Natural Disasters*, The Second Mallet Milne Lecture, SECED, Institution of Civil Engineers, London.

Dudley, E., 1987. 'Disaster mitigation: strong houses or strong institutions?', *Disasters*, **12**, (2), 111–121.

Dudley, E. and Haaland, A., 1993. *Communicating for Building Safety*, Intermediate Technology Publications, London.

Durkin, M.E., 1996. 'Casualty pattern in the 1194 Northridge California earthquake', *Eleventh World Conference on Earthquake Engineering, Paper 979*, Elsevier, Amsterdam.

Durkin, M.E. and Ohashi, H., 1988. 'Casualties, survival, and entrapment in heavily damaged buildings', *9th World Conference on Earthquake Engineering*, Tokyo.

Earthquake Research Institute, 1973. *Public Education Project on the Subject of Earthquakes and Earthquake-resistant Construction*, Ministry of Reconstruction and Resettlement, Republic of Turkey.

EEFIT, 1986. *The Mexico Earthquake of 23.9.1985: A Field Report by EEFIT*, The Institution of Civil Engineers, London.

EEFIT, 1988. *The Chile Earthquake of 3 March 1985: A Field Report by EEFIT*, The Institution of Civil Engineers, London.

EEFIT, 1994. *The Northridge Earthquake of 17th January 1994: A Field Report by EEFIT* (ed. Blakeborough, A. *et al.*), Institution of Structural Engineers, London.

EEFIT, 1999. *The Umbria Marche Earthquakes of 26 September 1997: A Field Report by EEFIT* (ed. R. Spence), Earthquake Engineering Field Investigation Team, Institution of Structural Engineers, London.

EEFIT, 2002a. *The Bhuj Earthquake of 26 January, 2001: A Field Report by EEFIT* (ed. G. Madabhushi), Earthquake Engineering Field Investigation Team, Institution of Structural Engineers, London.

EEFIT, 2002b. *The Kocaeli Earthquake of 17 August, 1999: A Field Report by EEFIT* (ed D. D'Ayala), Earthquake Engineering Field Investigation Team, Institution of Structural Engineers, London.

EERI, 1984. *The Anticipated Tokai Earthquake: Japanese Prediction and Preparedness Activities*, Publ. No. 84–05 (ed. C. Scawthorn), Earthquake Engineering Research Institute, 2620 Telegraph Avenue, Berkeley, CA 94704, USA.

EERI, 1985. *Urban Earthquake Hazards Reduction, Proceedings of US–Japan Workshop at Stanford University, July 1984*, Publ. No. 85–03, Earthquake Engineering Research Institution, 2620 Telegraph Avenue, Berkeley, CA 94704, USA.

EERI, 1986. *Reducing Earthquake Hazards: Lessons Learned from Earthquakes*, Publ. No. 86–02, Earthquake Engineering Research Institution, 2620 Telegraph Avenue, Berkeley, CA 94704, USA.

EQE, 2001. *The EQE Earthquake Home Preparedness Guide* (available from www.eqe.com).

Ergünay, O. and Erdik, M., 1984. 'Turkish experience on the earthquake performance of rural stone masonry buildings', *International Conference on Natural Hazards Mitigation Research and Practice, Small Buildings and Community Development*, New Delhi.

Esteva, L., 1982. 'Fundamental concepts of seismic risk analysis', *Seismic Risk Assessment and Development of Model Code for Seismic Design*, Vol. C of *Earthquake Risk Reduction in the Balkan Regions*, RER/79/014, UNESCO, Paris.

Fawcett, W. and Oliveira, C., 2000. 'Casualty treatment after earthquake disasters: development of a regional simulation model', *Disasters*, **24**, 271–287.

FEMA, 1997. *NEHRP Guidelines for the Seismic Rehabilitation of Buildings*, FEMA-273, US Federal Emergency Management Agency (www.fema.org).

FEMA, 1998. *Handbook for the Seismic Evaluation of Buildings a prestandard*, FEMA-310, US Federal Emergency Management Agency (www.fema.org).

FEMA, 1999. *HAZUS, 1999, Earthquake Loss Estimation Methodology*, US Federal Emergency Management Agency (www.fema.org).

Fielden, B.M., 1987. *Between Two Earthquakes: Cultural Property in Seismic Zones*, ICCROM, Via di San Michele 13, 00153 Roma, Italy.

Foster, H.D., 1980. *Disaster Planning: The Preservation of Life and Property*, Springer Verlag, New York.

Fournier d'Albe, E.M., 1982. 'An approach to earthquake risk management', *Engineering Structures*, **4**, (July), 147–152.

Frankel, A.D., Mueller, C.S., Barnhard, E.V., Leyendecker, E.V., Wesson, R.L., Harmsen, S.C., Klein, F.W., Perkins, D.M., Dickman, N.C., Hanson, S.L. and Hopper, M.G., 2000. 'USGS National Seismic Hazard Maps', *Earthquake Spectra*, **16**, 1, 1–19.

Freeman, J.R., 1932. *Earthquake Damage and Earthquake Insurance*, McGraw-Hill, New York.

Freeman, P.K., 2000. 'Catastrophe risk: a model to evaluate the decision process for developing countries', Unpublished Dissertation, University of Vienna.

Freeman, P.K. and Kunreuther, H., 1997. *Managing Environmental Risk Through Insurance*, The AEI Press, Washington, DC.

Fryer, L.S. and Griffiths, R.F., 1986. *Worldwide Data on the Incidence of Multiple-Fatality Accidents*, United Kingdom Atomic Energy Authority.

Gallagher, R.P., Reasenberg, P.A. and Poland, C.D., 1999. *ATC Tech Brief 2: Earthquake aftershocks – entering damaged buildings*, Applied Technology Council, Redwood City, CA.

Geller, R.J., 1997. 'Earthquake prediction: a critical review', *Geophysical Journal International*, **131**, 425–450.

Georgescu, E., 1988. 'Assessment of evacuation possibilities of apartment multistoried buildings during earthquakes or subsequent fires in view of earthquake preparedness', *9th World Conference on Earthquake Engineering*, Tokyo.

Grandori, G., 1982. 'Cost benefit analysis in earthquake engineering', *Seventh European Conference on Earthquake Engineering*, Athens, Vol. 7, 71–136.

Grandori, G. and Benedetti, D., 1973. 'On the choice of acceptable seismic risk', *Earthquake Engineering and Structural Dynamics*, **2**, 3–9.

Grandori, G., Garavaglia, E. and Molina, C., 1988. 'A new attenuation law of seismic intensity', *Proceedings of the Ninth European Conference on Earthquake Engineering*, Moscow.

Grünthal, G. (ed.), 1998. *European Macroseismic Scale 1998 (EMS 1998)*, Council of Europe, Cahiers du Centre Européen de Geodynamique et du Seismologie, Vol. 15.

Grünthal, G. (ed.), 1999. 'Seismic hazard assessment for Central, North and Northwest Europe: GSHAP Region 3', *Annali di Geofisica*, **42**, (6), 999–1011.

Grünthal, G., Bosse, C., Sellami, S., Mayer-Rosa, D. and Giardini, D., 1999. 'Compilation of the GSHAP regional seismic hazard for Europe, Africa and the Middle East', http://seismo.ethz.ch/GSHAP.

Gueri, M. and Alzate, H., 1984. 'The Popayan earthquake: a preliminary report on its effects on health', *Disasters*, **8**, (1), 18–20.

Gülkan, P. *et al.*, 1999. *Revision of the Turkish Development Law No 3194 and its attendant regulations with the objective of establishing a new building construction supervision system inclusive of incorporating technical disaster resistance-enhancing measures*, 3 vols, Turkish Ministry of Public Works and Settlement.

Gülkan, P. and Ergünay, O., 1984. 'Legislative aspects of mitigation of earthquake-induced losses', *International Symposium on Earthquake Risk in Less Industrialized Areas*, Zurich.

Gutenberg, B. and Richter, C.F., 1954. *Seismicity of the Earth and Associated Phenomena*, Princeton University Press, Princeton, NJ.

Hadfield, P., 1991. *Sixty Seconds That Will Change the World: The Coming Tokyo Earthquake*, Sidgwick & Jackson, London.

Hansen, R.D. and Soong, T.T., 2001. *Seismic Design with Supplemental Energy Devices*, MNO-8, Earthquake Engineering Research Institute, Berkeley, CA.

Hawkins, C.J. and Pearce, D.W., 1971. *Capital Investment Appraisal*, Macmillan, London.

Housner, G.W., 1989. *Coping with Natural Disasters*, The Second Mallet Milne Lecture, SECED, Institution of Civil Engineers, London.

Housner, G.W. (Chair), 1992. *The Economic Consequences of a Catastrophic Earthquake: Proceedings of a Forum August 1 & 2, 1990*, Committee on Earthquake Engineering, National Research Council. National Academy Press, Washington, DC.

IAEE, 1986. *Guidelines for Earthquake-Resistant Non-Engineered Construction*, International Association for Earthquake Engineering, Kenchiku Kaikan 3rd Floor, 5-26-20, Shiba, Minato-ku, Tokyo 108, Japan.

IAEE, 1988. *Earthquake Regulations – A World List*, International Association for Earthquake Engineering, Kenchiku Kaikan 3rd Floor, 5-26-20, Shiba, Minato-ku, Tokyo 108, Japan.

ICBO, 1988. *The Uniform Building Code*, International Conference of Building Officials, Whittier, CA.

ICBO, 2000. *2000 International Building Code* CD-ROM, International Conference of Building Officials, 5360 Workman Mill Road, Whittier, CA 90601-2298, USA (www.ocbo.org).

IFRC, 1996. *World Disasters Report 1996*, Oxford University Press, Oxford.

IFRC, 2001. *World Disasters Report 2001*, International Federation of Red Cross and Red Crescent Societies, Box 372, CH-121 Geneva 19, Switzerland (www.ifrc.org).

INSARAG, 2001. *International Search and Rescue Response Guidelines*, International Search and Rescue Advisory Group (INSARAG), www.reliefweb.int/insarag.

INTERTECT, 1971. *Refugee Camps and Camp Planning Series*, 4 Reports, INTERTECT International Disaster Specialists, PO Box 565502, Dallas, TX 75356, USA.

INTERTECT (ed.), 1981. *Proceedings of the International Workshop on Earthen Buildings in Seismic Areas*, Albuquerque, NM.

Isikara, A.M. and Vogel, A. (eds), 1980. *Multidisciplinary Approach to Earthquake Prediction, Proceedings of the International Symposium on Earthquake Prediction in the North Anatolian Fault Zone*, Istanbul, Friedr. Vieweg & Sohn, Braunschweig/Wiesbaden.

ISO, 1996. *Managing Catastrophe Risk*, Insurance Services Office, Inc., New York.

ISO, 1999. *Financing Catastrophe Risk: Capital Market Solutions*, Insurance Services Office, Inc., New York.

Itoh, S., 1985. 'The present scope of earthquake hazards mitigation strategies in Japan', *Proceedings of US–Japan Workshop on Urban Earthquake Hazards Reduction*, EERI Publ. No. 85-03, Earthquake Engineering Research Institution, 2620 Telegraph Avenue, Berkeley, CA 94704, USA.

Jackson, R., 1960. *Thirty Seconds at Quetta*, Evans Brothers, London.

Jaffe, M., Butler, J. and Thurow, C., 1981. *Reducing Earthquake Risks: A Planner's Guide*, Report No. 364, American Planning Association, Planning Advisory Service, 1313E. 60th St., Chicago, IL 60637, USA.

Jara, M., Hernandez, C., Garcia, R. and Robles, F., 1989. 'Typical cases of repair and strengthening of concrete buildings', *Earthquake Spectra*, **5**, (1), 175–194.

JNV, 1987. *Cómo Hacer Nuestra Casa de Adobe*, Junta Nacional de la Vivienda, Ecuador, in collaboration with United Nations Development Programme (UNDP), project ECU-87-004.

Johnson, L.A., Coburn, A.W. and Rahnama, M., 2000. 'Damage survey approach to estimating insurance losses', Chapter 12 in *Kocaeli, Turkey Earthquake Reconnaissance Report* (ed. Youd, T.L.), *Earthquake Spectra*, Supplement A to Vol. 16, 281–293.

Joyner, W.B. and Boore, D.M., 1981. 'Peak horizontal acceleration and velocity from strong-motion records including records from the 1979 Imperial Valley, California Earthquake', *Bulletin of the Seismological Society of America*, **71**, (6), 2011–2038.

Kagami, H. and Okada, S. 1986. 'Evaluation of earthquake risk regionality based on disaster sequences model: a study for 212 municipalities in Hokkaido Japan', *Proceedings Seventh Japanese Earthquake Engineering Symposium*, 213–218.

Kaila, K.L. and Madhava, Rao, 1975. 'Seismotectonic maps of the European area', *Bulletin of the Seismological Society of America*, **65**, (6), 1721–1732.

Karník, V., 1971. *Seismicity of the European Area, Part II*, Reidel, Dordrecht.

Karník, V., 1982. 'New trends in the management of earthquake risk', *UNDRO News*, March, Office of United Nations Disaster Relief Coordinator, Geneva, Switzerland.

Kendrick, T.D., 1956. *The Lisbon Earthquake*, Methuen, London.

Key, D., 1988. *Earthquake Design Practice for Buildings*, Thomas Telford, London.

Kircher, C.A., Nassar, A.A., Kustu, O. and Holmes, W.T., 1997. 'Development of building damage functions for earthquake loss estimation', *Earthquake Spectra*, **13**, (4), 663–682.

Kobayashi, H. and Kagami, H., 1972. 'A method for local seismic intensity zoning maps on the basis of subsoil conditions', *Proceedings Second International Conference on Microzoning*, 513–528.

Kreimer, A. and Munasinghe, M. (eds), 1991. *Managing Natural Disasters and the Environment*, Selected Materials from the Colloquium on the Environment and Natural Disaster Management, The World Bank, 1818 H Street, N.W., Washington, DC 20433, USA.

Krimgold, F., 1974. *The Role of International Aid for Pre-Disaster Planning in Developing Countries*, Avdelningen for Arkitektur, KTH, Stockholm.

Krimgold, F., 1987. *Developments in Post-Disaster Search and Rescue Technologies*, Lecture to the Martin Centre for Architectural and Urban Studies, University of Cambridge, 17 June.

Krimgold, F., 1989. 'Earthquake Casualty Estimation and Response Modeling', *International Workshop on Earthquake Injury Epidemiology*, The John Hopkins University, July 1989, pp. 17–25.

Kunreuther, H. and Roth, R.J. (eds), 1998. *Paying the Price: The Status and Role of Insurance Against Natural Disasters in the United States*, Joseph Henry Press, Washington, DC.

Kuran, U., 1986. 'Description and evaluation of documentary source material related to 1855 Bursa earthquake', Unpublished manuscript, Earthquake Research Center, Ministry for Housing and Public Works, Republic of Turkey.
Ladinski, V., 1989. '26 years of earthquake mitigation programmes in Skopje, Yugoslavia', *Course on Earthquakes and Related Hazards: Mitigation and Preparedness Planning*, University of Cambridge.
Lagorio, H., 1991. *Earthquakes: An Architects Guide to Non-structural Seismic Hazards*, John Wiley & Sons, New York.
Lechat, M.F., 1989. 'Corporal Damage as Related to Building Structure and Design: The Need for an International Survey', *International Workshop on Earthquake Injury Epidemiology*, The John Hopkins University, July 1989, pp. 1–16.
Leyendecker, E.V., Frankel, A.D. and Rukstales, K.S., 2000. 'Development of maximum considered earthquake ground maps', *Earthquake Spectra*, **16**, (1), 21–40.
Leslie, J.D.L., 1984. 'Think before you build: an earthquake reconstruction project in Yemen', *International Symposium on Earthquake Relief in Less-Industrialised Areas*, Zurich.
Lichtenstein, S., *et al.*, 1978. 'Judged frequency of lethal events', *Journal of Experimental Psychology: Human Learning and Memory*, **4**, (6).
LINUH, 1976. *Balinese Earthquake Manual*, B.I.C., Pusat Infirmasi Teknik Pembangunan, Bali, Indonesia.
Lubkowski, Z. and Duian, X., 2001. 'EN 1998 Eurocode 8: Design of Structures for Earthquake Resistance', *Civil Engineering*, **144**, (2), 55–60.
Mahin, S.A., 1986. *ATC 3-06: Tentative Provisions for the Development of Seismic Regulations for Buildings*, Prepublication manuscript, Department of Civil Engineering, University of California, Berkeley.
Marcellini, A., Daminelli, R., Francheschina, G. and Pagani, M., 1999. 'Some aspects of seismic microzonation', *Proceedings of the Advanced Study Course, Kefallinia*, European Commission, Brussels.
Maskrey, A., 1989. *Disaster Mitigation: A Community-Based Approach*, Oxfam Development Guidelines, No. 3.
McClure, F.E., 1984. 'Development and implementation of the University of California seismic safety policy', *Proceedings of the Eighth World Conference on Earthquake Engineering*, San Francisco, Vol. 7, 859.
McGuire, R.K., 1978. *FORTRAN Computer Program for Seismic Risk Analysis*, US Geological Survey Open File Report 78-1007.
McKay, M., 1978. 'The OXFAM/world neighbors housing education programme in Guatemala', in *Disasters and the Small Dwelling* (ed. I. Davis), Pergamon Press, Oxford.
Medvedev, S.V., 1965. *Engineering Seismology*, trans. from Russian, Israel Program for Scientific Translation, Jerusalem.
Medvedev, S.V., 1968. 'The international scale of seismic intensity', in *Seismic Zones of the USSR*, Nauka, Moscow.
Mileti, D., 1999. *Designing Future Disasters*, Joseph Henry Press, Washington, DC.
Mileti, D., Hutton, J. and Sorensen, J., 1981. *Earthquake Prediction Response and Options for Public Policy*, Institute of Behavioral Studies, University of Colorado, Boulder.
Milliman, J.W. and Roberts, R.B., 1985. 'Economic issues in formulating policy for earthquake hazard mitigation', *Policy Studies Review*, **4**, (4), 645–654.
Mucciarelli, M., Bettinali, F., Zaninetti, A., Vanini, M., Mendez, A. and Galli, P., 1996. 'Refining Nakamura's techniques: processing techniques and innovative instrumentation', *Seismology in Europe, Proceedings of the 25th General Assembly of the European Seismological Commission*, Reykjavik, 411–416.
Muir-Wood, R., 1981. 'Hard times in the mountains', *New Scientist*, **91**, 1270.

Munich Re, 1999. *Topics 2000: Natural Catastrophes – the Current Position*, Munich Re Group, Koniginstrasse 107, 80802 Munchen, Germany.
Murakami, H., 1996. 'Chances of occupant survival and SAR operations in the buildings collapsed by the Great Hanshin Earthquake, Japan, *Eleventh World Conference on Earthquake Engineering, Paper 852*, Elsevier, Amsterdam.
Murphy, J.R. and O'Brien, L.J., 1977. 'The correlation of peak ground acceleration amplitude with seismic intensity and other physical parameters', *Bulletin of the Seismological Society of America*, **67**, (3), 877.
Musson, R. and Van Rose, S., 1997. *Earthquakes: Our Trembling Planet*, British Geological Survey, London.
NCEER, 1989. *Proceedings of Conference on Reconstruction After Urban Earthquakes: An International Agenda to Achieve Safer Settlements in the 1990s*, National Center for Earthquake Engineering (NCEER), University of Buffalo, Buffalo, NY.
NEDA, 1990. *Reconstruction and Development Program Following the Earthquake of 16 July 1990*. National Economic and Development Authority, Republic of the Philippines, November.
Needham, J., 1971. *Science and Civilisation in China*, Vol. 4, Part 3, *Civil Engineering and Nautics*, Cambridge University Press, Cambridge.
New Zealand National Society for Earthquake Engineering, 1985. *Earthquake Risk Buildings: Recommendations and Guidelines for Classifying, Interim Securing and Strengthening*, NZNSEE, Wellington.
NLA, 1987. *Earthquake Disaster Countermeasures in Japan*, National Land Agency, Prime Minister's Office, Government of Japan, Tokyo.
Noji, E.K., 1989. 'Use of quantitative measures of injury severity in earthquakes', *International Workshop on Earthquake Injury Epidemiology*, Baltimore, MD.
NRC, 1975. *Earthquake Prediction and Public Policy*, Panel on the Public Policy Implications of Earthquake Prediction, Advisory Committee on Emergency Planning, Commission on Socio-Technical Systems, National Research Council, National Academy of Sciences, Washington, DC.
Ohta, Y., Ohashi, H. and Kagami, H., 1986. 'A semi-empirical equation for estimating occupant casualty in an earthquake', *Proceedings of the Eighth European Conference on Earthquake Engineering*, Lisbon, Vol. 2.3, 81–88.
Okada, S., Pomonis, A., Coburn, A.W., Spence, R.J.S. and Ohta, Y., 1991. *Factors Influencing Casualty Potential in Buildings Damaged by Earthquakes*, Collaborative Report, Hokkaido University, University of Cambridge, University of Tokyo.
Oliver, P., 1981. 'The cultural context of shelter provision', in *Disasters and the Small Dwelling* (ed. I. Davis), Pergamon Press, Oxford.
Page, R.A., Blume, J.A. and Joyner, W.B., 1975. 'Earthquake shaking and damage to buildings', *Science*, **189**, (4203).
PAHO, 1981. *A Guide to Emergency Health Management After Natural Disaster*, Scientific Publ. No. 407, Pan American Health Organization, Regional Office of the World Health Organization, 525 Twenty-third Street, N.W., Washington, DC 20037, USA.
PAHO, 1982. *Epidemiological Surveillance After Natural Disaster*, Scientific Publ. No. 420, Pan American Health Organization, Regional Office of the World Health Organization, 525 Twenty-third Street, N.W., Washington, DC 20037, USA.
Parker, R., 1982. 'The Guatemalan housing education program: successes and failures: lessons learnt', *A Final Evaluation for World Neighbors*, Oaxaca, Mexico.
Pate-Cornell, M.E., 1985. 'Costs and benefits of seismic upgrading of some buildings in the Boston area', *Earthquake Spectra*, **1**, (4), 721–740.
Paz, M., 1994. *International Handbook of Earthquake Engineering*, Chapman and Hall, London.

Penelis, G. and Kappos, A., 1997. *Earthquake-Resistant Concrete Structures*, E and FN Spon, London.

Platt, C.M. and Shepherd, R., 1985. 'Some cost considerations of the seismic strengthening of pre-code building', *Earthquake Spectra*, **1**, (4), 695–720.

Pomonis, A., Coburn, A.W., Okada, S. and Spence, R.J.S., 1991. *Reinforced Concrete Buildings Damaged by Earthquakes*, Working Report III of Project on Human Casualties in Building Collapse, The Martin Centre for Architectural and Urban Studies, University of Cambridge.

Pomonis, A., Coburn, A.W. and Spence, R.J.S., 1991. *Casualty Estimation Model for Masonry Buildings*, Working Report II of Project on Human Casualties in Building Collapse, The Martin Centre for Architectural and Urban Studies, University of Cambridge.

Postpischl, D. (ed.), 1985. *Atlas of Isoseismal Maps of Italian Earthquakes*, Progetto Finalizzato Geodinamica, CNR, Bologna, Italy.

Quetta Municipality Building Code, 1940.

Razani, R., 1981. 'Seismic protection of unreinforced masonry and adobe low-cost housing in less developed countries, policy issues and design criteria', *Disasters and the Small Dwelling* (ed. I. Davis), Pergamon Press, Oxford.

Razani, R., 1984. 'Earthquake disaster area reconstruction experience in Iran', *International Symposium on Earthquake Risk in Less Industrialized Areas*, Zurich.

Razani, R. and Nielsen, N.N., 1984. 'Valuation of human life in seismic risk analysis', *Proceedings of the Eighth World Conference on Earthquake Engineering*, San Francisco, Vol. 7, 947–954.

Report on the ad-hoc panel meeting of experts on up-dating of the MSK-64 seismic intensity scale, Jena, 10–14 March 1980, 1981. *Gerlands Beiträge Geophysik*, Leipzig.

Richter, C.F., 1958. *Elementary Seismology*, Freeman, San Francisco.

Rikitake, T., 1976. *Earthquake Prediction*, Elsevier, Amsterdam.

Rikitake, T., 1981. 'Practical approach to earthquake prediction and warning', *Current Research in Earthquake Prediction I*, 1–56.

RMS, 1995. *What if the 1906 Earthquake Strikes Again? A San Francisco Bay Area Scenario*, Topical Issue Series, May, Risk Management Solutions, 7015 Gateway Boulevard, Newark, CA 94560, USA.

Rosenblueth, E. (ed.), 1980. *Design of Earthquake Resistant Structures*, Pentech Press, London.

Rosenblueth, E., 1991. 'Public policy and seismic risk', *Nature and Resources*, **27**, (1), 10–18.

Sakai, S., 1990. 'Modelling search and rescue in building collapse', *International Conference on Impact of Natural Disasters*, UCLA, Los Angeles.

Sandi, H., 1982. 'Seismic vulnerability and seismic intensity', *Proceedings of the Seventh European Conference on Earthquake Engineering*, Athens, Vol. 2, 431.

Sarin, R.K., 1983. 'A social decision analysis of the earthquake safety problem: the case of Los Angeles existing buildings', *Risk Analysis*, **3**, (1), 35–50.

Sauter, F.F. and Shah, H.C., 1978. *Estudio de Seguro Contra Terremoto*, Instituto Nacional de Seguros, San Jose, Costa Rica.

Scawthorn, C., 1984. 'The locational approach to seismic risk mitigation: application to San Francisco', *Eighth World Conference on Earthquake Engineering*, San Francisco, Vol. 7, 939.

Scawthorn, C. (ed.), 1986. *Techniques for Rapid Assessment of Seismic Vulnerability*, American Society of Civil Engineers, New York.

Scawthorn, C., Bouhafs, M. and Blackburn, F.T., 1988. 'Demand and provision for post-earthquake emergency services: case study of the San Francisco Fire Department', *Proceedings Ninth World Conference on Earthquake Engineering*, Vol. 8, 1065–1070.

Scawthorn, C., Imemura, H. and Yamada, Y., 1981. 'Seismic damage estimation for low- and mid-rise buildings in Japan', *Journal of Earthquake Engineering and Structural Dynamics*, **9**, 93–115.

Schultz, C.H., Koenig, K.L. and Noji, E., 1996. 'A medical disaster response to reduce immediate mortality after an earthquake', *New England Journal of Medicine*, **15**, (February), 438–444.

Seed, H.B., Ugas, C. and Lysmer, J., 1974. *Site-dependent Spectra for Earthquake Resistant Design*, Report No. EERC 74-12, Earthquake Engineering Research Center, University of California, Berkeley.

Shiono, K. and Krimgold, F., 1989. 'A computer model for the recovery of trapped people in a collapsed building', *International Workshop on Earthquake Injury Epidemiology*, Baltimore, MD.

Sieh, K. and LeVay, S., 1998. *The Earth in Turmoil: Earthquakes, Volcanoes and their Impact on Mankind*, Freeman, New York.

Singh, S.K., Astiz, L. and Havskov, J., 1981. 'Seismic gaps and recurrence periods of large earthquakes along the Mexican subduction zone: a re-examination', *Bulletin of the Seismological Society of America*, **71**, (3), 827.

Singh, S.K., Lermo, J., Dominguez, T., Ordaz, M., Espinosa, J.M., Mena, E. and Quaas, R., 1988. 'A study of amplification of seismic waves in the valley of Mexico with respect to a hill zone site', *Earthquake Spectra*, **4**, (4), 653–674.

Sinha, S. 1990. *Mini-cement: a Review of Indian Experience*, Intermediate Technology Publications, London.

Skeet, M., 1977. *Manual for Disaster Relief Work*, Churchill Livingstone, Edinburgh.

Smith, D.G.E., 1988. *Guidelines for the Earthquake-resistant Design of Low-Rise Buildings*, Scott Wilson Kirkpatrick and Partners, Scott House Basing View, Basingstoke, RG21 2JG, UK.

Soong, T.T. and Spencer, B.F., 2000. 'Active, semi-active and hybrid control of structures', *World Conference on Earthquake Engineering, Paper 2834*, Auckland.

Spence, R.J.S., 1981. 'The vulnerability to earthquakes of houses of low-strength masonry', *International Workshop on Earthen Buildings in Seismic Areas*, University of New Mexico, Albuquerque, May.

Spence, R.J.S., 1983. 'A note on earthquake vulnerability in Quetta (Pakistan)', *Disasters*, **7**, (2), 91–93.

Spence, R.J.S., 1986. 'Earthquake risk assessment: a review of methods with particular reference to rural housing in eastern Turkey', *Middle East and Mediterranean Regional Conference on Earthen and Low Strength Masonry Buildings in Seismic Areas*, Ankara.

Spence, R.J.S., 1988. 'Earthquake risk in Campania region: risk levels and alternative mitigation policies', *Seminar on Tools and Methods for Seismic Risk Analysis*, LUPT, University of Naples.

Spence, R.J.S., 1990. 'A note on the beta distribution in vulnerability analysis', Martin Centre Working Paper, Department of Architecture, University of Cambridge.

Spence, R. (ed.), 1998a. *Reducing Earthquake Risk to Structures and Monuments in the EU, Conference Proceedings*, EU Environment and Climate Programme.

Spence, R., 1998b. 'The Umbria earthquake and its consequences', *Magdalene College Magazine*, New Series, **42**, 32–39.

Spence, R., 1999. 'Intensity, damage and loss in earthquakes', in *Seismic Damage to Masonry Buildings* (ed. A. Bernardini), 27–40, Balkema, Rotterdam.

Spence, R, 2000. 'Recent earthquake damage in Europe and its implications for loss estimation methodologies', Chapter 7, 77–90, in *Implications of Recent Earthquakes for Seismic Risk*, Imperial College Press, London.

Spence, R. and Adams, R., 1999. 'Geological hazards: earthquake', in *Natural Disaster Management* (ed. J. Ingleton), 52–54, Tudor Rose, UK (www.ndm.co.uk).

Spence, R.J.S. and Coburn, A.W., 1984. 'Traditional housing in seismic areas', in *The International Karakoram Project* (ed. K.J. Miller), Vol. 1, Cambridge University Press, Cambridge.

Spence, R.J.S. and Coburn, A.W., 1987a. *Reducing Earthquake Losses in Rural Areas*, Final Report of Project R3662, Vulnerability of Low-Income Houses in Earthquake Areas, to Overseas Development Administration, HM Government, UK, The Martin Centre for Architectural and Urban Studies, University of Cambridge.

Spence, R. and Coburn, A.W., 1987b. 'Earthquake protection – an international task for the 1990's', *The Structural Engineer*, **65A**, (8), August.

Spence, R. and Coburn, A.W., 1992. 'Strengthening buildings of stone masonry to resist earthquakes', *Meccanica*, **26**, (3), September.

Spence, R., Coburn, A. and Dudley, E., 1989. *Gypsum Plaster: its manufacture and use*, Intermediate Technology Publications, London.

Spence, R., Coburn, A.W., Pomonis, A. and Sakai, S., 1992a. 'Correlation of ground motion with building damage: the definition of a new damage-based seismic intensity scale', *Tenth World Conference on Earthquake Engineering*, Madrid.

Spence, R.J.S., Coburn, A.W., Sakai, S. and Pomonis, A., 1991a. 'A parameterless scale of seismic intensity for use in seismic risk analysis and vulnerability assessment', in SECED (ed.) *Earthquake, Blast and Impact: Measurement and Effects of Vibration*, Elsevier Applied Science, Amsterdam.

Spence, R.J.S., Coburn, A.W., Sakai, S. and Pomonis, A., 1991b. *Reducing Human Casualties in Building Collapse: First Report*, The Martin Centre, Department of Architecture, Cambridge University.

Spence, R.J.S., Coburn, A.W., Sakai, S. and Pomonis, A., 1992b. *Reducing Human Casualties in Building Collapse: Methods of Optimising Disaster Plans to Reduce Injury Levels*, Project Final Report, Science and Engineering Research Council, The Martin Centre for Architectural and Urban Studies, University of Cambridge.

Spence, R.J.S. and Cook, D., 1983. *Building Materials in Developing Countries*, John Wiley & Sons, Chichester.

Spence, R. and D'Ayala, D., 1999. 'Damage assessment and analysis of the 1997 Umbria-Marche earthquakes', *Structural Engineering International*, March, 229–233, Zurich.

Spence, R., D'Ayala, D., Oliveira, C. and Pomonis, A., 2000. 'The performance of strengthened masonry buildings in recent European earthquakes', *12th World Conference on Earthquake Engineering, Paper 1366*, Auckland.

Spence, R., del Re, D. and Thompson, A., 2002a. 'Performance of buildings', Chapter 3 in *The Bhuj Earthquake of 26.1.01: A Field Report by EEFIT* (ed. G. Madabhushi), Institution of Structural Engineers, London.

Spence, R., Peterken, O., Booth, E., Aydinoglu, N., Bommer, J. and Tabuchi, S., 2002b. 'Seismic loss estimation for Turkish catastrophe insurance, *Proceedings 7th US National Conference on Earthquake Engineering, Paper 722.*

Spence, R., Pomonis, A., Dowrick, D.J. and Cousins, J.A., 1998. 'Assessment of casualties in urban earthquakes', in *Seismic Design Practice into the Next Century* (ed. E. Booth), Balkema, Rotterdam.

Starr, C., 1969. 'Social benefit versus technological risk: what is our society willing to pay for safety?', *Science*, **165**, 1232–1236.

Steinbrugge, K.V., 1982. *Earthquakes, Volcanoes and Tsunamis: An Anatomy of Hazards*, Skandia America Group, 280 Park Avenue, New York, NY 10017, USA.

Steinbrugge, K.V., Algermissen, S.T. and Lagorio, H.J., 1984. 'Determining monetary losses and casualties for use in earthquake mitigation and disaster planning', *Eighth World Conference on Earthquake Engineering*, San Francisco, Vol. 7, 615–623.

Stephens, L.H. and Green, S.J., 1979. *Disaster Assistance: Appraisal, Reform and New Approaches*, New York University Press, New York.

Swiss Re, 2000. 'Natural catastrophes and manmade disasters in 2000', *Sigma*, No. 2 (annual series available from www.swissre.com/portal).

Tanabashi, R., 1960. 'Earthquake resistance of traditional Japanese wooden structure', *Second World Conference on Earthquake Engineering*, Tokyo, 151–163.

Tanaka, H. and Baxter, P., 2001. Personal communication.

Thompson, P., Cuny, C., Coburn, A.W., Cheretis, J. and Georgoussis, G., 1986. *Earthquake Damage Assessment and Recovery Issues in Kalamata Region, Peloponnese, Greece*, Report prepared by INTERTECT, Dallas, USA, in collaboration with The Earthquake Planning and Protection Organisation (OASP), Athens, and The Ministry of Environment, Planning and Public Works, Greece, for the Office of US Foreign Disaster Assistance, Agency for International Development, Washington, DC 20523, USA.

Tiedemann, H., 1982. 'Structural and non-structural damage related to building quality', *Proceedings of the Seventh European Conference on Earthquake Engineering*, Athens, Vol. 6, 27.

Tiedemann, H., 1984. 'A model for the assessment of seismic risk', *Proceedings of the Eighth World Conference on Earthquake Engineering*, San Francisco, Vol. 1, 199.

Tiedemann, H., 1984a. 'Economic consequences of earthquakes', *International Symposium on Earthquake Risk in Less Industrialized Areas*, Zurich, 63–68.

Tiedemann, H., 1986. 'Insurance and the mitigation of earthquake disasters', *UNDRO News*, Jan/Feb issue.

Tiedemann, H., 1989. 'Casualties as a Function of Building Quality and Earthquake Intensity', *International Workshop on Earthquake Injury Epidemiology*, The John Hopkins University, July 1989, pp. 420–434.

Tiedemann, H., 1992. *Earthquakes and Volcanic Eruption: A Handbook on Risk Assessment*, Swiss Reinsurance Company, Mythenquai 50/60, PO Box, CH-8022, Zurich, Switzerland.

Tobriner, S., 1982. *The Genesis of Noto – An Eighteenth Century Sicilian City*, A. Zwemmer, London.

Tobriner, S., 1984. 'A history of reinforced masonry construction designed to resist earthquakes 1755–1907', *Earthquake Spectra*, **1**, (1), 125–150.

Toro, G., Abrahamson, N. and Schneider, J., 1997. 'Model of strong ground motion from earthquakes in central and eastern North America', *Seismological Research Letters*, **68**, 41–57.

Toro, G.R. and McGuire, R.K., 1987. 'An investigation into earthquake ground motion characteristics in Eastern North America', *Bulletin of the Seismological Society of America*, **77**, (2), 468–489.

UN, 1970. *Skopje Resurgent*, United Nations, New York.

UN, 1975. *Low-Cost Construction Resistant to Earthquake and Hurricanes*, United Nations Sales No. E75 IV7, New York.

UN, 1988. Resolution 42/169 of the General Assembly of United Nations, 11 December.

UNCHS, 1989, *Human Settlements and Natural Disasters*, United Nations Centre for Human Settlements (Habitat), PO Box 30030, Nairobi, Kenya.

UNCHS, 1990. Small-scale Manufacture of Low-Cost Building Materials, United Nations Centre for Human Settlements (Habitat), PO Box 30030, Nairobi, Kenya.

UNCHS, 2001. *Cities in a Globalizing World: Global Report on Human Settlements*, Earthscan, London.

UNDP, 1991. *Disaster Mitigation*, UNDP/UNDRO Disaster Management Training Programme.

UNDRO, 1979. *Natural Disasters and Vulnerability Analysis: Report of Expert Group Meeting*, Office of United Nations Disaster Relief Coordinator (UNDRO), Palais des Nations, CH-1211 Geneva 10, Switzerland.

UNDRO, 1982. *Shelter After Disaster*, Office of United Nations Disaster Relief Coordinator (UNDRO), Palais des Nations, CH-1211 Geneva 10, Switzerland.
UNDRO, 1984. *Disaster Prevention and Mitigation: A Compendium of Current Knowledge*, 12 vols, Office of the United Nations Disaster Relief Coordinator (UNDRO), Palais des Nations, CH-1211 Geneva 10, Switzerland.
UNDRO, 1989, *Natural Disasters and Insurance, Proceedings of the 1st Meeting of the International Working Group sponsored by UNDRO in collaboration with UNESCO and The Geneva Association*, Office of United Nations Disaster Relief Coordinator (UNDRO), Palais des Nations, CH-1211 Geneva 10, Switzerland.
UNESCO, 1982. *Earthquake Risk Reduction in the Balkan Region, Final Report*, 5 vols, UNESCO in association with UNDRO, Project Number RER/79/014, United Nations Educational, Scientific and Cultural Organization (UNESCO), 7, place de Fontenoy, 75700 Paris, France.
UNHCR, 1999. *Handbook for Emergencies*, United Nations High Commissioner for Refugees, Geneva.
UNICEF, 1986. *Assisting in Emergencies: A Resource Handbook for UNICEF Field Staff*, Prepared by Ron Ockwell, United Nations Children's Fund, May.
UNIDO, 1984. *Building Construction under Seismic Conditions in the Balkan Region*, 7 vols, prepared by UNIDO in collaboration with UNDP, Project Number RER/79/015, United Nations Industrial Development Organization (UNIDO), Vienna International Centre, PO Box 300, A-1400 Vienna, Austria.
USACE, 1999. *Urban Search and Rescue Structures Specialist: Field Operation Guide*, US Army Corps of Engineers Readiness Support Center, San Francisco.
USGS, 2002. *US Design Hazard Maps for use with the Uniform Building Code* (http://geohazards.cr.usgs.gov/eq/design/ibc/IBC1615-lus.pdf).
Vaciago, G., 1989. *'Seismic microzonation as a practical urban planning tool'*, MSc Dissertation, Imperial College.
Ville de Goyet, C., 1976. 'Earthquake in Guatemala, epidemiologic evaluation of the relief effort', *Bulletin Pan American Health Organisation*, **10**, (2), 95–109.
Ville de Goyet, C., 2000. 'Stop propagating disaster myths', *The Lancet*, **356**, 762–764.
Warburton, G., 1991. *The Reduction of Vibrations*, The Mallet-Milne Lecture, SECED, Institution of Civil Engineers, London.
Warner, J., 1984. 'Important aspects of cementitious materials used in repair and retrofit', *Eighth World Conference on Earthquake Engineering*, San Francisco, Vol. 1, 493–499.
West, W.D., 1935. 'Preliminary geological report on the Baluchistan (Quetta) earthquake of May 31', *Records of the Geological Survey of India*, Vol. LXI.
Watabe, M., Mochizuki, K.T., Takahashi, T. and Hase, T., 1991. 'Microzonation on seismic intensity in Tokyo', *Proceedings, Fourth International Conference on Seismic Zonation*, Vol. 1, 691–701.
Westgate, K., 1981. 'Land-use planning, vulnerability and the low-income dwelling', in *Disasters and the Small Dwelling* (ed. I. Davis), Pergamon Press, London.
Whitman, R.V., Biggs, J.M., Brennan, J.E., III, Cornell, C.A., De Neufville, R.L. and Vanmarcke, E.H., 1975. 'Seismic design decision analysis', *Journal of the Structural Division*, ST5 ASCE, 1067–1084.
Whitman, R.V., Heger, F.J., Luft, R.W. and Krimgold, F., 1980. 'Seismic resistance of existing buildings', *Journal of the Structural Division*, ST7 ASCE 1573–1591.
Whitman, R.V., Reed, J.W. and Hong, S.T., 1973. 'Earthquake probability matrices', *Proceedings of the Fifth World Conference on Earthquake Engineering*, Vol. 1, 2531.
Wiegel, R.L., 1970. *Earthquake Engineering*, Prentice Hall, Englewood Cliff, NJ.
Wiggins, J.H. and Moran, C., 1970. *Earthquake Safety in the City of Long Beach Based on the Concept of Balanced Risk*, J.H. Wiggins Co., Redondo Beach, CA.

Winchester, P., 1992. *Power, Choice and Vulnerability: A Case Study in Disaster Mismanagement in South India*, James and James Science Publishers, London.
Wolfe, M.R., Bolton, P.A., Heikkala, S.G., Greene, M.M. and May, P.J., 1986. *Land Use Planning for Earthquake Hazard Mitigation: A Handbook for Planners*, Special Publ. 14, Natural Hazards and Application Information Center, Institute of Behavioral Science #3, Campus Box 482, University of Colorado, Boulder, CO 80309-0482, USA.
Wong, K.M., 1987. *Seismic Strengthening of Unreinforced Masonry Buildings*, Centre for Environmental Design Research, University of California, Berkeley, CA.
Woo, G., 1999. *The Mathematics of Natural Catastrophes*, Imperial College Press, London.
Wu Liang Yong, 1981. 'Reconstruction after the Tangshan earthquake', Seminar in Department of Architecture, Cambridge University, 4 March.
Yanev, P., 1974. *Peace of Mind in Earthquake Country: How to Save Your Home and Life*, Chronicle Books, San Francisco.
Zhang Quinnan, 1987. 'Urban earthquake disaster mitigation planning and information in China', *International Research and Training Seminar on Regional Development Planning for Disaster Prevention*, United Nations Centre for Regional Development, Tokyo.
Zuccaro, G. 1998. 'Seismic vulnerability of Vesuvian villages: structural distributions and a possible scenario', in *Reducing Earthquake Risk to Structures and Monuments in the EU, Conference Proceedings* (ed. R. Spence), EU Environment and Climate Programme.

Index

Note: Figures and Tables are indicated by *italic page numbers*, footnotes by suffix "n[X]" where "X" is the note number (e.g. "32n[21]" is note 21 at the foot of page 32)